Michael Krystek
Measurement Uncertainties

Also of interest

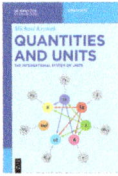

Quantities and Units
The International System of Units
Michael Krystek, 2023
ISBN 978-3-11-134405-8, e-ISBN (PDF) 978-3-11-134411-9

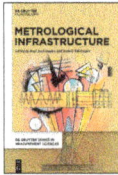

Metrological Infrastructure
Edited by Beat Jeckelmann, Robert Edelmaier, 2023
ISBN 978-3-11-071568-2, e-ISBN (PDF) 978-3-11-071583-5

in
De Gruyter Series in Measurement Sciences
Edited by Klaus-Dieter Sommer, Thomas Fröhlich
ISSN 2510-2974, e-ISSN 2510-2982

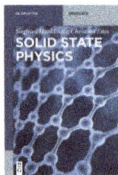

Solid State Physics
Siegfried Hunklinger, Christian Enss, 2022
ISBN 978-3-11-066645-8, e-ISBN (PDF) 978-3-11-066650-2

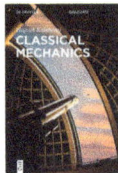

Classical Mechanics
Hiqmet Kamberaj, 2021
ISBN 978-3-11-075581-7, e-ISBN (PDF) 978-3-11-075582-4

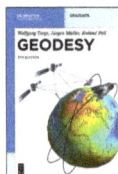

Geodesy
Wolfgang Torge, Jürgen Müller, Roland Pail, 2023
ISBN 978-3-11-072329-8, e-ISBN 978-3-11-072330-4

Michael Krystek

Measurement Uncertainties

——

Error Propagation, Probabilistic Modelling,
Statistical Methods

DE GRUYTER

Author

Dr. Michael P. Krystek studied physics and mathematics at the Technical University of Berlin, received his PhD in physics and his habilitation in metrology. He was a Senior Scientist at PTB (Physikalisch-Technische Bundesanstalt) and chairman of IEC/TC 25 Quantities and Units until 2022. He was project leader for the last revision of the ISO/IEC 80000 series of standards in ISO/TC 12 Quantities and Units, PTB expert in the BIPM-CCU during the revision process of the International System of Units (SI), and expert for ISO and IEC in the JCGM working groups for VIM and GUM.

ISBN 978-3-11-145343-9
e-ISBN (PDF) 978-3-11-145371-2
e-ISBN (EPUB) 978-3-11-145396-5

Library of Congress Control Number: 2024939080

Bibliographic information published by the Deutsche Nationalbibliothek
The Deutsche Nationalbibliothek lists this publication in the Deutsche Nationalbibliografie; detailed bibliographic data are available on the Internet at http://dnb.dnb.de.

© 2024 Walter de Gruyter GmbH, Berlin/Boston
Cover image: Michael Krystek

www.degruyter.com

Preface

A complete measurement result consists of the value of a measured quantity and the associated measurement uncertainty. The uncertainty allows an assessment of the quality of the measured quantity value.

The *Guide to the Expression of Uncertainty in Measurement (GUM)*, published about thirty years ago, has internationally standardized the evaluation of measured data. The publication of this document was motivated by the globalization of trade and the worldwide use of the International System of Units (SI), because only the use of standardized methods ensures that measurement results obtained in different countries and at different times can be easily compared.

Before the introduction of the GUM, the so-called "theory of errors" was used to estimate the uncertainty of measurement. However, this method has the disadvantage that only the actual measured data are used to calculate a best estimate of the measurand and its uncertainty. Any additional information, such as e. g. prior knowledge about the measurand and the characteristics of the measurement device, or available information about influence factors not directly measured, could not be taken into account in the evaluation. In particular, the treatment of systematic errors caused problems.

Since the publication of the GUM, it has often been said that the document is too complicated for the user. This complaint is partly true and partly false. For those who simply wish to apply the rules of the GUM, this guide provides clear instructions on how to do so. The problem seems to be more one of the *understanding* of the rules that are stated in the GUM. This book attempts to fill this gap, not by giving "guidelines to the guide", but by providing the reader with additional information not contained in the GUM and explaining its background.

When applying the methods of the GUM to practical problems, the user is usually faced with two main difficulties, namely the mathematical modelling of the measurement and the correct application of the evaluation method B, which, unlike the more familiar evaluation method A, is based on non-statistical methods. Once these two difficulties have been overcome, the evaluation of the measurement data and the calculation of the uncertainty of measurement can easily be carried out using various existing programmes. Therefore, this book concentrates mainly on these two aspects of data evaluation.

The first chapter serves only as an introduction and tries to motivate the reader to look more closely at the data evaluation methods introduced by the GUM. The second chapter then presents some basic metrological concepts, which are explained in detail. These terms are usually familiar to metrologists, but some of them have undergone a change of meaning in the past, which is reflected in new standards and guidelines.

Mathematical modelling of measurement requires considerable experience and each new model is usually challenging for the metrologist. In order to assist the user in the development of mathematical models, instructions and general rules have been compiled in the third chapter. By following these recommendations, some difficulties

https://doi.org/10.1515/9783111453712-001

can be avoided, especially when dealing with large models. Linear and quadratic approximations as described in the GUM and their limitations are also discussed in detail in this chapter.

The correct application of the methods of the GUM and the evaluation of the results obtained requires a basic understanding of probability theory and statistics. Without this, these methods cannot be fully applied to complex measurement tasks. This is particularly true for evaluation method B, which is based on probability distributions and an interpretation of probability that may be unfamiliar to many users. In order to understand the meaning of the rules and procedures specified in the GUM, some additional knowledge is required. Providing this knowledge is the aim of chapters four and five, which form the main part of this book.

The final chapter compares the methods of the GUM with the traditional method of measurement data evaluation to enable the reader to identify differences and similarities. This comparison is quite critical and shows that there is still room for improvement in the GUM.

This book is a revised translation of the second edition of a German version first published in 2012, which has been used in teaching for more than ten years. In addition, two appendices have been added which were not included in the first German edition. The first appendix deals with the calculation of multivariate measurement uncertainty, the second with the application of BAYESian methods to systematic measurement errors. Both appendices are written as tutorials, are self-contained and can be read independently of the main body of the book.

I would like to take this opportunity to thank the team at De Gruyter for their efficient co-operation in the production of this book. My special thanks go to Ms. Karin Sora and Dr. Damiano Sacco for their kind decision to include this book in the De Gruyter programme.

Berlin, April 2024   Michael Krystek

Contents

Preface —— V

1   **Introduction** —— 1
1.1   The importance of measurement uncertainty —— 1
1.2   The nature of measurement uncertainty —— 3

2   **Fundamental terms of metrology** —— 7
2.1   Quantities, quantity values and units —— 7
2.2   Measurement —— 14
2.3   The true value of a measured quantity —— 16
2.4   Principles, methods, procedures —— 21
2.5   Accuracy, trueness, precision, resolution —— 23
2.6   Measurement errors —— 28
2.7   Measurement uncertainty —— 32

3   **Measurement models** —— 39
3.1   Modelling methods —— 39
3.2   Model equations —— 44
3.3   Submodels —— 50
3.4   Modelling strategies —— 52
3.5   Linear approximation —— 55
3.6   Quadratic approximation —— 69
3.7   Graphical modelling —— 74

4   **Basics of probability theory** —— 89
4.1   Probability concepts —— 89
4.2   Events and outcomes —— 106
4.3   Mathematical probability —— 115
4.4   Conditional probability —— 118
4.5   Rules of probability calculations —— 123
4.6   The theorem of BAYES and LAPLACE —— 129
4.7   Stochastic independence —— 132
4.8   Random quantities —— 136
4.9   Probability distribution functions —— 140
4.10   Probability density functions —— 150
4.11   Transformations of random quantities —— 156
4.12   Expectations —— 159
4.13   Variances and standard deviations —— 166
4.14   Multivariate random quantities —— 172

4.15   Multivariate distribution functions —— 175
4.16   Multivariate density functions —— 179
4.17   Marginal distributions —— 185
4.18   Multivariate expectations —— 193
4.19   Covariances and correlations —— 199
4.20   Approximate estimations —— 206
4.21   Central limit theorem —— 209

5   **Statistical methods** —— 211
5.1   Populations and random sampling —— 211
5.2   Statistics —— 213
5.3   Estimators —— 215
5.4   Method of moments —— 221
5.5   Maximum likelihood method —— 224
5.6   Least squares method —— 234
5.7   BAYESian methods —— 240
5.8   Interval and region estimation —— 255

6   **Measurement uncertainty concepts** —— 273
6.1   The traditional method —— 273
6.2   The methods of the GUM —— 280

A   **From univariate to multivariate uncertainty** —— 287
A.1   Introduction —— 287
A.2   Univariate uncertainty calculations —— 287
A.3   Multivariate uncertainty calculation —— 292
A.4   Matrix representation —— 295
A.5   Generalization —— 297
A.6   Coverage regions —— 298
A.7   Summary of multivariate calculations —— 300
A.8   Examples —— 302

B   **Dealing with systematic measurement errors** —— 309
B.1   Introduction —— 309
B.2   Preliminary remarks —— 310
B.3   Ignoring an existing systematic error —— 311
B.4   BAYES' theorem and marginalization —— 312
B.5   The principle of maximum entropy —— 313
B.6   A procedure to handle systematic errors —— 316
B.7   The influence of temperature on length —— 318
B.8   The cosine error in length measurement —— 321
B.9   The influence of form deviations on the distance —— 323

B.10   Noise as a systematic error — **328**

C   **Bayesian linking of key comparisons** — **333**
C.1   Introduction — **333**
C.2   The scenario — **335**
C.3   The information available — **336**
C.4   Establishing key comparison reference values — **337**
C.5   Degrees of equivalence — **341**
C.6   Conformity tests — **341**
C.7   Examples — **342**
C.8   Uncorrelated measurement results — **345**
C.9   Conclusion — **347**

List of Definitions — **351**

List of Propositions — **355**

List of Examples — **357**

List of Figures — **361**

List of Tables — **363**

List of Symbols — **365**

Bibliography — **371**

Index — **373**

1 Introduction

> I am prepared to concede that all data
> have some uncertainty, and should
> therefore, if possible, be confirmed by
> other data.
>
> *(Bertrand Russell, 1940)*

Measurements are made to determine the value of a quantity. In doing so, we will always obtain deviations of greater or lesser magnitude from the value of the quantity that actually exists. This is inevitable, even if we are as careful as possible. In principle, we cannot determine the true value of a measured quantity, it remains a fiction. In order to be able to judge the reliability of measurements and to be able to compare them, it is therefore always necessary to state the uncertainties of the measurements. This requirement has long been understood in the field of scientific and technical metrology, but it is also becoming increasingly important in everyday measurements.

1.1 The importance of measurement uncertainty

Measurements are ubiquitous in everyday life, although most people are unaware of this fact. Examples include the measurement of time by our watches, the measurement of speed by the speedometer in our cars, the measurement of volume at the pumps in a petrol station, the determination of the weight[1] of items in a supermarket, or the measurement of the consumption of water, gas and electricity in our homes. In medicine, we measure body temperature or the concentration of certain substances in blood or urine to determine the state of a person's health. We measure the concentration of pollutants in water, air or soil, the exposure to ultraviolet rays or radioactive substances, the concentration of alcohol in drinks to raise taxes, or in the air we breathe to detect a violation of regulations, and so on. The modern social order is inconceivable without metrology. Metrology has become a science in itself and is becoming increasingly important.

There has been a need to measure and weigh since the dawn of civilization. Objective and accurate measurement has been an essential aspect of human society since pre-Christian times. Early measurements were mainly based on astronomical observations, the proportions of the human body or comparisons with objects in the environment. Ancient units of measurement, such as time units like year and day, length units like foot and yard, or power units like horsepower, still remind us of the beginnings of metrology.

The comparability of measurements was initially ensured by the use of arbitrary material measures, such as the length of a foot or an ell of the ruling sovereign. In the

---

[1] To be precise, we should speak of the determination of a mass, but in everyday life we do not usually make a distinction between mass and weight.

https://doi.org/10.1515/9783111453712-001

Middle Ages, representations of material measures were placed in places easily accessible to everyone, such as the outer walls of churches or town halls. Certain properties of objects could be quantified by comparison, thus ensuring fair exchange and trade, at least locally. Measurement uncertainty, as we know it today, was irrelevant.

Advances in science and technology demanded ever more precise measurements. Only a quantitative description of natural phenomena based on measurements with the smallest possible uncertainties allows the mathematical formulation of the laws of nature and their reliable application in science and technology. On the other hand, new scientific knowledge enabled the development of metrology and an increase in the accuracy of measurements.

As improved measurement methods were applied, it soon became apparent that measurement results showed more or less pronounced scatter, the causes of which had to be analysed and quantified.

The transition from piecework by craftsmen and manufacturers to mass production during industrialization was made possible by reliable measurement results. Today, the globalization of trade and production requires a high degree of confidence in the comparability of measurement results and, in addition, an internationally agreed system of units. This is impossible without quantitative information on the accuracy of measurements. The statement of the measurement uncertainty associated with each measurement value, as is common practice today, serves this purpose and has therefore become mandatory.

The importance of a quantitative statement of the accuracy of measurements is illustrated by a few arbitrarily chosen examples.

Measurements need to be compared with limits in many areas. In industrial metrology, for example, it is used to verify compliance with the designer's specifications, in medicine to help decide whether a disease exists, or in legal metrology to enable compliance with regulations. In all these cases, the uncertainty of measurement can be used to decide whether a measurement result is within specified limits with a given probability, or whether the requirements are not met. If a measured value is very close to one of the specified limits, there is a high risk that the measured property may not meet the specified requirements. In this case, the measurement uncertainty associated with the measurement result is an important tool to obtain a realistic estimate of this risk and to allow a reasonable decision to be made.

In the event of product liability, objective evidence must be presented in court to demonstrate that the methods used to manufacture and qualify the product in question were appropriate. In the absence of such evidence, taking into account the measurement uncertainty, doubts about the measurement results used to assess the essential characteristics of the product cannot be resolved with certainty.

In order to have confidence in the accuracy of measurement results, all measurement instruments and systems must be checked for compliance with their specifications. In this case, it is also essential to use the measurement uncertainty to make a decision.

In medical and pharmaceutical research, the efficacy and, if necessary, the harmfulness of new drugs must be demonstrated. This is usually done by performing a series of tests. Without knowledge of the measurement uncertainty, it is not possible to judge the significance of the results obtained from these tests.

In the natural sciences, new theories are always verified by measuring their predictable phenomena. The results of new measurements may reveal inconsistencies with existing theories, rendering them inaccurate or inadequate. Spectacular examples of this were, at the beginning of the last century, the general theory of relativity, with which A. EINSTEIN was able to explain the observed deviations in the orbit of the planet Mercury, which could not be reconciled with the classical mechanics of I. NEWTON, or the assumption of quantization of energy by which M. PLANCK was able to explain the deviations from the radiation law of WIEN, which could be verified by precise measurements carried out by scientists at the Physikalisch-Technische Reichsanstalt (today the Physikalisch-Technische Bundesanstalt, PTB, the German national metrology institute) and which are now considered to be the beginnings of quantum mechanics. In both cases, the accuracy of the measurement results played an important role.

Metrology is one of the most important fields of technology. The demand for accurate measurements in our modern society is constantly increasing, and measurement uncertainty serves as a measure of this accuracy.

1.2   The nature of measurement uncertainty

The examples given in the previous section show that measurement uncertainty does not have a negative connotation or indicate a flaw in a measurement. Instead, it allows an assessment of the quality of a measurement. Therefore, measurement uncertainty must be considered as an essential part of a complete measurement result. But what exactly do we mean when we say that measurement uncertainty is associated with a measurement result? What are we really uncertain about? What exactly is the nature of measurement uncertainty? To answer these questions, we must first make it clear that, before we make a measurement, we already have some knowledge of the possible values of the quantity we are interested in. If this were not the case, we would not be able to make a measurement at all. Choosing an appropriate measuring instrument generally requires that we already know something about the value of the quantity to be measured. No measurement is possible if we are completely ignorant. But we always have some prior knowledge.

For example, a nurse will choose a clinical thermometer rather than a kitchen thermometer to measure a patient's body temperature because she knows the range of the temperature to be measured. However, she is uncertain about the exact value of the temperature. Therefore, she carries out the measurement.

On the basis of their specific knowledge, physicists design complex and often very expensive measurement systems and modify them whenever necessary in the light of new information, because the search for scientific knowledge is always an iterative process. New measurements are made on the basis of the information gained from previous measurements.

The purpose of any measurement is, therefore, to obtain additional information and thereby to reduce as far as possible the uncertainty about the value of the quantity being measured. However, we are never completely ignorant of this value, our knowledge is only imperfect.

The result of a measurement is a value of a quantity. But this value cannot be absolutely exact, as we can easily recognize. The main reason for this is that we can never know all the influences that affect a measurement.

The nurse, based on her knowledge of the temperature range to be expected, will certainly use an appropriate thermometer, but how can she be sure of the accuracy of the indication of this thermometer? There is an uncertainty in the calibration of a clinical thermometer. But even if this were not the case, and the thermometer could be considered ideal, the temperature reading on the scale cannot be absolutely accurate. And there are other reasons why the body temperature of the patient cannot be accurately determined, e. g. that no one can be sure that the thermodynamic equilibrium has been reached, or that the conductive heat transfer at the point of measurement may have been inadequate.

Every measurement is imperfect and subject to influences which are generally not precisely quantifiable or, in some cases, not even fully known. These influences may be intrinsic to the measured object (for example, the non-ideal geometric shape of a brick does not allow its exact dimensions to be determined), to the measuring instrument (e. g. a zero point deviation, a linearity error or an imperfection in the graduation of the scale), to environmental influences (e. g. variations in ambient temperature, air pressure and humidity) or even by the operator of the measurement system (e. g. an inaccuracy or error in reading the display, or even a certain, albeit unconscious, bias with regard to the expected measurement value). Finally, the choice of an appropriate measurement method or the stability of a reference material or standard used to calibrate the measurement system is relevant to the correct evaluation of the measured data.

Another reason for a measurement uncertainty is that we are often not sure what we are actually measuring. A vague specification of the quantity to be measured and an imperfect modelling of the measurement are also contributors to measurement uncertainty that are often ignored.

It is a well known fact that repeated measurements of a quantity, even under constant conditions, will generally not yield the same measured value. We observe that the measured values are subject to irregular variations, even when we make every effort to keep the measuring conditions as constant as possible and try to bring all known systematic influences under control.

Despite many attempts in the past, there is still no satisfactory explanation for these variations. The term "random" is commonly used to describe these variations, but its meaning is not really well understood.

Leaving aside this philosophical consideration, we can restrict ourselves to a more or less precise description of the phenomenon itself. For this purpose, a considerable and powerful mathematical discipline has been developed over the last few centuries—the theory of probability—which makes it possible to express the principles that the variations of the measured values seem to follow. Today, this theory serves as the fundamental basis for evaluating measurement data and calculating the associated measurement uncertainties in order to assess their accuracy.

Measurement uncertainty characterizes our lack of knowledge about the value of a quantity. It does not simply describe random variations in the measured values. Two distinct contributions to measurement uncertainty have been identified: random and systematic effects. However, there is still an ongoing debate about the correct approach to dealing with these effects.

2 Fundamental terms of metrology

> To measure a quantity means to
> represent it by a number which states
> how often the underlying unit is
> contained in the measured quantity.
>
> *(Friedrich Kohlrausch, 1887)*

This chapter deals with the basic concepts of metrology as they are essential for the evaluation of measurement results. The definitions and terms used are those of the corrected 3rd edition of the *International Vocabulary of Metrology — Basic and General Concepts and Associated Terms (VIM)*[1] [2, 3], published in 2007 and currently in force. However, we will not always adhere to its definitions and may add our own perspectives. When discussing the "true" value of a quantity, we will also refer to the *Guide to the Expression of Uncertainty in Measurement (GUM)*[2] [5] as well as some related terms in part one of the German standard DIN 1319 [6], which was a predecessor of the GUM.

2.1 Quantities, quantity values and units

In colloquial language, the term "quantity" has several different meanings. The Merriam-Webster dictionary gives the following examples:
- The wine is made in small quantities.
- The boss is worried about quantity as well as quality.
- The family buys food in quantity.

In metrology, however, the term "quantity" is understood in the sense of "physical quantity", e.g. as a quantifiable property of a physical object in the sense of mathematics. This corresponds largely to the Merriam-Webster definition of "the aspect in which an object[3] is measurable in terms of greater, lesser, or equal, or of increasing or decreasing magnitude" or "the subject of a mathematical operation". It can also be said that an object serves as a *carrier* of a property.

An object is usually a carrier of several properties at the same time, e. g. length, mass, volume, temperature, hardness, colour, smell, etc. Not all of these properties are quantifiable, because some of them we cannot compare in terms of "more", "less" or "equal" and assign a number to the property that is not just used as a label. The

---

1 A corrigendum was published in 2008 [1].
2 A corrected version has been published in 2008 [4].
3 The dictionary uses the more general term "thing".

https://doi.org/10.1515/9783111453712-002

non-quantifiable properties, such as e. g. colour or smell, are called nominal properties[4] (VIM:2008, 1.30). They are not quantities.

According to the International Vocabulary of Metrology (VIM), an object does not necessarily have to be of a material kind, i. e. a body or a substance. It can also be a phenomenon, i. e. a process or a state, such as radiation (light, heat or ionizing radiation) or an electromagnetic field. Based on this understanding, the VIM gives the following definition of the term "quantity" (VIM:2008, 1.1):

Definition 2.1 (Quantity — VIM)

A quantity is a property of a phenomenon, body or substance, where the property has a magnitude that can be expressed as a number and a reference.

There is a problem with this definition. Although we agree with it insofar as "a quantity is a property" that "has a magnitude", we disagree with the statement "a magnitude … can be expressed as a number and a reference", because it does not exclude nominal properties such as e. g. Mohs hardness. Comparability alone is not sufficient to define a quantity. In metrology, only properties that are measurable can be considered quantities. Measurable properties are those where there is a relationship between each pair of characteristic values that can be assigned a real number (for more information see [7]). Therefore, deviating from the VIM definition, we give the following definition of "quantity":

Definition 2.2 (Quantity)

A quantitative property is called quantity if there exists a ratio[5] for each pair of its characteristic values to which a real number can be assigned.

Both definitions of "quantity" initially imply a restriction to scalars. But there are quantities which are vectors or tensors. In these cases, however, their components can be treated as scalars.

The symbols for quantities are generally italicised. Recommended symbols are published in the ISO/IEC 80000 series of standards *Quantities and units*.

It is important to note that quantity symbols are not uniquely associated with specific quantities. A given symbol may denote different quantities.

The generic term "quantity" can be hierarchically subdivided into several subtopics, as shown in Fig. 2.1 using the quantity *length* as an example. All these quantities are of the same kind.

A quantity has a value which is called "quantity value" and is defined as:

---

4 The value of a nominal property should not be confused with *nominal value*.
5 Note that this ratio is *not* a quotient, which is the result of the division of two numbers, but rather a relation between two magnitudes.

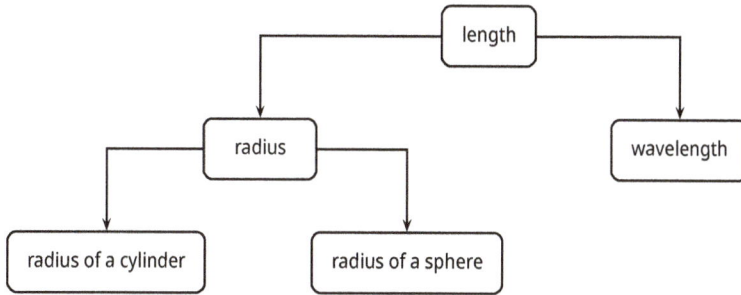

Fig. 2.1: Hierarchic subdivision of a quantity.

Definition 2.3 (Quantity value)

The representation of a characteristic value of a quantity as a product of a number and a unit is called quantity value, i. e.

$$|Q| = \{Q\}[Q]$$

applies, where $|Q|$ denotes the quantity value of the quantity Q and $[Q]$ its unit. The number $\{Q\}$ is called the numerical value of the quantity.

The definition of the term "quantity value" given in the *International Vocabulary of Metrology* (VIM:2008, 1.19) is:

Definition 2.4 (Quantity value — VIM)

A quantity value is a number and reference together expressing the magnitude of a quantity.

The reference in this definition can be either a measurement unit or a measurement procedure or a reference material or a combination of all. The VIM provides examples to illustrate how this applies (VIM:2008, 1.19):

Example 2.1

1. The length of a given rod: 5.34 m or 534 cm
2. The mass of a given body: 0.152 kg or 152 g
3. The curvature of a given arc: $112 \, \mathrm{m}^{-1}$
4. The Celsius temperature of a given sample: $-5 \, °C$
5. The electric impedance of a circuit element at a given frequency: $(7 + 3\,\mathrm{j})\,\Omega$, where j denotes the imaginary unit
6. The refractive index of a given sample of glass: 1.32
7. The Rockwell C hardness of a given sample: 43.5 HRC
8. The mass fraction of cadmium of a given sample of copper: $3 \, \mu g/kg$ or $3 \cdot 10^{-9}$
9. The molality of Pb^{2+} in a given sample of water: $1.76 \, \mu mol/kg$

10. The amount-of-substance concentration of lutropin in a given sample of human blood plasma (WHO International Standard 80/552 used as a calibrator): 5.0 IU/l, where "IU" stands for "WHO International Unit"
11. A force, for example specified in Cartesian co-ordinates:

$$F = \begin{pmatrix} F_x \\ F_y \\ F_z \end{pmatrix} = \begin{pmatrix} 31.5 \text{ N} \\ 43.2 \text{ N} \\ 17.0 \text{ N} \end{pmatrix}$$

This example shows that, according to the definition stated in the VIM, a quantity value—depending on the type of reference—can be either

– a product of a number and a measurement unit (items 1–5, 8 and 9),
– a number and a reference to a measurement procedure (item 7), or
– a number and a reference material (item 10).

It is up to the reader to decide how useful it is to use a measurement method or reference material instead of the SI unit of the underlying quantity to determine the quantity value. The VIM definition appears to be an inadmissible extension of the concept of quantity to include properties that can be compared but for which there is no relation to which a number can be uniquely assigned.

The measurement unit "one" is generally not stated for a quantity of dimension number[6] (see items 6 and 8 in the example 2.1). The number can also be complex (see item 5 in the example 2.1).

Often a quantity value can be expressed in more than one way (see items 1, 2 and 8 in the 2.1 example). In the case of vector or tensor quantities, each component has a quantity value (see item 11 in the example 2.1).

Strictly speaking, we must always distinguish between a quantity and a quantity value. For example, instead of saying "the speed is 50 m/s", we should say "the value of the speed is 50 m/s", because the term "speed" refers to a quantity, not a quantity value. In everyday life, however, we do not distinguish between a quantity and its value, because everyone understands from the context what is actually meant. For the sake of simplicity, we will follow this loose usage in most cases, and for this reason we will often use the same symbol for quantities and quantity values. For example, the symbol X can mean both the quantity X itself and its quantity value. However, to avoid misun-

---

6 In the VIM, the term "quantities of dimension one" is still used. However, it is better to speak of "quantities of dimension number", because the values of quantities of this dimension are pure numbers. This concept corresponds to the fact that the quantity values of e. g. "quantities of the dimension length" are length values and those of "quantities of the dimension mass" are mass values. In the VIM, the term "quantity of dimension one" is used to reflect the fact that in ISO standards the symbolic representation for the dimension of such quantities is currently denoted by the symbol "1" (see also ISO 80000-1:2009, 3.8, Note 3). However, this concept is mathematically incorrect because "1" is not a dimension but a number. A more detailed discussion of the role of quantity dimensions in the SI can be found in [8].

derstandings, we will add appropriate information in cases where it is not clear what is meant.

As can be seen in the example 2.1 above, a quantity value consists of a numerical value and a measurement unit. The term "measurement unit" is defined as (VIM:2008, 1.9, modified):

Definition 2.5 (Measurement unit)

A measurement unit is a real scalar quantity value, defined and adopted by convention, with which a value of any other quantity of the same kind can be compared to express the ratio of the two quantity values as a number.

In the VIM, a measurement unit is defined as a *quantity*, not a *quantity value*. There is currently no consensus on which definition is correct.

Sometimes measurement units of quantities of the same dimension are denoted by the same unit name and symbol, even though the quantities are not of the same kind. For example, "joule per kelvin" and J/K are the unit name and symbol of both effective heat capacity and entropy, even though these two quantities cannot generally be considered to be of the same kind.

In some cases, however, a special unit name is used for clarification, which is restricted to use with a quantity of a special kind. For example, the unit "second to the power of minus one" (s^{-1}) is called "hertz" (Hz) when used for frequencies and "becquerel" (Bq) when used for radionuclide activities.

The unit of all quantities of the dimension number, which are pure numbers, is the number one, symbol "1". In some cases this unit has special names, like e. g. radian and steradian, which are used for the plane angle and the solid angle, respectively, or they are expressed by quotients, like "millimol per mol" (mmol/mol, which is equal to 10^{-3}) and "microgram per kilogram" (µg/kg, which is equal to 10^{-9}).

Instead of the term "unit", the short term "unit" is often used. For a quantity, this short term is often combined with the quantity name, such as "mass unit" or "unit of mass".

In principle, units are arbitrary in their definition, and their regulation and use are subject only to the convention in force at any given time, which may change at any time. Examples of such units can still be found in Anglo-Saxon countries with their old units of length and volume, such as e. g. feet and gallons, but also in aviation, for example, when reporting flight altitudes and speeds, or in maritime navigation when specifying distances. But it does not make sense for everyone to adopt different units of measurement, or for these units to depend on place and time, as was the case in the past, when units of measurement were arbitrarily chosen by local rulers and could be changed by them at any time. With the increasing globalization of trade, the need for global agreement has become more important. Today, measurement units are usually based on the *International System of Units (SI)*, which comprises the names and symbols of the units, including a set of prefixes and their names and symbols, together with

Tab. 2.1: Names and symbols of the seven base units and their dimensions.

Base quantity		Base unit	
Name	Symbol of the dimension	Name	Symbol of the unit
length	L	metre	m
mass	M	kilogram	kg
time (duration)	T	second	s
electric current	I	ampere	A
thermodynamic temperature	Θ	kelvin	K
amount of substance	N	mol	mol
luminous intensity	J	candela	cd

rules for their use, adopted by the *General Conference on Weights and Measures (CGPM)*, founded in 1875 and substantially revised in 2019 (see [7]). The names and symbols of the seven base units, together with the corresponding symbols of their dimensions, are given in Tab. 2.1.

A full description and explanation of the *International System of Units (SI)* is given in the current edition of the SI brochure published by the *Bureau International des Poids et Mesures (BIPM)* in Paris. This brochure can be downloaded free of charge from the BIPM website.

The second component of a quantity value, besides the measurement unit, is its numerical value, which is defined as (VIM:2008, 1.20):

Definition 2.6 (Numerical quantity value)
A numerical quantity value is a number in the expression of a quantity value, other than any number serving as the reference.

For quantities of dimension number, the reference is a measurement unit, which is itself a pure number and, of course, is not considered part of the numerical quantity value.

Example 2.2
In an amount-of-substance fraction equal to 3 mmol/mol, the numerical quantity value is 3 and the unit is mmol/mol. The unit mmol/mol is numerically equal to the number 10^{-3} or 0.001, respectively, but this number is not part of the numerical quantity value, which remains 3.

For quantities with a measurement unit (i. e. quantities other than ordinal quantities), the numerical value $\{Q\}$ of a quantity Q can be expressed by $\{Q\} = |Q|/[Q]$, where $|Q|$

is the value of the quantity and $[Q]$ is the measurement unit.[7] However, if a quantity unit has a defined symbol, as is the case for all base units in the SI, then that symbol *without* the square brackets must be used instead of $[Q]$. This rule is often ignored when labelling axes in graphs and when using units in tables.

Example 2.3

For the mass quantity m with a quantity value of 5.7 kg, the numerical quantity value is $\{m\} = (5.7\,\text{kg})/\text{kg} = 5.7$; the same quantity value can be expressed as 5700 g in which case the numerical quantity value is $\{m\} = (5700\,\text{g})/\text{g} = 5700$.

This example shows in detail that the *numerical value* of a quantity depends on the chosen quantity unit. However, the *quantity value* itself does *not* change when its representation is changed by using different units (the two mass values 5.7 kg and 5700 g represent *the same* quantity value).

The relation between physical quantities is given by physical laws (theories), which are usually written in the form of so-called quantity equations.

A quantity equation is defined by (VIM:2008, 1.22):

Definition 2.7 (Quantity equation)

A quantity equation is a mathematical relation between quantities in a given system of quantities, independent of the measurement units.

Example 2.4

1. The equation for the kinetic energy

$$T = \frac{m}{2}v^2,$$

   where T denotes the kinetic energy, v the speed, and m the mass.
2. The equation for the speed of a particle

$$\boldsymbol{v} = \frac{\mathrm{d}\boldsymbol{r}}{\mathrm{d}t},$$

   where $\boldsymbol{v}$ denotes the speed vector of the considered particle, $\boldsymbol{r}$ its position vector, and t the time.
3. The equation for the amount of substance of a component during an electrolysis

$$n = \frac{It}{zF},$$

---

7 There is no consensus in the metrological literature on how to denote the value of a quantity. Here we follow the convention in mathematics of denoting the "size" of a mathematical object, such as the length of a vector or the cardinality of a set, by absolute lines.

where n denotes the amount of substance of the component, I the electric current, t the duration of the electrolysis, z the valency of the ions of the component, and F the FARADAY constant.

4. The equation for the heat during the heating of a body

$$Q = cm\Delta T\,,$$

where Q denotes the transported heat, m the mass of the body, c its effective heat capacity, and ΔT the observed temperature change.

We will return to quantity equations when we look at the mathematical description of measurement by model equations.

2.2 Measurement

A measurement is a sequence of planned operations to compare a quantity value with a measurement unit and includes counting by convention. The term "measurement" is defined as (VIM:2008, 2.1):

Definition 2.8 (Measurement)
A measurement is a process of experimentally obtaining one or more quantity values that can reasonably be attributed to a quantity.

Measurement presupposes a description of the quantity commensurate with the intended use of a measurement result, a measurement procedure, and a calibrated measuring system operating according to the specified measurement procedure, including the measurement conditions.

In metrology, measurand and quantity are often used as synonymous terms, because the term "measurand" is defined as (VIM:2008, 2.3):

Definition 2.9 (Measurand)
A measurand is a quantity intended to be measured.

Note the difference between the two terms "measurand" and "measured quantity". The former term indicates what we intend to measure, while the latter indicates what we actually measure.

The specification of a measurand generally requires the knowledge of the kind of quantity, a detailed description of the state of the phenomenon, body, or substance as carrier of the quantity, including any relevant components, and the chemical entities involved.

The measurement, including the measuring system and the conditions under which the measurement is carried out, might change the phenomenon, body, or substance such that the quantity being measured can differ from the measurand as defined. In this case, an adequate correction is necessary.

Example 2.5
The potential difference between the terminals of a battery decreases when using a voltage meter with a significant internal conductance to perform the measurement. The open circuit potential difference can be calculated from the internal resistances of the battery, the voltage meter and, of course, the wires connecting them.

Example 2.6
The length of a steel rod in equilibrium with the ambient temperature of 23 °C is different from the length at the specified standard temperature of 20 °C, which is usually the measurand. In this case, a correction is necessary.

When performing a measurement, we cannot assume that the measurement operation can be performed in an ideal way. Therefore, we cannot expect to get the exact value of the respective quantity as a result of the measurement. We will rather get what is called a "measurement result". This term is defined as (VIM:2008, 2.9):

Definition 2.10 (Measurement result)
A measurement result is a set of quantity values being attributed to a measurand together with any other available relevant information.

A measurement result generally contains "relevant information" about the set of quantity values being attributed to a measurand, such that some may be more representative of the measurand than others. This can be expressed in the form of a probability density function.

However, a measurement result is generally not given in the form of a probability density function, but is expressed as a single measurement value and an associated measurement uncertainty. If the measurement uncertainty is considered to be negligible for some purpose, the measurement result may be expressed only by a single quantity value. This is a common way in daily life and some other fields to express a measurement result.

In the traditional literature and in the second edition of the VIM, the term "measurement result" was defined as a value attributed to a measurand and explained to mean an indication, an uncorrected result, or a corrected result, depending on the respective context. Nowadays, however, this view is considered obsolete.

A measurement result consists of a *set of quantity values*. This concept leads to the term "measured value" (VIM:2008, 2.10):

Definition 2.11 (Measured value)
A measured value is a quantity value representing a measurement result.

For a measurement involving replicate indications, each indication can be used to provide a corresponding measured quantity value. This set of individual measured quantity values may also be used to calculate a resulting measured quantity value, such as an average or median, usually with a smaller associated measurement uncertainty.

In the case that a range of the true quantity values, which are believed to represent the measurand, is not small in comparison with the measurement uncertainty, a measured quantity value is often an estimate of an average or median of the set of these quantity values.

In the GUM, the terms "result of measurement" and "estimate of the value of the measurand" or just "estimate of the measurand" are used instead of the term "measured quantity value".

2.3   The true value of a measured quantity

In the previous section we have used the term "true quantity value" in connection with the definition of the term "measured quantity value". The term "true value" is still frequently used today in metrology without there being a sufficiently clear idea of its actual meaning.

When people are asked about the meaning of the term "true value", it seems certain that they think that the term should at least be understood as an actually existing value of a quantity. But this thinking already reveals a logical misconception. A quantity of a thing, be it a body, a substance or a phenomenon, does not have a unique value. Neither the scale nor the unit of a physical quantity is given by nature. Quantity values are assigned to a property by us according to a convention, and will change each time the convention is changed. So what does the word "true" in the term "true value" actually mean?

To determine the exact meaning of a term, it is common to consult an encyclopaedia or relevant standards. Unfortunately, this is not very helpful in this case because there are several different definitions of "true value" in the standards. However, it is important to note that these definitions are not contradictory, but rather reflect the different perspectives of experts in the different areas of application covered by each standard. In the following, we will take a closer look at the definitions in the different standards and comment on them.

In the German standard DIN 55350-13:1987 [9] we find the definition:

The true value is the actual property value under the conditions prevailing during its determination.

NOTE 1   Often the true value is an ideal value, because it can only be determined if all systematic errors of the result could be avoided, or it results from theoretical considerations.

NOTE 2 *The true value of a theoretical (mathematical) characteristic is also called "exact value". The exact value will, however, not always be the result of determination by a numerical calculation method. For example, the exact value of an area of a circle with diameter d is given by $d^2\pi/4$.*

According to this definition, the *true value* is equal to the *actual value* of a quantifiable property, i. e. representable by a numerical value and thus measurable. That is to say, a quantity at a particular place and time under the conditions prevailing at that exact location and that instant of time.

Therefore, even without taking the first note into account, it is already clear in this case that the "true value" is only understood as an *ideal value*, even if *all systematic errors of the result could be avoided* (i. e. if we were able to perform an ideal measurement), because already the complete knowledge of the *prevailing conditions* is impossible.

From this point of view, the second part of the first note—"or it results from theoretical considerations"—is not immediately comprehensible, because even in case of purely *theoretical considerations* not all *prevailing conditions* can completely be taken into account. Here it seems to be meant that cases exist where the "true value"—by reason of generally accepted physical laws—is independent of any preconditions, as e. g. the value of the speed of light in vacuum which today according to EINSTEIN's theory of relativity is supposed to be constant (particularly convincing experimental evidence has been provided by D. SADEH [10]) with a value fixed exactly[8] at $c = 299\,792\,458\ \mathrm{m{\cdot}s^{-1}}$ by the *Committee on Data for Science and Technology (CODATA)* [11]. Similarly, the absolute zero point temperature $T = 0\,\mathrm{K}$ (corresponding to $-273.15\,°\mathrm{C}$) are fixed by convention (see e. g. [11]).

On the other hand, the *value of a theoretical (mathematical) characteristic*, called "exact value" in the second note, is a matter of a deductively inferable value of mathematics, where the *prevailing conditions* are solely the laws of mathematics. In geometry, for example, the mathematical constant π, but also the angular sum of a plane triangle are such true values.

The second note also points out that the use of *a numerical calculation method* does not necessarily result in *the exact value*. However, the example given requires a more detailed explanation in order to understand its exact meaning. What is meant here is not the possible rounding error that can occur in numerical calculations, but the fact that the mathematical constant π cannot be represented by a rational number or a decimal number with a finite number of digits, so that any approximation of π will inevitably lead to an inaccurate result. This problem is unavoidable in principle and occurs with any numerical representation of irrational numbers. In practice, however, this is not a

---

8 The value of the speed of light in vacuum was determined by measurements and subsequently fixed by convention as an exact value, i. e. without measurement uncertainty.

real problem, since we can always approximate any real number by a rational number as exactly as we wish.

The German standard DIN 55350-13:1987 has already provided us with some pieces of the puzzle. Next we look at an international standard. The definition of the term "true value" in ISO 3534-2:2006 [12] is identical in content to the definition given in the German standard DIN 55350-13:1987, but less detailed. It reads (ISO 3534-2:2010, 3.2.5):

> The true value is the value which characterizes a quantity or quantitative characteristic perfectly defined in the conditions which exist when that quantity or quantitative characteristic is considered.
>
> NOTE 1   The true value of a quantity or quantitative characteristic is a theoretical concept and, in general, cannot be known exactly.

The note to this definition explicitly states that the "true value" *generally cannot be known exactly*. It is only a theoretical concept that does not and cannot have any practical implications. However, this conclusion can also be drawn from the definition in the German standard DIN 55350-13:1987 and therefore does not lead to any further insight.

The definition of the "true value" as given in ISO 3534-2:2006 is problematic, because within this standard there is no definition of the term "quantitative characteristic", i. e. the whole definition is obscure. Moreover, since *every* quantity is also a quantitative characteristic [13] (*viz.* to some extent a subset of quantitative characteristics), the question arises as to what other quantitative characteristics might be meant by this ISO definition.

The definitions of the "true value" given in DIN 55350-13:1987 and ISO 3534-2:2006 have in common that they define this term as something real and existing, but not precisely known. In contrast, the very short definition given in DIN 1319-1 [6] follows a slightly different philosophy. This definition is (DIN 1319-1, 1.3):

> The true value (of a measurand) is the value of the measurand as objective of the evaluation of measurements of the measurand.

In this standard, it is not said—as in the previous definitions—what the meaning of the "true value" is, but rather for what purpose it is used, i. e. that the aim of any measurement is to determine a value called "true value".

A more philosophical question is whether it makes sense to consider the value obtained from a measurement as the "true" value. In this context, the following assumptions are likely to have been made implicitly:
a)   Exactly one "true value" exists for each measurand.
b)   The "true value" of a measurand is a fixed quantity value.
c)   If an ideal measurement would exist, the "true value" could be determined.

If these assumptions were actually true, the true value could be considered as an unknown parameter of the evaluation model. It is then the *objective of the evaluation of the measurements of the measurand* to determine from the measurement data a best estimate of this unknown parameter (i. e. of the measurand) which is as close as possible to the true value.

However, the concept of a "true" value to be assigned to a quantity presupposes that the measurand is well-defined, i. e. that the evaluation model is complete. But this is, strictly speaking, never possible, because to define a measurand completely, an infinite amount of information would be needed, since *all* influence quantities, whether known or unknown, would have to be taken into account in the modelling. But since this is impossible, the measurement of the quantity of interest can only be approximately described by a model, i. e. in principle each model will lead to a different "true" value.

We now turn to the definition of the "true value" as given in the *Guide to the Expression of Uncertainty in Measurement (GUM)* [5], published in 1995 and still valid. The definition is (GUM, B.2.3):

> *The true value of a quantity is the value consistent with the definition of a given particular quantity.*
>
> *NOTE 1   This is a value that would be obtained by a perfect measurement.*
>
> *NOTE 2   True values are by nature indeterminate.*
>
> *NOTE 3   The indefinite article "a", rather than the definite article "the", is used in conjunction with "true value" because there may be many values consistent with the definition of a given particular quantity.*

At the time of publication of the GUM, this definition was in accordance with the 2nd edition of the *International Vocabulary of Basic and General Terms in Metrology (VIM)* [3], which was published before the GUM (see VIM:1993, 1.19). This takes into account that depending on the definition of the measurand there may be different true values—one for each definition of the measurand—which can then all be considered ideal.

In the 3rd edition of the *International Vocabulary of Basic and General Terms in Metrology (VIM)*, this issue is further clarified in the notes on the definition of the "true value" (VIM:2008, 2.11):

Definition 2.12 (True quantity value)

The true quantity value is the quantity value consistent with the definition of a quantity.

NOTE 1   In the error approach to describing measurement, a true quantity value is considered unique and, in practice, unknowable. The uncertainty approach is to recognize that, owing to the inherently incomplete amount of detail in the definition of a quantity, there is not a single true quantity value but rather a set of true quantity values consistent with the definition. However, this set of values is, in principle and in practice, unknowable. Other approaches dispense altogether with the concept of true quantity value and rely on the concept of metrological compatibility of measurement results for assessing their validity.

NOTE 2 In the special case of a fundamental constant, the quantity is considered to have a single true quantity value.

NOTE 3 When the definitional uncertainty associated with the measurand is considered to be negligible compared to the other components of the measurement uncertainty, the measurand may be considered to have an "essentially unique" true quantity value. This is the approach taken by the GUM and associated documents, where the word "true" is considered to be redundant.

The so-called error approach mentioned in Note 1 of this definition is based on the same principles as the "theory of errors" introduced by C. F. GAUSS in 1821 and used in metrology until the implementation of the GUM in 1995, while the so-called uncertainty approach is based on the methods first published in the GUM.

The fundamental constants mentioned in Note 2 are the constants of nature, i. e. special physical quantities, where it is presupposed that their quantity values neither spatially nor temporally change. They are the unknown parameters of the physical laws (theories). The theories, however, cannot establish these values, they rather must be determined by measurements. This exceptional position of the fundamental constants explains why they are considered to have a single true value.

Where appropriate, the value of a fundamental constant may be fixed by convention at the most recently agreed value, as has been done in the past for the speed of light in vacuum, for example, and recently also for other fundamental constants at the time of the major revision of the SI in 2019. In such a case, the value can obviously no longer have any uncertainty.

It is immediately apparent from Note 3 of the definition 2.12 that we can completely dispense with the true value when calculating the measurement uncertainty according to the methods of the GUM, provided that the definitional uncertainty is negligibly small. A more detailed justification of this position can be found in Appendix D of the GUM [5]. The term "definitional uncertainty" will be discussed in more detail in section 2.7.

All in all, the standards do not give a satisfactory answer to the question of what the term "true value" really means. The reason, of course, is that the problem is not really a technical one but a philosophical one. Its solution depends on the definition of "truth", which has plagued philosophers for at least the last three thousand years. So the best we can do is to follow the suggestion of the GUM and consider the word "true" to be redundant in metrology.

2.4 Principles, methods, procedures

The experimental determination of the value of a measurand can often be carried out in various ways. The basis of each measurement is a "measurement principle" defined as (VIM:2008, 2.4):

Definition 2.13 (Measurement principle)
A measurement principle is a phenomenon which serves as a basis of a measurement.

The phenomenon can be of a physical, chemical or biological nature, but it is important that a natural phenomenon is used which can always and everywhere be reproduced in the same way.

Most quantities cannot be measured directly. However, if there is a unique scientific relationship—known empirically and quantitatively—between the measurand and another quantity that is more readily accessible to direct measurement, then this relationship can be used to determine the value of the measurand from the measured data.

Example 2.7
1. The thermoelectric effect applied to the measurement of temperature.
2. The energy absorption applied to the measurement of amount-of-substance concentration.
3. The lowering of the concentration of glucose in blood of a fasting rabbit applied to the measurement of insulin concentration in a preparation.

In addition to the measurement principle a particular "measurement method" must be specified. This term is defined as (VIM:2008, 2.5):

Definition 2.14 (Measurement method)
A measurement method is a generic description of a logical organization of operations used in a measurement.

Measurement methods may be qualified as
– substitution measurement method,
– differential measurement method, and
– null measurement method;
or
– direct measurement method, and
– indirect measurement method.

We give an example for each of the measurement methods to make their differences more comprehensible:

Example 2.8

A *substitution measurement method* is, for example, the measurement of a mass using a spring balance and a set of calibrated weights. In this case the length variation of a spring caused by the load of a calibrated weight of known mass and by the load of the object to be measured is used as measurement principle. If the spring constant is known, the unknown mass of the object to be measured can be calculated from the measured length values, provided that the length of the spring changes linearly with the force exerted by the masses in the gravity field of the Earth.

Example 2.9

A *differential measurement method* is, for example, the measurement of the volume of an irregularly shaped body (e. g. a stone) by determining the volume of the displaced liquid during a complete submersion of the body in a liquid. In this case the constancy of mass is used as the measurement principle.

Example 2.10

A *null measurement method* is, for example, the measurement of a mass using a beam balance and a set of calibrated weights. If the beams of the balance are exactly of the same length and the balance is in a state of equilibrium, the mass values of the calibrated weight and the object to be measured are equal. In this case, the measurement principle is that the sum of the torques caused by the masses of the calibrated weight and the object is zero.

Example 2.11

A *direct measurement method* is, for example, the measurement of a length using a ruler or a tape measure. In this case, a direct comparison is made between the object to be measured and the scale of the measuring device.

Example 2.12

An *indirect measurement method* is, for example, the measurement of a temperature with a liquid expansion thermometer. In this case the volume expansion of the liquid with temperature is used as measurement principle. The length of the liquid column is then (approximately) proportional to the temperature.

We can see that the differentiation of the measurement methods in the first group emphasizes the procedural aspect of a measurement, while in the second group the key aspect is whether a measured value can be obtained directly from an indication or whether it has to be calculated subsequently.

The practical application of one or more measurement principles together with a measurement method is called a "measurement procedure". This term is defined as (VIM:2008, 2.6):

Definition 2.15 (Measurement procedure)
A measurement procedure is a detailed description of a measurement according to one or more measurement principles and to a measurement method, based on a measurement model and including any calculation to obtain a measurement result.

A measurement procedure is usually documented in sufficient detail to enable an operator to perform and evaluate a measurement. This should include the model of the measurement as well as any equations and algorithms needed to calculate the measurement result. In addition, it is advisable not to forget information on relevant literature or tables. Such information may be useful later if a verification of the evaluation procedure becomes necessary.

The achievable precision of a measurement is usually determined by the choice of the measurement procedure. A measurement procedure may include a statement of a target measurement uncertainty to be met, i. e. a measurement uncertainty specified as an upper bound and specified on the basis of the intended use of the measurement result (VIM:2008, 2.34). Thus, for economic reasons, any measurement should always be performed *as good as necessary*, not *as good as possible*.

2.5 Accuracy, trueness, precision, resolution

It often happens that the terms "accuracy", "trueness", "precision" and "resolution" are either confused or used in the wrong context. Therefore, in this section we will briefly discuss these terms.

Measurement accuracy

The term "measurement accuracy" is defined as (VIM:2008, 2.13):

Definition 2.16 (Measurement accuracy)
Measurement accuracy is the closeness of agreement between a measured quantity value and a true quantity value of a measurand.

Measurement accuracy is not a quantity and cannot be given a numerical quantity value. Only a qualitative statement is possible, as e. g. that a measurement is more accurate than another measurement when it has a smaller measurement error or measurement uncertainty.

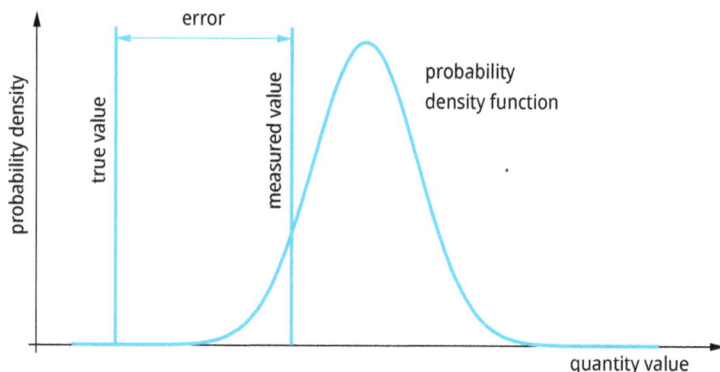

Fig. 2.2: On the definition of measurement accuracy.

Measurement accuracy is sometimes understood as the closeness of agreement between measured values attributed to the measurand. However, this view is not correct, because the definition of measurement accuracy does not allow such a statement (see also Fig- 2.2 for clarification). On the other hand, it would be admissible to consider measurement accuracy as the closeness of agreement between a measured quantity value and a reference quantity value assigned by convention, because a reference quantity value is a quantity value that is used as a basis for comparison with values of quantities of the same kind (VIM:2008, 5.18). However, this presupposes that the selected reference value can reasonably be regarded as a substitute for the "true value", which is in principle unknown.

Measurement trueness

The term "measurement trueness" is defined as (VIM:2008, 2.14):

Definition 2.17 (Measurement trueness)

Measurement trueness is the closeness of agreement between the average of an infinite number of replicate measured quantity values and a reference quantity value.

Measurement trueness is not a quantity and therefore cannot be expressed numerically. It is inversely related to the term "systematic measurement error", but not to the term "random measurement error". Measures for closeness of agreement are given in the series of standards ISO 5725.

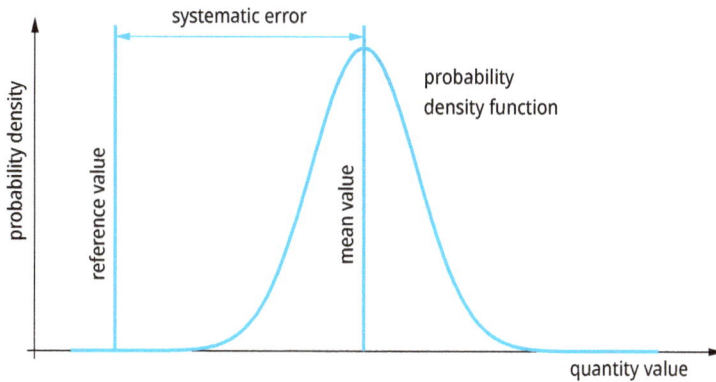

Fig. 2.3: On the definition of measurement trueness.

Measurement precision

The term "measurement precision" is defined as (VIM:2008, 2.15):

Definition 2.18 (Measurement precision)
Measurement precision is the closeness of agreement between indications or measured quantity values obtained by replicate measurements on the same or similar objects under specified conditions.

Measurement precision is usually expressed numerically by measures of imprecision, such as standard deviation, variance, or coefficient of variation under specified measurement conditions.

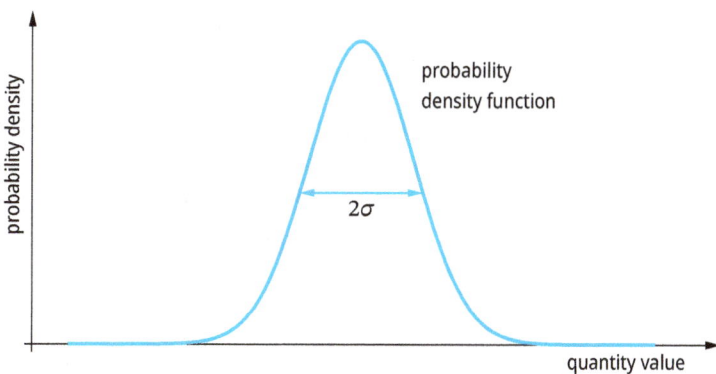

Fig. 2.4: On the definition of measurement precision which is specified here as inversely proportional to the standard deviation σ.

Measurement trueness and measurement precision, although not being quantities, are related to different influences. This is illustrated in Fig. 2.5 using the scatter of the hits on a target as an example. The centre (aiming point) of the target corresponds to the reference quantity value. It can be seen that a small scatter of the hits (measured values) corresponds to a large measurement precision, independent of the measurement trueness which can only describe the systematic measurement error of the hits with respect to the aiming point.

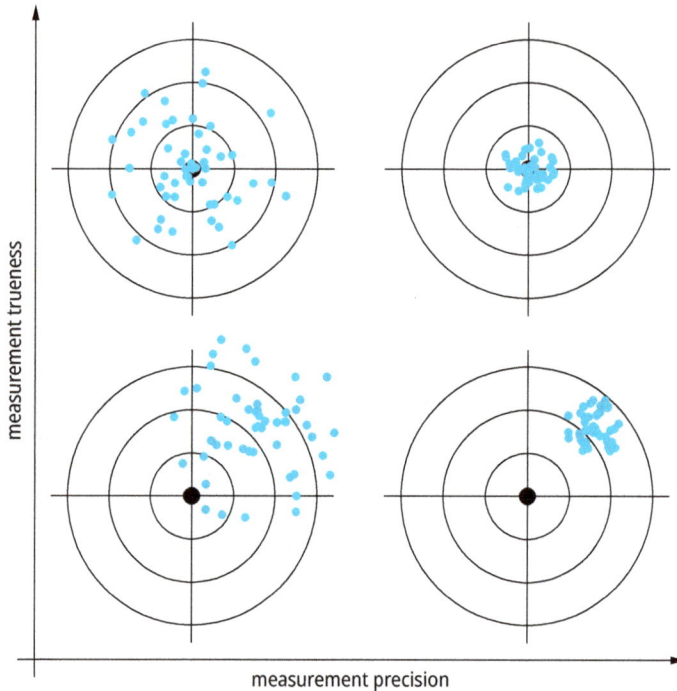

Fig. 2.5: Comparison of measurement trueness and measurement precision.

The term "measurement precision" is used to define the terms "measurement repeatability" (VIM:2008, 2.21), "intermediate measurement precision" (VIM:2008, 2.23), and "measurement reproducibility"(VIM:2008, 2.25). The respective "specified conditions" are "repeatability conditions of measurement", "intermediate precision conditions of measurement", and "reproducibility conditions of measurement".

Repeatability condition of measurement

A repeatability condition of measurement is (VIM:2008, 2.20):

Definition 2.19 (Repeatability condition of measurement)
A repeatability condition of measurement is a condition of measurement, out of a set of conditions that includes the same measurement procedure, same operators, same measuring system, same operating conditions and same location, and replicate measurements on the same or similar objects over a short period of time.

A condition of measurement is a repeatability condition only with respect to a specified set of repeatability conditions.

Intermediate precision condition of measurement

An intermediate precision condition of measurement is (VIM:2008, 2.22):

Definition 2.20 (Intermediate precision condition of measurement)
An intermediate precision condition of measurement is a condition of measurement, out of a set of conditions that includes the same measurement procedure, same location, and replicate measurements on the same or similar objects over an extended period of time, but may include other conditions involving changes.

Changes may include new calibrations, calibrators, operators and measurement systems. A specification of conditions should include, to the extent practicable, the changed and unchanged conditions.

Reproducibility condition of measurement

A reproducibility condition of measurement is (VIM:2008, 2.24):

Definition 2.21 (Reproducibility condition of measurement)
A reproducibility condition of measurement is a condition of measurement, out of a set of conditions that includes different locations, operators, measuring systems, and replicate measurements on the same or similar objects.

Different measurement systems may or may not use different measurement procedures. A specification should state the changed and unchanged conditions where practical.

Resolution

The term "resolution" is defined as (VIM:2008, 4.14):

Definition 2.22 (Resolution)

The resolution is the smallest change in a quantity being measured that causes a perceptible change in the corresponding indication.

The resolution can depend on, for example, friction or noise (internal or external). It may also depend on the value of the quantity being measured.

The resolution of a displaying device is the smallest difference between its indications that can meaningfully be distinguished (VIM:2008, 4.15).

Conclusion

We have seen that, by definition, only the resolution and the measurement precision are quantifiable. The resolution directly enters into the calculation of the measurement uncertainty as one of its components. In contrast, as we will see later, the measurement precision can be seen as a measure that is inversely proportional to the measurement uncertainty, provided that no systematic measurement error occurs, or that systematic effects have been corrected for.

Finally, it must be emphasized that neither measurement trueness nor measurement accuracy can be quantified because, by definition, they are not quantities. Unfortunately, statements that ignore this simple fact are still frequently made in the scientific and technical literature and at academic meetings. The negative consequences are that students are confused by different interpretations of these terms and may learn incorrect concepts. In principle, it would be better to use the term "measurement uncertainty" (see section 2.7)—which is a quantity—instead of the vague term "measurement accuracy".

2.6   Measurement errors

It has been known for centuries that the value of a quantity cannot be determined with arbitrary precision. C. F. GAUSS, P. S. LAPLACE and their contemporaries have already dealt with what they called "observational errors", also called "errors in measurement" by later authors, and today referred to as "measurement errors". Since then, the term "measurement error" has undergone many changes, both in naming and in definition.

In the German standard DIN 1319-1:1995 [6] the following definition can be found (DIN 1319-1:1995, 3.5):

The measurement error is the deviation of a value, obtained from measurements and assigned to the measurand, from the true value.

In this definition the measurement error still refers only to the *true value* of the measurand. In contrast, the currently valid definition of the term "measurement error" is
(VIM:2008, 2.16):

Definition 2.23 (Measurement error)
The measurement error is the measured quantity value minus a reference quantity
value.

This new definition is more general than the old one, since it does not only refer to the
true value of the measurand, which is in principle unknown, but also to a value of a
reference quantity, which may or may not be the true value (VIM:2008, 5.18, Note 1).

The difference between the new definition of the measurement error and the old one
is that instead of using the true value of the measurand, we can also use an arbitrarily
agreed value as the reference value when calculating a measurement error. This has the
advantage that, unlike the true value, we actually know the reference value. On the other
hand, we have to accept that a reference value usually has also an associated measurement uncertainty. The consequence is that the measurement error contains—in addition
to the measurement uncertainty contribution from the measured value itself—yet another contribution to the measurement uncertainty which comes from the reference
value alone. Therefore, whenever possible, we will generally try to choose the reference
value so that its measurement uncertainty is small enough to be negligible compared to
the measurement uncertainty of the measurand under consideration.

The definition of the term "reference quantity value" is (VIM:2008, 5.18):

Definition 2.24 (Reference quantity value)
A reference quantity value is a quantity value used as a basis for comparison with
values of quantities of the same kind.

A reference quantity value can be a true quantity value of a measurand, in which case
it is unknown, or a conventional quantity value, in which case it is known. A reference
quantity value with its associated measurement uncertainty is usually provided with
reference either to
a) a material, e. g. a certified reference material,
b) a device, e. g. a stabilized laser,
c) a reference measurement procedure, or
d) a comparison of measurement standards.

According to the *International vocabulary of metrology (VIM)* the concept of "measurement error" can be used both (VIM:2008, 2.16, Note 1)
a) when there is a single reference quantity value to refer to, which occurs if a calibration is made by means of a measurement standard with a measured quantity value

having a negligible measurement uncertainty or if a conventional quantity value is given, in which case the measurement error is known, and

b) if a measurand is supposed to be represented by a unique true quantity value or a set of true quantity values of negligible range,[9] in which case the measurement error is not known.

The statement b) in this note is unclear. Since according to the definition of the true value (VIM:2008, 2.11, Note 1) *a true quantity value is considered unique*, the meaning of the phrase *a set of true quantity values of negligible range* is at least questionable from a mathematical point of view.

The term "measurement error" should not be confused with the terms "production error" or "mistake". A measurement error has nothing in common with the incorrect operation of a machine or measuring instrument, a wrong or missing number in measurement data, or a programming error in software used to evaluate measurement data. Such mistakes and errors are generally not considered to be measurement errors.

Before the introduction of the concept of measurement uncertainty, the measurement error was sometimes called *"error in measurement"* and the method of calculating its value was known as "calculation of errors".

Two different types of measurement error are usually distinguished, namely random measurement error and systematic measurement error. The term "random measurement error" is defined as (VIM:2008, 2.19):

Definition 2.25 (Random measurement error)
The random measurement error is the component of the measurement error that in replicate measurements varies in an unpredictable manner.

A possible reference quantity value for a random measurement error is the average that would result from an infinite number of repeated measurements of the same measurand. However, such an average is almost as good an ideal value as the true value of a measurand. In practice, therefore, we are satisfied with a mean value which results from a sufficiently large number of repeated measurements of the same measurand under, as far as possible, unchanging conditions.

The random measurement errors of a set of repeated measurements form a distribution which can be characterized by its expectation, generally assumed to be zero, and its variance. The meaning of the terms "distribution", "expectation" and "variance" will be clarified by the respective definitions given in chapter 4 when we deal with the basics of probability theory.

---

9 The range of an interval $[a, b]$ is defined as the difference $(b - a)$ of the boundary values of the interval and is denoted by the expression $r[a; b]$. [3, 2]

The total measurement error consists of the random measurement error and the systematic measurement error. The term "systematic measurement error" is defined as (VIM:2008, 2.17):

Definition 2.26 (Systematic measurement error)
The systematic measurement error is the component of the measurement error that in replicate measurements remains constant or varies in a predictable manner.

This definition of the systematic measurement error also includes the possibility of a drift, a hysteresis, or a predictably periodically changing systematic effect, such as the effect of variations in temperature control. Systematic measurement deviations are always strictly deterministic, even if the underlying physical law is not definitely known.

A possible reference quantity value for a systematic measurement error is a true quantity value (only in principle, because it is unknown), or a measured quantity value of a measurement standard with negligible measurement uncertainty, or a conventional quantity value.

It is not always possible to make a clear distinction between systematic and random errors. Sometimes systematic errors are not immediately recognized as such. It is possible that a measurement error is initially considered to be random due to insufficient knowledge of the measurement process, but that additional influences of a more systematic nature arise from a better understanding of the measurement, and we must then speak of a systematic measurement error.

A systematic measurement error and its causes may be completely unknown or partially known. However, it is very important to understand that it is by no means possible to find out whether systematic effects exist or not by repeated measurements, even performed under unchanged conditions.

A correction can be applied in order to compensate for a known systematic measurement error. The term "correction" is defined as (VIM:2008, 2.53):

Definition 2.27 (Correction)
A correction is a compensation for an estimated systematic effect.

The information given in this definition, that only a compensation for an *estimated systematic effect* is possible, is essential because a systematic measurement error can never be known completely, but only a (best) estimate can be determined. Therefore, strictly speaking, it is impossible to completely correct for the influence of a systematic effect.

To compensate for a systematic measurement error, we can, for example, measure a standard and compare the measured quantity value with the quantity value stated in its calibration certificate. The difference between these two values can be used as an estimate of the systematic measurement error and can therefore be used to compensate for the underlying systematic effect.

The compensation can be of various types, such as a summand, a factor or a tabular value. Usually the estimate of the respective systematic measurement error—also called "measurement bias" (VIM:2008, 2.18)—with a changed sign is used for the compensation. But then it has to be noted that the measurement uncertainty associated with this estimate is not zero and therefore has to be taken into account in the calculation of the measurement uncertainty of the *corrected* result.

A known systematic measurement error shall be corrected in order not to unduly increase the measurement uncertainty. The GUM assumes that the result of a measurement has been corrected for all recognized significant systematic effects (GUM, 3.2.4). This recommendation dates back to C. F. GAUSS [14].

Finally, it should be noted that a small measurement uncertainty does not imply the existence of a small systematic measurement error, even if a correction has been applied. A measurement result may have a very small uncertainty, but there may still be a large systematic measurement error, which is unknown, because, for example, one or more influence quantities of a systematic nature have taken effect, which have remained unrecognized, may have been overlooked, or, for whatever reason, have not been included in the evaluation model of the measurement.

2.7 Measurement uncertainty

In this section we finally turn to the concept of "measurement uncertainty". We have already mentioned in the introduction that the measurement uncertainty can be used to assess the quality of a measurement result and there is consensus that the measurement uncertainty should be as small as possible. However, we still need to discuss how the term "measurement uncertainty" should be defined, because in principle there are different options. It is always a matter of convention which agreement is reached on the definition of measurement uncertainty. It is advisable, however, that there should be only one definition worldwide in order to ensure the comparability of measurement results.

A definition established by convention is usually subject to change and is adapted over time to reflect new knowledge and requirements. Unfortunately, expert discussions do not necessarily lead to a generally accepted result. C. F. GAUSS has already clearly shown that an ideologically biased discussion does not make sense in such cases,[10] when

---

10 Quodsi quis hanc rationem pro arbitrio, nulla cogente necessitate, electam esse obiciat, libenter assentiemur. Quippe quaestio haec per rei naturam aliquid vagi implicat, quod limitibus circumscribi nisi per principium aliquatenus arbitrarium nequit.

... utrum enim error duplex aeque tolerabilis putetur quam simplex bis repetitus, an aegrius, et proin utrum magis conveniat, errori duplici momentum duplex tantum, an maius, tribuere, quaestio est ne qua per se clara, neque demonstrationibus mathematicis decidenda, sed libero tantum arbitrio remittenda. [14]

he was concerned with whether the measure he favoured was better suited to describe errors of observation than the measure initially used by P. S. LAPLACE.

Even today, the discussion about the "correct" approach does not seem to be completely over. In the introduction to the *International Vocabulary of Metrology (VIM)*, the approach commonly used in the past (called the "Error Approach") is compared with the approach used today (called "Uncertainty Approach"). There it states (VIM:2008, Introduction):

> *No preference is given in this third edition to any of the particular approaches.*
> ⋮
> *The objective of measurement in the Error Approach is to determine an estimate of the true value that is as close as possible to that single true value. The deviation from the true value is composed of random and systematic errors. The two kinds of errors, assumed to be always distinguishable, have to be treated differently. No rule can be derived on how they combine to form the total error of any given measurement result, usually taken as the estimate. Usually, only an upper limit of the absolute value of the total error is estimated, sometimes loosely named "uncertainty".*
> ⋮
> *The objective of measurement in the Uncertainty Approach is not to determine a true value as closely as possible. Rather, it is assumed that the information from measurement only permits assignment of an interval of reasonable values to the measurand, based on the assumption that no mistakes have been made in performing the measurement. Additional relevant information may reduce the range of the interval of values that can reasonably be attributed to the measurand. However, even the most refined measurement cannot reduce the interval to a single value because of the finite amount of detail in the definition of a measurand. The definitional uncertainty, therefore, sets a minimum limit to any measurement uncertainty. The interval can be represented by one of its values, called a "measured quantity value".*

Thus, the VIM seems to leave it to the user to decide which approach to apply. In practice, however, the uncertainty approach corresponds to the current convention, since the definition of measurement uncertainty given in the *Guide to the Expression of Uncertainty in Measurement (GUM)* reads (GUM, 2.2.3):

[But if someone would object that this assignment was arbitrarily made without compelling necessity, we willingly agree. This question after all contains by the nature of things something indefinite which in a sense can only definitely be limited by an arbitrary principle.

... whether twice the error is indeed considered as equally tolerable as the single one repeated twice, or as worse, and whether it is therefore more adequate to assign only a duplicated moment to a twofold error or a larger one is a question which is neither clear per se nor can be decided by a mathematical proof, but is left solely to absolute discretion.]

The uncertainty (of measurement) is a parameter, associated with the result of a measurement, that characterizes the dispersion of the values that could reasonably be attributed to the measurand.

NOTE 1 The parameter may be, for example, a standard deviation (or a given multiple of it), or the half-width of an interval having a stated level of confidence.

NOTE 2 Uncertainty of measurement comprises, in general, many components. Some of these components may be evaluated from the statistical distribution of the results of series of measurements and can be characterized by experimental standard deviations. The other components, which also can be characterized by standard deviations, are evaluated from assumed probability distributions based on experience or other information.

NOTE 3 It is understood that the result of the measurement is the best estimate of the value of the measurand, and that all components of uncertainty, including those arising from systematic effects, such as components associated with corrections and reference standards, contribute to the dispersion.

This definition is identical with the definition given in the 2nd edition of the VIM (see VIM:1993, 3.9). In the current 3rd edition it now reads (VIM:2008, 2.26):

Definition 2.28 (Measurement uncertainty)
The measurement uncertainty is a non-negative parameter characterizing the dispersion of the quantity values being attributed to a measurand, based on the information used.

NOTE 1 Measurement uncertainty includes components arising from systematic effects, such as components associated with corrections and the assigned quantity values of measurement standards, as well as the definitional uncertainty. Sometimes estimated systematic effects are not corrected for but, instead, associated measurement uncertainty components are incorporated.

NOTE 2 The parameter may be, for example, a standard deviation called standard measurement uncertainty (or a specified multiple of it), or the half-width of an interval, having a stated coverage probability.

NOTE 3 Measurement uncertainty comprises, in general, many components. Some of these may be evaluated by Type A evaluation of measurement uncertainty from the statistical distribution of the quantity values from series of measurements and can be characterized by standard deviations. The other components, which may be evaluated by Type B evaluation of measurement uncertainty, can also be characterized by standard deviations, evaluated from probability density functions based on experience or other information.

NOTE 4   In general, for a given set of information, it is understood that the measurement uncertainty is associated with a stated quantity value attributed to the
measurand. A modification of this value results in a modification of the associated
uncertainty.

In the following the differences between the old and the new definition of the term
"measurement uncertainty" will be discussed. These differences, although only small,
clearly show the changes in the understanding of measurement uncertainty that have
taken place in the 15 years between the 2nd and the 3rd editions of the VIM. The new
edition is more in line with the thinking underlying the current version of the GUM.

The new definition emphasizes that the measurement uncertainty *is a non-negative
parameter* (this was, of course, already the case without explicitly mentioning it) *characterizing the dispersion of the quantity values being attributed to a measurand, based on the
information used*, whereas it was sufficient in the old definition that it *characterizes the
dispersion of the values that could reasonably be attributed to the measurand.* Therefore,
the meaning of the requirement "reasonably" is now stated more precise, namely that
the measurement uncertainty should be *based on the information* available. In addition,
the subjunctive "could … be attributed" has been replaced by "being attributed", in order to avoid the vagueness of the previous statement. The reference to the *information
used* makes it clearer now that the measurement uncertainty *reflects the lack of exact
knowledge of the value of the measurand* (see GUM, 3.3.1).

In the new definition, the measurement uncertainty is no longer a *parameter, associated with the result of a measurement*, but is *associated with a stated quantity value
attributed to the measurand* (VIM:2008, 2.26, Note 4). This change is a sensible correction, because the complete measurement result consists of the stated quantity value *and*
the associated measurement uncertainty. It also makes clear that the measurement is
associated with the quantity *value*, not with the quantity itself. This should, by the way,
intuitively be clear, because a quantity itself is not uncertain, we are merely uncertain
about its value. Thus, it makes no sense to talk about, for example, the measurement
uncertainty of a voltage—as it unfortunately frequently happens—but only about the
measurement uncertainty of the *value* of a voltage of, for example, 1.5 V. Hence, it becomes understandable why a modification of a quantity value *results in a modification
of the associated uncertainty* (VIM:2008, 2.26, Note 4). A change of the evaluation model
may also lead to a changed quantity value and consequently inevitably to a different
associated measurement uncertainty. Therefore, if different evaluation models are used,
estimated quantity values and the associated measurement uncertainties are not directly
comparable.

In the new definition—as previously in the old one (VIM:1993, 3.9, Note 1)—the standard deviation is permitted as one possibility to determine the measurement uncertainty,
which, however, is now called *standard uncertainty* (VIM:2008, 2.26, Note 2). This corresponds to the definition of the *standard uncertainty of the result of a measurement
expressed as a standard deviation* as given in the GUM (GUM, 2.3.1). In contrast, the *half-*

width of an interval having a stated level of confidence (VIM:1993, 3.9, Note 1) has been replaced in the new definition by the *half-width of an interval, having a stated coverage probability* (VIM:2008, 2.26, Note 2). These words make it clear that the (expanded) measurement uncertainty corresponds *not* to a *confidence interval* in the statistical sense and *the term "confidence level" is not used in connection with this interval but rather the term "level of confidence"* (GUM, 6.2.2). It has been emphasized that the *terms "confidence interval" and "confidence level" are only applicable when certain conditions are met, including that all components of the uncertainty are obtained from Type A evaluations* (GUM, 6.2.2).

Note 2 of the old definition is essentially identical to Note 3 of the new definition. In addition, the new definition refers to *Type A evaluation of measurement uncertainty* as a *method of evaluation of uncertainty by statistical analysis of series of observations* (GUM, 2.3.2) and to *Type B evaluation of measurement uncertainty* as a *method of evaluation of uncertainty by means other than statistical analysis of series of observations* (GUM, 2.3.3). With this extension of the definition, consistency has been achieved between the VIM and the GUM.

Note 1 of the new definition is in agreement with Note 3 of the old definition inasmuch as both components resulting from systematic effects and components associated with corrections and assigned quantity values of measurement standards are now included in the determination of the measurement uncertainty. Moreover, the definitional uncertainty has been added as another component of the measurement uncertainty (VIM:2008, 2.26, Note 1). The term "definitional uncertainty" is defined as (VIM:2008, 2.27):

Definition 2.29 (Definitional uncertainty)
The definitional uncertainty is the component of measurement uncertainty resulting from the finite amount of detail in the definition of a measurand.

The definitional uncertainty is the smallest measurement uncertainty that can practically be achieved in any measurement of a given measurand. Any change in the descriptive details, however, leads to a different definitional uncertainty. In the GUM the term "intrinsic uncertainty" is used instead of the term "definitional uncertainty".

In the GUM, the definitional uncertainty is considered to be negligible in comparison to other components of measurement uncertainty. (VIM:2008, Introduction), although such disregard is not always acceptable.

For example, the geometric quantity "diameter" is not well-defined for every measurement problem, because it assumes the ideal circle as an evaluation model. In reality, however, there are always more or less pronounced deviations from circularity. Thus, the evaluation of a set of data points measured on a real circular object yields more than one possible diameter value, such as the distance between any two points on opposite sides of the circumference, the diameter of the largest inscribed or smallest circumscribed circle associated with the data set, or the diameter determined by the least squares method, to name just a few possibilities. The differences between these possible results of a diameter measurement can be significant contributors to the measurement uncertainty. Therefore, the addition of the definitional uncertainty in the new definition of measurement uncertainty is very useful, even though it is considered negligible in the GUM.

In the new definition of measurement uncertainty—in contrast to the old one—it is pointed out that estimated systematic effects that have not been corrected will lead to an additional contribution to the measurement uncertainty (VIM:2008, 2.26, Note 1). In general, corrections for systematic measurement errors should be applied. Only under *very special circumstances should corrections for known significant systematic effects not be applied to the result of a measurement* (GUM, 6.3.1, Note). However, it should be noted that in any case the *uncertainty* of the systematic measurement error contributes to the measurement uncertainty, even if the correction has been made as recommended by the GUM. This fact is immediately comprehensible because our knowledge of a systematic measurement uncertainty is always insufficient, as we have already pointed out in the previous section.

However, one point is still missing from the new definition of measurement uncertainty, namely *that the result of the measurement is the best estimate of the value of the measurand* (VIM:1993, 3.9, Note 3). It is therefore possible that the measurement uncertainty can be stated even in cases where the above statement is not valid. This is the case, for example, when no correction for a systematic measurement error has been applied and therefore the determined value of the measurand is *not* the best estimate. The dispersion of the values is then determined with respect to the uncorrected value of the measurand and is therefore characterized by an increased measurement uncertainty.

We have addressed all differences between the new and old definitions of measurement uncertainty in the VIM and GUM. In summary, the changes are mainly minor corrections and more precise wording, but the underlying philosophy has not changed.

3 Measurement models

The experiences are more or less
completely broken down into simpler,
more common elements and always
symbolized with a sacrifice of accuracy
for the purpose of communication.

(Ernst Mach, 1897)

The calculation of the value of a measurand and its uncertainty always requires an evaluation model that represents the knowledge of the influence of the known input quantities
on the output quantity to be determined. The model should include all relevant input
quantities, where individual input quantities may be described by sub-models. The models used in metrology are based almost solely on the laws of the natural sciences and can
therefore, in most cases, be easily described by one or more mathematical equations.

3.1 Modelling methods

In science and engineering, modelling by means of mathematical methods serves as an
adequate solution of problems and is often a prerequisite for systematic problem-solving.
Different modelling methods, adapted to the nature of the underlying tasks, are used
in various fields. It should be the aim of every good modelling to describe, as well as
possible, the aspects which are substantial for the solution of the respective problem,
however, in doing so *not* to take into account those aspects which are irrelevant to the
particular case under consideration.

The term model is used in various contexts with different meanings. A model can
be a copy of an already existing or still to be created original (e. g. a flight simulator
or a true-to-scale representation of a building) or it can serve as an attempt to explain
particular phenomena of the real world (e. g. models of physical theories, like the atomic
model of BOHR). Models are used, for example, in order to

- investigate, understand, or explain certain properties of complex relations (e. g.
  physical models),
- specify requirements for the manufacturing of a product (e. g. technical drawings
  or product specifications),
- illustrate the look and properties of an original (e. g. in architecture),
- verify that all essential properties of a complex system are described correctly and
  completely (e. g. the validation of a measurement system).
- perform actions which are either not possible with the original or cannot be performed any more (e. g. a simulation of the sequence of actions which possibly have
  caused an aircraft crash),

https://doi.org/10.1515/9783111453712-003

Models can and should never give a complete image of reality. They always include only certain characteristics and disregard others. The respective intended use of the model determines which characteristics have to be taken into account and which description is appropriate for that purpose. Some important items which should be taken into account during modelling are:
- local and global closeness,
- degree of detailing,
- degree of abstraction,
- static and dynamic performance,
- performance with respect to perturbations,
- system stability,
- mathematical complexity,
- numerical feasibility.

To develop a good model for a given problem is the most difficult task of measurement data evaluation and demands some skills and experiences. Unfortunately, there are no general rules of procedure for successful modelling. Often several comparable methods are available to choose from, i. e. there is not necessarily only *one* unambiguous modelling. However, depending on the intended use of the model, the information available about the system to be described, as well as the underlying mathematical methods of the modelling, constraints often arise which facilitate the selection of a suitable method.

During modelling a difference is made between a more theoretical and a more practical approach. The decision for or against a particular method of modelling depends on the information available about the system.

Usually experimental modelling is used whenever a system is either not well known or it is not necessary or reasonable to describe all its details *in extenso*. In this case, the modelling includes all concepts which lead to a suitable system description and are solely based on the measured response of a system to given input signals. Thereby it is irrelevant whether the system performance can be described by specific equations or algorithms, or whether already existing measured data serve as a basis for the model. However, given that tabulated measured data can cause additional problems (e. g. discontinuities not present in reality or missing differentiability), a mathematical description will generally be preferable. In this case, a single equation or system of equations is given and the unknown model parameters are determined by an appropriate optimization method so that the deviations of the modelled data from the observed data are as small as possible.

An experimental model can always be adjusted to available measured data as closely as desired (for example by choosing the model order, if a polynomial approach is used). But it is very important to ensure that the quality of approximation of the model to the measured data is not carried too far, and thus existing *random* measurement errors become part of the model. However, in the case of *systematic* measurement errors this might be desired.

An experimental model can be *extrapolated* only in a few cases, i. e. it is normally not usable outside the range covered by the measured data. The strength of such models lies mainly in the *interpolation* of the measured data. Moreover, for any other system the entire optimization procedure necessary to determine the model parameters must be repeated in most cases, even if the systems are similar. The model parameters are often not interpretable, i. e. in many cases an understanding of the effective function of a system is not immediately possible.

Experimental modelling is sometimes called black-box modelling, because it is almost exclusively based on the measured response of a system to prescribed input signals, without the need of any knowledge concerning the inner structure or functionality of the system itself.

Theoretical or physical modelling describes a system on the basis of generally accepted physical theories. In this case all existing subsystems are usually described separately and subsequently combined to a more complex system. This usually leads to a better understanding of the system, since every quantity appearing in the model has a real physical meaning, but requires significantly more knowledge.

Theoretical modelling generally yields computationally demanding models, because physical descriptions are available for many effects which, however, in most cases lead to analytically not solvable problems and thus generally require approximations or a numerical treatment. Typical examples are dynamical systems, flow simulations, or the calculation of thermal or elastic material properties, which generally need time-consuming calculations.

Physical modelling enables a high reusability of models. In most cases no or only a few measured data are necessary in order to determine the model parameters. The information used is limited to characteristic quantities, such as specified geometrical dimensions, material properties, or physical constants, which can be looked up in appropriate tables or scientific publications. Since a system description based on physical laws provides a high degree of transparency, this type of modelling is also known as white-box modelling, or in theoretical physics as modelling *ab initio*.

An essential difference between experimental and theoretical modelling can exist in the structure of the system under consideration. When using experimental modelling, the structure of the system is usually predefined and thus its dynamical characteristics are already determined in advance. On the other hand, when theoretical modelling is used, the structure of the model is only formed by the modelling process itself. In this case, improper simplifications of complex systems may sometimes cause problems with computability. In order to assure a simplified but nevertheless physical reasonable system description, the quantity equations used in the model have to obey the known laws of physics and, in addition, to fulfil certain mathematical constraints.

A restriction solely to theoretical or experimental modelling does not always lead to a satisfactory result. Therefore, a combination of both methods is often used. Such a method is sometimes called grey-box modelling. The main idea of this modelling is to use prior knowledge about a system in form of physical laws, as well as suitable empirical model parameters, which enable an optimal approximation of the whole system model to available measured data. These parameters may or may not be interpretable either physically or technically.

Grey-box modelling involves two conceptually different approaches. One possibility is to use physical modelling as a starting point and then consider certain quantities as model parameters that allow an optimal approximation of the whole model to existing measured data. The parameter values obtained in this way will generally not match the actual values of the quantities. In most cases, however, they describe the real system much better than would be the case without adjustment of the selected quantities. Furthermore, regardless of a better mathematical description, it is often possible to predict how a particular quantity will affect the system. In the case of such a system optimization, the actual values of the quantities are usually good initial values. This can be crucial for the success of the chosen optimization method, especially when non-linear models of high complexity are used.

In the case of a modular system, grey-box modelling provides an alternative to describing a particular subsystem using a black-box approach. This practice is often advisable when complex relationships between one or more parts of a large system need to be simplified in order to make a large model manageable.

Which modelling method should be applied in a particular case depends on one hand on the required precision of the model, but on the other hand also on the complexity of the underlying physical relations. If the physical description becomes very extensive, as in cases of complex systems, the degree of abstraction generally increases and thus the emphasis of the modelling process is usually shifted more in the direction of black-box modelling.

In order to illustrate the course of actions during modelling, we consider the model of a semiconductor diode. In this case, a mere theoretical modelling of the characteristics of the diode is not advisable, because the underlying physical processes are very complex.

If we restrict ourselves to modelling just one particular characteristic of a semiconductor diode, such as e. g. its direct current properties, we only need a simple non-linear model.

Example 3.1 (SHOCKLEY diode model)

The SHOCKLEY model of a semiconductor diode facilitates a very simple description of the current-voltage characteristic of such a diode. This model is already sufficient for most direct current applications.

The model equation named after W. B. SHOCKLEY, an American inventor and physicist,[1] uses an exponential function to describe the I-U characteristic of a semiconductor diode. This model provides—except of the so-called "reverse break-

---

[1] W. B. SHOCKLEY was the manager of a research group at *Bell Labs* that included J. BARDEEN and W. BRATTAIN. In 1956 they were jointly awarded the Nobel Prize in Physics for "their researches on semiconductors and their discovery of the transistor effect".

down"—a good description of the direct current characteristics of a semiconductor diode.

In the forward voltage range (positive voltage) we notice the exponential rising of the current with increasing voltage (see Fig. 3.1). In contrast, in the reverse voltage range (negative voltage) the reverse-blocking current asymptotically approaches the so-called "reverse bias saturation current" I_S.

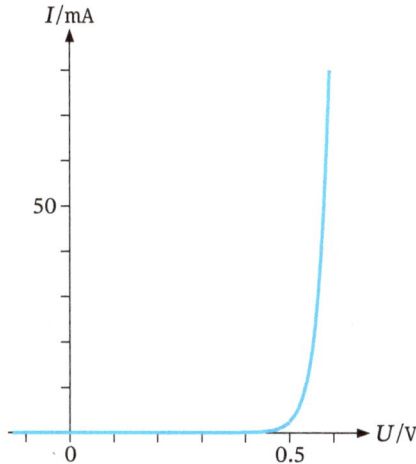

Fig. 3.1: Typical I-U characteristic of a semiconductor diode.

Although the SHOCKLEY model belongs to the ideal diode models, it is able to describe very well the reality of the direct current characteristic of a semiconductor diode in a wide range. For a certain voltage U the current I through the diode is approximately given by the SHOCKLEY equation

$$I = I_S \left[\exp\left(\frac{U}{n\,U_T}\right) - 1 \right]$$

where U_T denotes the thermal voltage (at 20 °C approximately 25 mV), n the emission coefficient (range 1 ... 4), and I_S the reverse bias saturation current (at 20 °C about 100 nA for germanium and 10 pA for silicon). The actual value of the reverse bias saturation current I_S, besides its already mentioned dependency on the semiconductor material, depends also on the area of the depletion layer, as well as on the temperature, and increases strongly with increasing temperature.

The thermal voltage is given by

$$U_T = \frac{k\,T}{e}$$

where k denotes the BOLTZMANN constant, e the elementary electric charge, and T the absolute temperature of the semiconductor.

In this example the emission coefficient and the reverse bias saturation current are typical model parameters, whose values have to be determined by measuring the characteristic of the respective diode type, in order to get a sufficiently good approximation of the diode model to reality. In contrast, the thermal voltage is given by a mere physical model.

It is very common to use hybrid models, like the SHOCKLEY diode model, in electronics to represent complex devices, like e. g. operational amplifiers or A/D-converters. Such models are usually provided by the manufacturer of the devices and may be found in the respective data sheet.

3.2 Model equations

In measurement data evaluation usually models are used which are given in form of one or more mathematical equations. These model equations can either be linear or non-linear.

In order to keep things simple, we shall restrict ourselves in this chapter to model equations where one single output quantity depends on several input quantities. This is also the approach used in the *Guide to the Expression of Uncertainty in Measurement (GUM)*. However, the extension of the mathematical model to more than one output quantity is straightforward.

We start our considerations with linear model equations. To begin with, we will look at two simple examples.

Example 3.2 (Resistive voltage drop in an electric power line)
An electric current is flowing though an electric power line with resistance R. According to OHM's law the resistive voltage drop U in the line is given by the equation

$$U = RI.$$

Here the resistivity R is considered as a model parameter whose value has been determined by another measurement and is thus assumed to be known.

Example 3.3 (Ceiling of an aeroplane)
An aeroplane is taking of at an aerodrome whose elevation above sea level is given by the quantity h_0. After take off the aeroplane climbs with the constant rate of climb v_s. After the time t it reaches the altitude

$$h = h_0 + v_s t$$

above sea level. The aerodrome elevation and the aeroplane's rate of climb are considered as model parameters and thus assumed to be known.

The examples 3.2 and 3.3 are special cases of a linear model. If only one input quantity and only one output quantity occur in the model, a linear model is mathematically given by an equation of the form

$$Y = aX + b,$$ (3.1)

where X denotes the input quantity and Y the output quantity of the model. The quantities a and b denote constant model parameters.

Beside the linear models with only one input and one output quantity there exist also linear models with only one output quantity which have more than one input quantity. Here we consider the following two examples.

Example 3.4 (Mass of a mixture of liquids)

Two liquids of different mass density are mixed. The volumes of the liquids are denoted by V_1 and V_2 and their respective mass densities by ρ_1 and ρ_2, respectively. Because of the law of conservation of mass, the mass m of the mixture is given by the equation

$$m = \rho_1 V_1 + \rho_2 V_2.$$

The quantities ρ_1 and ρ_2 are considered as model parameters whose values are assumed to be known. They could have been measured separately, for example, or they could have been looked up in an appropriate manual.

Example 3.5 (RICHMANN's mixture rule)

When several different liquids of the same volume, but of different temperature, are mixed the mixture temperature is determined by the mixture rule

$$T_{\mathrm{m}} = \sum_{i=1}^{n} a_i T_i,$$

stated in 1750 by G. W. RICHMANN,[2] where n denotes the number of liquids in the mixture, and T_i the temperature of the i-th liquid component. The coefficients a_i are given by

$$a_i = \frac{c_i \rho_i}{\sum\limits_{k=1}^{n} c_k \rho_k}, \qquad i = 1, \dots, n,$$

where c_i denotes the specific thermal capacity and ρ_i the mass density of the i-th liquid component. The coefficients a_i ($i = 1, \dots, n$) are assumed to be known, because they can be calculated from known quantities which can be found in a suitable handbook.

---

2 The mixing rule he discovered experimentally became the starting point for the later discovery of specific and latent heat.

Examples 3.4 and 3.5 are cases of the application of a general linear model with one output quantity and several input quantities. For such a model we can state the following definition:

Definition 3.1 (Linear model)
A linear model is mathematically given by the model equation

$$Y = \sum_{k=1}^{n} a_k X_k + b,\qquad(3.2)$$

where the n input quantities are denoted by X_k ($k = 1, \ldots, n$) and the output quantity by Y. The coefficients a_k ($k = 1, \ldots, n$) and b are constant model parameters that are assumed to be known.

The simpler linear model according to equation (3.1) can obviously be obtained from equation (3.2) for the case $n = 1$.

Unfortunately, linear models are exceptional cases in practice. Therefore, we now turn to non-linear model equations. To begin with, we are going to look again at some simple examples.

Example 3.6 (Area of a circular disk)
We consider a circular disk of diameter d. Its area is given by

$$A = \frac{\pi}{4} d^2,$$

i. e. the area is proportional to the square of the diameter, where the constant of proportionality $\pi/4$ is a known model parameter.

Example 3.7 (Volume of a straight circular cylinder)
We consider a straight circular cylinder of diameter d and height h. Its volume is given by

$$V = \frac{\pi}{4} d^2 h,$$

i. e. the volume is proportional to the product of the square of the diameter and the height of the cylinder, where the constant of proportionality $\pi/4$ is a known model parameter.

Example 3.8 (Mass density of a liquid)
The mass density of a liquid can be calculated from its mass m and its volume V according to the equation

$$\rho = \frac{m}{V}.$$

Since the volume generally changes with temperature, the mass density is depending on temperature. This temperature dependence could still be taken into account by extending the model equation, but here we are only interested in a simplified model for a particular fixed temperature, and so we do not consider the influence of the temperature.

Example 3.9 (Electric power loss in a wire)

An electric current I is flowing through a wire of length ℓ, diameter d, and specific resistance ρ. The electric power loss is then given by

$$P = \frac{4\rho\ell}{\pi} \left(\frac{I}{d}\right)^2 . \tag{3.3}$$

The power loss leads to a heating of the wire. In order to minimize this heating, the wires used in electric power lines are made of a material with a high specific conductivity (today usually copper or, in special cases, even silver) and their diameter is chosen according to the maximum current I_{max} to be expected, so that the ratio I_{max}/d is kept as constant as possible.

In this model the thermal expansion of the wire and the temperature dependence of the specific resistance is not taken into account, although this would in fact be necessary because of the heating of the wire caused by the electric power loss.

All these examples lead to non-linear models. For this type of model we can give the following general definition:

Definition 3.2 (Non-linear model)

A non-linear model is given by the model equation

$$Y = f(X_1, \dots, X_n),$$

where the n input quantities are denoted by X_k ($k = 1, \dots, n$) and the output quantity by Y. The relation between the input quantities and the output quantity is given by a non-linear function denoted by f. This function may also depend on an arbitrary number of constant model parameters.

We note that the example 3.1 of the SHOCKLEY model of a semiconductor diode considered at the end of the previous section also leads, as expected, to a non-linear model equation.

At the end of this section we shall deal with another type of model equation, called *implicit* model equation,[3] which is sometimes occurring in practice. For this purpose, we will look at the following example.

---

3 The *Guide to the Expression of Uncertainty in Measurement (GUM)* only deals with *explicit* model equations. It is even not mentioned that other possibilities exist.

Example 3.10 (Van der Waals equation)

The approximate equation of state for real gases (van der Waals equation) is given by

$$p - \frac{nRT}{V - nb} + \frac{n^2 a}{V^2} = 0 \, ,$$

where p denotes the pressure, V the volume, T the thermodynamic temperature, and n the number of moles of the gas. The universal gas constant R, the cohesive pressure a, and the co-volume b are known model parameters, which can be obtained from the scientific literature. However, only R can be regarded as a constant of nature, while the constants a and b are only valid for a particular gas at any given time.

If we consider the volume V, the pressure p, and the temperature T as input quantities to be determined by a measurement, the number n of moles of the gas as output quantity of the model is given by the equation

$$abn^3 - aVn^2 + (bp + RT)V^2 n - pV^3 = 0 \, . \tag{3.4}$$

This is a cubic equation for the unknown number n and therefore has three possible solutions. But we are only interested in a real positive solution, because the number of moles is a positive real number.

In this example the model equation (3.4) is not solved for the quantity n we are looking for. This is an example of an implicit model. In general, we can give the following definition:

Definition 3.3 (Implicit model)

An implicit model is given by the model equation

$$f(Y, X_1, \dots, X_n) = 0 \, ,$$

where the n input quantities are denoted by X_k ($k = 1, \dots, n$) and the output quantity by Y. The relation between the input quantities and the output quantity is given by a function, denoted by f, which is uniquely solvable for Y. This function may also depend on an arbitrary number of constant model parameters.

In this definition it is essential that it should be possible to solve the implicit model equation uniquely for the output quantity. This requirement is by no means self-evident, as there are implicit equations where this is not possible. In such a case, however, the model equation would be useless because neither the value of the output quantity nor the associated measurement uncertainty could be calculated straightforwardly from such a model.

The requirement of uniqueness of an existing solution of an implicit model equation $f(Y, X_1, \dots, X_n) = 0$ can, if necessary, always be assured by additional constraints. However, whether a solution for Y is actually possible can be decided by the *implicit function theorem* explained in every good calculus textbook. In most of the cases in which we are interested in the evaluation of measurement data, we can simplify this theorem to the following form:

Proposition 3.1 (Solvability of an implicit model equation)

Let $f(Y, X_1, \ldots, X_n) = 0$ be a continuously differentiable implicit model equation. The quantities $X_1, \ldots, X_n$ are known to take the values $x_1, \ldots, x_n$. If now $Y = y$ is a solution of the equation $f(Y, x_1, \ldots, x_n) = 0$ and if

$$\left(\frac{\mathrm{d}f(Y, x_1, \ldots, x_n)}{\mathrm{d}Y} \right)_{Y=y} \neq 0 \,,$$

then the implicit model equation $f(Y, X_1, \ldots, X_n) = 0$ is uniquely solvable for the quantity Y.

We demonstrate the application of this proposition by a simple example.

Example 3.11 (Semicircle in a plane)

A semicircle in the (x, y) plane above the x-axis, with its centre at the origin of the co-ordinate system, is given by the implicit equation

$$f(Y, X, R) = Y^2 + X^2 - R^2 = 0 \,, \quad \text{with constraint} \quad Y \geq 0 \,, \tag{3.5}$$

where X and Y denote the co-ordinates[4] of an arbitrary point located on the semicircle, and R its radius.

We assume the value of the quantity R, denoted by r, to be known. In addition, one particular value of the quantity X, denoted by x, is also assumed to be known. Then the value $y = \sqrt{r^2 - x^2}$ is the solution of the equation $f(Y, x, r) = 0$ under the given constraint. Moreover, we obtain

$$\left(\frac{\mathrm{d}f(Y, x, r)}{\mathrm{d}Y} \right)_{Y=y} = 2\sqrt{r^2 - x^2}$$

for the derivation at this point. This derivation can take the value zero, if and only if $x = \pm r$. Therefore, if the condition $-R < X < +R$ is fulfilled, equation (3.5) is uniquely solvable for Y.

If evidence can be provided that a given model equation is uniquely solvable for the output quantity Y, this does not mean, however, that we are actually able to write the model in the form of an explicit equation of the output quantity. An implicit equation can generally not be solved analytically for one of its variables. Usually we can only obtain a numerical solution.

---

4 Note that R, X and Y denote quantities, not quantity values, because equation (3.5) is a quantity equation, like all model equations.

Practitioners usually argue that since a model reflects the laws of nature underlying the respective measurements, it is not worth the trouble of mathematically proving the unique solvability of an implicit model equation. This argument, however, is not substantive, since any model is, of course, generally *not* a complete representation of the phenomena that actually occur in nature, i. e. there exists a definite possibility that a model could turn out to be wrong, or at least inappropriate, and that this fact will go unnoticed.

The possibility that a model could be mathematically inappropriate is one of the reasons to require a validation of each complex model before using it for the evaluation of measured data. This requirement is not only reasonable for implicit model equations but quite generally accepted. A validation might not be necessary, provided the correctness of a particular model equation has already been approved in practical applications.

3.3 Submodels

A complex model can almost always be represented more clearly if submodels are used. This allows a complex task to be broken down into easier-to-handle subtasks and also makes the calculation easier to understand. This can significantly reduce the risk of incorrect calculations. In addition, we make it easier for others to understand our thoughts, thereby enabling them to correct our calculations for any errors that may exist. In order to demonstrate the application of submodels, we use once more the model given in example 3.9.

> **Example 3.12 (Electric power loss in a wire; using submodels)**
> When an electric current I is flowing through a wire with resistance R, an electric power loss
>
> $$P = RI^2 \tag{3.6}$$
>
> arises, which leads to a heating of the wire.
> The resistance of a wire of length ℓ, cross-sectional area A, and specific resistance ρ can be calculated from the equation
>
> $$R = \rho \frac{\ell}{A}. \tag{3.7}$$
>
> Here we have not yet specified the actual cross-sectional shape of the wire.
> If the wire has a circular cross-section of diameter d, its cross-sectional area is given by (see example 3.6)
>
> $$A = \frac{\pi}{4} d^2. \tag{3.8}$$
>
> From equations (3.6), (3.7) and (3.8) we get again equation (3.3).

Equations (3.6) to (3.8) are the submodel equations of the model already stated in example 3.9. It is noticeable that exactly the same model was obtained, but in a more easily understandable form and at the same time the model equations of the submodels are simpler. We can also decide what to do if, for example, the wire has a rectangular cross-section rather than a circular one.

The utilization of submodels is very useful, in order to derive model equations from known physical phenomena or technical facts. We shall demonstrate this by deriving the model given in example 3.5.

Example 3.13 (Derivation of RICHMANN's mixture rule)

When several liquids with different temperatures are mixed, an equilibrium with an average temperature will be reached after a certain time. This average temperature, called mixture temperature, is created, because those portions of the liquid with a temperature higher than the mixture temperature release heat to the mixture, while those portions with a lower temperature absorb heat from the mixture.

We consider a mixture of n liquids with temperatures T_k ($k = 1, \ldots, n$), which represent the respective heat quantities Q_k ($k = 1, \ldots, n$) of the liquids. Since, due to the law of conservation of energy, stating that heat can neither be generated nor disappear during the mixture, the sum of all heat quantities, respectively released to or absorbed from the mixture, has to be zero, i.e. we have to require

$$\sum_{k=1}^{n} Q_k = 0. \tag{3.9}$$

The heat quantity Q_k absorbed or released by a certain amount of liquid is proportional to its mass m_k, as well as to the difference of its temperature T_k and the mixture temperature, with a constant of proportionality c_k being equal to the specific heat capacity of the respective liquid portion. We thus obtain the equations

$$Q_k = m_k c_k (T_k - T_m), \qquad k = 1, \ldots, n. \tag{3.10}$$

where T_m denotes the resulting mixture temperature. We can therefore conclude, that we get a positive quantity of heat for a hot liquid portion (i.e. if $T_k > T_m$), which will thus be released to the mixture, while we get a negative heat quantity for a cold liquid portion (i.e. if $T_k < T_m$), which will thus be absorbed from the mixture. But if the temperature of a liquid portion is equal to the temperature of the mixture (i.e. if $T_k = T_m$), its heat quantity is zero, i.e. heat will neither be released to nor absorbed from the mixture by this particular liquid portion.

If we now insert equation (3.10) into equation (3.9) and subsequently solve for the mixture temperature T_m, we obtain the model equation

$$T_m = \sum_{i=1}^{n} a_i T_i, \tag{3.11}$$

with the coefficients

$$a_i = \frac{m_i c_i}{\sum\limits_{k=1}^{n} m_k c_k}, \qquad i = 1, \ldots, n.\tag{3.12}$$

The masses of the liquid portions can be calculated by means of the equations

$$m_i = \rho_i V_i, \qquad i = 1, \ldots, n.\tag{3.13}$$

from the respective volumes V_i and mass densities ρ_i.

Equations (3.11), (3.12) and (3.13) are equivalent to the model equations given in example 3.5, provided that the volumes of the liquid portions are equal.

We have shown by this example, how the model equations yielding RICHMANN's mixture rule can be obtained from well known physical laws. In doing so we have used several easily understandable submodels. From the derivation of the model equations, we can also see which mixing rules would result if other conditions were present, e. g. if the constitution of all liquids involved is equal but the volumes of the liquids are not equal.

Since the utilization of submodels is generally advantageous, this method should be used as much as possible to derive model equations, but also subsequently during the calculation of the measurement results and their associated uncertainties. A modular structure of the model facilitates also the reusability of submodels when new models need to be derived or existing models have to be adapted to changing conditions.

Over time, the decomposition of models into submodels yields a set of model equations for recurring subtasks, such as modelling of the thermal expansion in length or volume measurements, the consideration of the change in e. g. the specific electrical conductivity or any other material characteristics with temperature, the models of analogue or digital measurement systems, or the modelling of certain frequently occurring systematic measurement errors, such as zero point deviation, linearity error, drift, or hysteresis.

The advantage of a modularization during modelling is not only that we do not have to rethink proven methods over and over again, but also that we can reuse proven and verified submodels at any time. This makes any modelling process faster and less error-prone.

3.4 Modelling strategies

The main problem when evaluating measurement data is often the specification of a suitable model. Once the model is known, the actual calculations are usually left to a specially designed software package.

Modelling is not possible without a conscious involvement of the human mind, i. e. not by a purely mechanical way through a more unconscious action—like, for example, walking—without a deeper consideration of the phenomena or influences involved in the process being modelled.

But the human mind is unable to understand complex systems that are influenced by many factors. Nonlinear relationships demand too much of the human imagination, although most of the time we are not aware of this fact. Therefore, predictions about the behaviour of complex systems are almost always doomed to failure. Even brilliant brains are generally not able to comprehend more than simple cause-effect relations (see e. g. [15]). That also explains why the behaviour of complex software is sometimes surprising, even if it is a self-coded program.

Based on this insight we can conclude that specifying a good model is by no means a routine job. On the contrary, modelling is more an art than a science. We learn it best—like calculating or a craft—by doing it over and over again. It is also recommended to ask experienced colleagues for advice and assistance. This is even advisable for experts in the respective field of the intended application of the model, because to know something about a matter is certainly good, but to know too much about it can also be a hindrance, because one sometimes deals too much with marginal bagatelles. In addition, a continuous exchange of experiences with colleagues helps not to overlook essential aspects of a complex problem.

Modelling complex systems usually requires a competent knowledge of mathematics. Sometimes it is necessary to involve mathematicians with knowledge of practical and numerical mathematics. Complex models can often only be developed and verified through the collaboration of various experts.

Before actually starting with modelling it often makes sense to study the relevant specialist literature, because it is generally better to learn from the experience of others. In certain fields collections of model examples already exist (see e. g. [16, 17]), which can either directly be used or accordingly be modified, in order to suit a particular purpose.

In some cases it may also be reasonable to have a look at similar problems which have already been addressed in other areas. You may come across ideas that you can apply to your own modelling. Models of different disciplines often have a very similar mathematical structure. This knowledge has been utilized in the past in modelling of various systems with the aid of electrical circuits (analogue computers). Unfortunately, such special knowledge has already largely been forgotten today. A review or a discussion with senior colleagues might be helpful in restoring some of those skills, because the usage of analogies is still applicable in the world of digital computers.

The design of a model is a process which can be subdivided into several steps. The temptation to specify the whole model in a single step should be resisted. It is better to start with a simple model, which initially only describes the essential relations determined by the chosen measurement procedure (ideal model), and subsequently gradually refine this model until the required accuracy of approximation to reality has been accomplished. But it has always to be kept in mind that the model should describe the reality only to a certain extent. The more parameters (e. g. material constants, etc.) which need to be estimated or determined in advance (e. g. by consulting appropriate handbooks), the higher the risk that small deviations will be magnified, that the model will become unrealistic, and that it will no longer be able to fulfil the task assigned to it. A more complex model does not necessarily increase the accuracy of the approximation, but it always increases the number of parameters.

The quality of a model cannot be determined *a priori*, but it should be considered as a rule that a simpler model is better than a more complex one,[5] as long as it serves the intended purpose. This requirement, however, does not preclude the possibility that more than one suitable model may exist for a particular task, and often does, i. e. models are not necessarily unique for a given problem.

Sometimes it is necessary to model very complex systems. In this case a systematic approach becomes inevitable and the use of submodels is indispensable. Here we should also start initially with an ideal cause-effect relation, but for each submodel separately. Subsequently, all essential influence quantities, starting with the most important one, are taken into account step-by-step. The individual submodels may be developed by various experts or, alternatively, existing submodels can be used, which shall be adjusted to predefined conditions, if necessary. After the design of the submodels is completed, they are assembled, in order to yield the final model, whereupon any significant feedback (i. e. the "cause" is influenced by the "effect") between the submodels needs also to be taken into account, as far as applicable. Finally, the whole model needs to be validated. This is best done through a computer simulation. It is important to ensure that *all* components of the model are tested and that the model as a whole can be implemented.

It is difficult to judge the quality of a model. In extreme cases, models can be either too complex or too simple. If they are too complex, many model parameters need to be estimated, and it may be impossible to provide the necessary information. Such models are usually developed by experts who want to include all known aspects of a problem. They often go too far. At the other extreme are models that are too simple. They can lead to incorrect results when evaluating measurement data, because they do not adequately represent the reality by ignoring essential features. Oversimplified models are often the result of insufficient knowledge of the measurement procedure and the effective influence factors, or simply because one does not want to put in too much effort. In the latter case, one is satisfied with a very general model, leaving out important details that differentiate between different measurement procedures. As usual, an optimal model can be found somewhere between the two extremes.

It is not unusual for a model to be developed iteratively, as is common in the natural sciences. Here, the information gained from measurements leads to a better understanding of reality and thus to a model that is better adapted to the new knowledge. Measurements, evaluation of the measured data and refinement of the model occur cyclically, until eventually—at least temporarily—no new knowledge can be gained. Then the model has evolved into a scientific theory, valid until discrepancies arise that force

---

5 Law of parsimony, usually attributed to W. v. OCCAM, and therefore also known as "Occam's razor".

a better adapted model of reality. Every model can only be valid in the context of the available information, which will never be complete.

3.5   Linear approximation

Under certain conditions, nonlinear models can be linearized, i. e. they can, in a first approximation, be expressed sufficiently well by linear model equations. In the *Guide to the Expression of Uncertainty in Measurement (GUM)* only linear or linearized models are used. The reason is that measurement uncertainty calculations become particularly simple if linear or linearized model equations are available. Therefore, we will deal in this section with two directly connected questions, namely, whether and how model equations can be linearized, and which conditions have to be taken into account when doing so.

We start our considerations with the simplest case and assume that the output quantity Y depends only on one single input quantity X, i. e. we deal with the non-linear model equation $Y = f(X)$. Our goal is to approximate this equation as best as possible by the linear equation $Y = aX + b$, where a and b are real constants that need to be determined.

We can, of course, not expect that for every value of X a linearization of the model equation with a negligibly small deviation is always possible. Therefore, we restrict the linear approximation of the model to the neighbourhood of a particular point on the curve represented by the equation $Y = f(X)$. This point, with the co-ordinates X_0 and $Y_0 = f(X_0)$, will be called operating point[6] in the following and denoted by P.

In order to determine the constants a and b of the model equation, linearized at the operating point, we use the approach

$$f(X) = aX + b + R(X),\qquad(3.14)$$

where $R(X)$ denotes the remainder, being the deviation of the linear approximation from the original model equation at a point denoted by X.

It is reasonable to require that at the operating point the linear approximation $Y = aX + b$ gives the same result as the original model equation $Y = f(X)$, i. e. we require $R(X_0) = 0$. If we set $X = X_0$ in equation (3.14) and use the stated requirement, we obtain

$$b = f(X_0) - aX_0.\qquad(3.15)$$

Substituting this result into the equation (3.14) gives for the relative deviation, related to the distance $\Delta X = X - X_0$ from the operating point measured along the X axis,

$$r(X) = \frac{R(X) - R(X_0)}{X - X_0} = \frac{f(X) - f(X_0)}{X - X_0} - a,\qquad(3.16)$$

---

6 In mathematics this point is also called the *centre* of the approximation.

where we again made use of the requirement $R(X_0) = 0$. The first term on the right side of this relation is the slope of a secant of the function $f(X)$ through the operating point P and another point Q in its immediate neighbourhood on the curve described by $Y = f(X)$, where we assume the difference $\Delta X = X - X_0$ to be small. This situation is depicted in Fig. 3.2.

Now we imagine that the deviation ΔX becomes smaller and smaller. Then the point Q is moving along the curve closer and closer to the point P, until finally the two points coincide and at the same time the secant turns into the tangent of the function $f(X)$ at the point P. If the function is differentiable at this point, the slope of the tangent is the first derivative

$$f'(X_0) = \lim_{X \to X_0} \frac{f(X) - f(X_0)}{X - X_0} \tag{3.17}$$

at this point. With this result we obtain from equation (3.16)

$$r(X_0) = R'(X_0) = f'(X_0) - a\,, \tag{3.18}$$

i. e. the relative deviation at the operating point is equal to the first derivative of the remainder at this particular point. This derivative, however, only exists, if the function $f(X)$ is differentiable at the operating point.

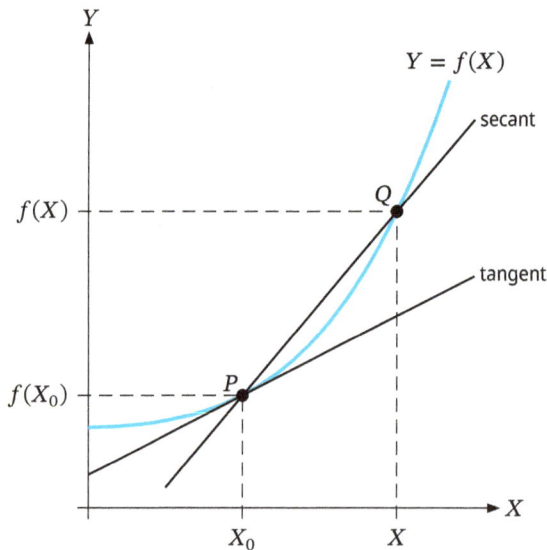

Fig. 3.2: Linearization of a non-linear equation $Y = f(X)$.

In order to obtain, as good as possible, a linear approximation at the operating point, we now require, in addition to $R(X_0) = 0$, that the relative deviation at the operating point is also zero, i. e. $r(X_0) = 0$. Thus, equation (3.18) yields

$$a = f'(X_0).$$
(3.19)

If we insert this result, as well as the result for the constant b from equation (3.15), into equation (3.14), we obtain

$$f(X) = f(X_0) + f'(X_0)(X - X_0) + R(X).$$

If the remainder $R(X)$ is negligible within a certain range of interest around the operating point, then the function $Y = f(X_0) + f'(X_0)(X - X_0)$ is a sufficiently good approximation of the non-linear model equation $Y = f(X)$ in that particular range, and we have achieved our goal.

Now we will examine under which conditions it is a good approximation to replace the non-linear model by the tangent at the operating point. To do this, we insert the result (3.19) into equation (3.16), which leads to the expression

$$r(X) = \frac{f(X) - f(X_0)}{X - X_0} - f'(X_0)$$
(3.20)

for the relative deviation. The first term on the right side of this equation is the slope of the secant passing through the operating point and another point denoted by X, and the second term is the slope of the tangent at the operating point. Hence, the relative deviation is equal to the difference of these two slopes. This difference is small, if the function $f(X)$ is only slightly curved and the respective point under consideration is not too far away from the operating point.

In order to get an impression of the magnitude of the deviation occurring by virtue of the linearization of a non-linear model function, we give two examples for the calculation of the remainder $R(X)$.

Example 3.14
We consider the quadratic function $f(X) = aX^2$. Then we have

$$\frac{f(X) - f(X_0)}{X - X_0} = \frac{aX^2 - aX_0^2}{X - X_0} = \frac{a(X + X_0)(X - X_0)}{X - X_0} = a(X + X_0).$$

According to equation (3.17) this yields

$$f'(X_0) = 2aX_0.$$

If we insert these results into equation (3.20) and subsequently multiply with the factor $(X - X_0)$, we obtain the remainder

$$R(X) = a(X - X_0)^2.$$

Example 3.15

We consider the reciprocal function $f(X) = a/X$. Then we have

$$\frac{f(X) - f(X_0)}{X - X_0} = \frac{1}{X - X_0}\left(\frac{a}{X} - \frac{a}{X_0}\right) = -\frac{a}{XX_0}.$$

According to equation (3.17) this yields

$$f'(X_0) = -\frac{a}{X_0^2}.$$

If we substitute these results into equation (3.20) and subsequently multiply with the factor $(X - X_0)$, we obtain the remainder

$$R(X) = \frac{a}{XX_0^2}(X - X_0)^2.$$

We notice that in both examples the remainder is proportional to the quadratic expression $(X - X_0)^2$. It can be demonstrated that this is generally the case (see e. g. [18]). Thus, the deviation introduced by linearizing a non-linear model function increases quadratically with the distance of the point under consideration from the operating point. This explains why a linearized model can only give useful results in a small neighbourhood of the operating point, i. e. a linear approximation is generally only locally applicable.

We summarize our considerations by the following proposition:

Proposition 3.2 (Linearization of a non-linear function)

A non-linear function $Y = f(X)$, differentiable at the point X_0, can be approximated there by the linear function

$$Y_{\text{lin}} = f(X_0) + f'(X_0)(X - X_0), \tag{3.21}$$

provided that the deviation represented by the remainder

$$R(X) = \left[\frac{f(X) - f(X_0)}{X - X_0} - f'(X_0)\right](X - X_0) \tag{3.22}$$

is negligibly small in the neighbourhood of X_0.

We demonstrate the application of this proposition by another example, but in this case we will explicitly state the operating point.

Example 3.16

In this example we are looking for a linear approximation of the non-linear function $f(X) = 1/(1 + X)$ at the operating point $X_0 = 0$. We first calculate the general expressions and subsequently insert $X_0 = 0$.

The slope of the secant through the operating point is given by

$$\frac{f(X) - f(X_0)}{X - X_0} = -\frac{1}{(1 + X)(1 + X_0)}. \tag{3.23}$$

Using relation (3.17) leads to the slope

$$f'(X_0) = -\frac{1}{(1+X_0)^2} \tag{3.24}$$

of the tangent at the operating point. With this result we obtain from equation (3.21) the linear approximation

$$Y_{lin} = \frac{1 + 2X_0 - X}{(1+X_0)^2}.$$

Inserting the results (3.23) and (3.24) into equation (3.22) yields the expression

$$R(X) = \frac{(X - X_0)^2}{(1+X)(1+X_0)^2}$$

for the remainder. Using now $X_0 = 0$ leads finally to the linear approximation

$$Y_{lin} = 1 - X,$$

with the remainder

$$R(X) = \frac{X^2}{1+X}.$$

It is easy to prove that summing Y_{lin} and $R(X)$ gives $f(X)$, as it should be.

In order to get an impression of the quality of the linear approximation for values in the neighbourhood of the operating point, we use $X = \pm 0.01$. For these values we get, within the range $-0.01 \leq X \leq 0.01$, an order of magnitude $R(X) \leq 10^{-4}$ for the remainder, i. e. the approximation is sufficiently good.

So far we have only looked at the linearization of model equations where the output quantity depends solely on a single input quantity. However, this is the exception rather than the rule. In general, the output quantity Y depends on several input quantities $X_1, \dots, X_n$, i. e. in most cases we are dealing with—usually non-linear—model equations of the form $Y = f(X_1, \dots, X_n)$. Therefore, it is necessary to think about the linearization of such model equations.

The aim is to approximate the model equation $Y = f(X_1, \dots, X_n)$ at a given point as well as possible by the linear function $Y = c_0 + c_1 X_1 + \cdots + c_n X_n$, where the constants $c_0, \dots, c_n$ have to be determined.

For the same reason as before, we restrict the linearization of the model equation to the neighbourhood of a point given by $X_1 = \xi_1, X_2 = \xi_2, \dots, X_n = \xi_n$ which we again call operating point. We start with the approach

$$f(X_1, \dots, X_n) = c_0 + \sum_{i=1}^{n} c_i X_i + R(X_1, \dots, X_n), \tag{3.25}$$

where the remainder $R(X_1, \dots, X_n)$ describes the deviation of the linear approximation from the original model function.

It is again reasonable to require that the linear approximation at the operating point leads to the same result as the original model equation, i. e. we require $R(\xi_1, \dots, \xi_n) = 0$. With this requirement

$$c_0 = f(\xi_1, \dots, \xi_n) - \sum_{i=1}^{n} c_i \xi_i$$

results from equation (3.25). If we insert this result into equation (3.25), we obtain

$$f(X_1, \dots, X_n) = f(\xi_1, \dots, \xi_n) + \sum_{i=1}^{n} c_i(X_i - \xi_i) + R(X_1, \dots, X_n). \tag{3.26}$$

We set in this equation $X_i = \xi_i$ $(i = 1, \dots, n)$ for all co-ordinates, except X_k for an arbitrary index k, in order to determine the constants c_i $(i = 1, \dots, n)$. This yields

$$f(\xi_1, \dots, X_k, \dots, \xi_n) = f(\xi_1, \dots, \xi_n) + c_k(X_k - \xi_k) + (X_k - \xi_k)r(X_k),$$

with

$$r(X_k) = \frac{R(\xi_1, \dots, X_k, \dots, \xi_n) - R(\xi_1, \dots, \xi_n)}{X_k - \xi_k}, \tag{3.27}$$

where we have used the requirement $R(\xi_1, \dots, \xi_n) = 0$.

By comparison of equations (3.16) and (3.27) we deduce, that the expression $r(X_k)$ describes the relative deviation of the linear approximation from the original model equation, when moving away from the operating point along a direction given by increasing or decreasing values of X_k, while the values of all other co-ordinates X_i remain unchanged.

Solving equation (3.27) for the relative deviation $r(X_k)$, we obtain

$$r(X_k) = \frac{f(\xi_1, \dots, X_k, \dots, \xi_n) - f(\xi_1, \dots, \xi_n)}{X_k - \xi_k} - c_k. \tag{3.28}$$

The first term on the right side of this equation is the slope of a straight line through the operating point P and another point Q in its neighbourhood on the curve of intersection with the (Y, X_k) plane, described by $Y = f(\xi_1, \dots, X_k, \dots, \xi_n)$, where we again assume the difference $\Delta X_k = X_k - \xi_k$ to be small. The straight line through the points P and Q can be considered as secant of the function $f(\xi_1, \dots, X_k, \dots, \xi_n)$ solely depending on X_k. This corresponds to a similar situation as depicted in Fig. 3.2, if therein X is replaced by X_k and X_0 is replaced by ξ_k.

Now we imagine that the deviation ΔX_k becomes smaller and smaller and thus the point Q moves further and further along the curve of intersection towards the point P, until both points finally coincide and, at the same time, the secant turns into the tangent of the function $f(\xi_1, \dots, X_k, \dots, \xi_n)$ at the operating point P. If the function is differentiable at this point, the slope of the tangent is given by the partial derivative[7]

---

7 The partial derivative of a function depending on more than one quantity is calculated in the same way as an ordinary derivative, but the values of all quantities are kept constant, except for the respective quantity under consideration.

with respect to X_k, i. e. by

$$
\left[\frac{\partial f(X_1, \dots, X_n)}{\partial X_k}\right]_{X_1=\xi_1,\dots,X_n=\xi_n} =
$$

$$
\left[\lim_{X_k\to\xi_k} \frac{f(\xi_1, \dots, X_k, \dots, \xi_n) - f(\xi_1, \dots, \xi_k, \dots, \xi_n)}{X_k - \xi_k}\right]_{X_1=\xi_1,\dots,X_n=\xi_n}.
$$

(3.29)

Using this result we obtain from equations (3.27) and (3.28)

$$
r(\xi_k) = \left[\frac{\partial R(X_1, \dots, X_n)}{\partial X_k}\right]_{X_1=\xi_1,\dots,X_n=\xi_n} =
$$

$$
\left[\frac{\partial f(X_1, \dots, X_n)}{\partial X_k}\right]_{X_1=\xi_1,\dots,X_n=\xi_n} - c_k,
$$

(3.30)

i. e. the relative deviation concerning a change of X_k at the operating point is equal to the first partial derivative of the remainder at this point with respect to X_k. This derivative, however, exists only, if the function $f(X_1, \dots, X_n)$ is partial differentiable with respect to X_k at the operating point.

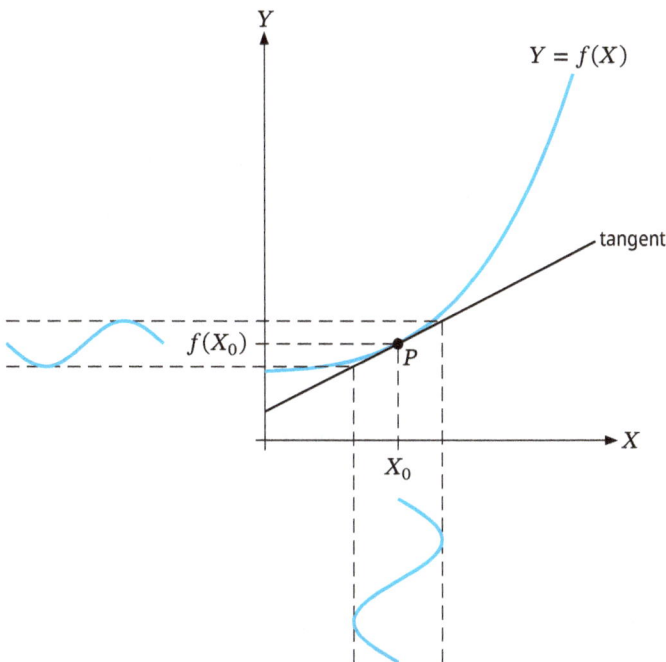

Fig. 3.3: Sensitivity of the value of the output quantity Y to small changes in the value of the input quantity X.

In order to obtain, as well as possible, a linear approximation at the operating point, we now require in addition to $R(\xi_1, \ldots, \xi_n) = 0$ that the relative deviation at the operating point is also zero, i. e. $r(\xi_k) = 0$. Thus, equation (3.30) yields

$$c_k = \left[\frac{\partial f(X_1, \ldots, X_n)}{\partial X_k} \right]_{X_1=\xi_1,\ldots,X_n=\xi_n} . \tag{3.31}$$

Since during our considerations we have used an arbitrary index k, equation (3.31) is applicable for the calculation of the coefficients c_k for all indices k.

The coefficients c_k $(k = 1, \ldots, n)$ are called sensitivity coefficients in the *Guide to the Expression of Uncertainty in Measurement (GUM)*, because they can be interpreted as a measure of the sensitivity of the value of the output quantity with respect to small changes of the value of the respective input quantity. This is illustrated in Fig. 3.3 for a periodically changing input quantity.

The larger the slope of the tangent of the respective input quantity under consideration at the operating point, the greater is the effect of a change of its value on the value of the output quantity. Therefore, the sensitivity coefficients may be considered as a kind of amplification factors.

The linear approximation of the function $Y = f(X_1, \ldots, X_n)$ is given by equation (3.26), with sensitivity coefficients to be calculated according to relation (3.31). For the deviation of the approximation, described by the remainder, we obtain from equation (3.26) the expression

$$R(X_1, \ldots, X_n) = f(X_1, \ldots, X_n) - f(\xi_1, \ldots, \xi_n) - \sum_{i=1}^{n} c_i(X_i - \xi_i) .$$

We are now able to generalize proposition 3.2 to functions which depend on an arbitrary number of input quantities:

Proposition 3.3 (Linearization of an explicit model function)
An explicit model function $Y = f(X_1, \ldots, X_n)$, which is differentiable at the point $X_1 = \xi_1, \ldots, X_n = \xi_n$, can be approximated at this point by the linear function

$$Y_{\text{lin}} = f(\xi_1, \ldots, \xi_n) + \sum_{i=1}^{n} c_i(X_i - \xi_i), \tag{3.32}$$

with the coefficients

$$c_k = \left[\frac{\partial f(X_1, \ldots, X_n)}{\partial X_k} \right]_{X_1=\xi_1,\ldots,X_n=\xi_n} , \qquad k = 1, \ldots, n, \tag{3.33}$$

provided that the deviations represented by the remainder

$$R(X_1, \ldots, X_n) = Y - Y_{\text{lin}} \tag{3.34}$$

are negligibly small in the neighbourhood of $X_1 = \xi_1, \ldots, X_n = \xi_n$.

In order to demonstrate how the calculations work, we apply this proposition to a simple non-linear function, namely a weighted sum of squares. Such functions often occur in practical applications.

Example 3.17

We consider the non-linear function

$$f(X_1, \ldots, X_n) = \sum_{i=1}^{n} a_i X_i^2 \,. \tag{3.35}$$

By substitution into relation (3.33) and application of equation (3.29) we obtain the coefficients

$$c_k = \left[\frac{\partial f(X_1, \ldots, X_n)}{\partial X_k} \right]_{X_1=\xi_1, \ldots, X_n=\xi_n} =$$

$$\left[\lim_{X_k \to \xi_k} \frac{f(\xi_1, \ldots, X_k, \ldots, \xi_n) - f(\xi_1, \ldots, \xi_k, \ldots, \xi_n)}{X_k - \xi_k} \right]_{X_k=\xi_k} =$$

$$\left[\lim_{X_k \to \xi_k} a_k \frac{X_k^2 - \xi_k^2}{X_k - \xi_k} \right]_{X_k=\xi_k} = \left[\lim_{X_k \to \xi_k} a_k \frac{(X_k + \xi_k)(X_k - \xi_k)}{X_k - \xi_k} \right]_{X_k=\xi_k} =$$

$$\left[\lim_{X_k \to \xi_k} a_k (X_k + \xi_k) \right]_{X_k=\xi_k} = 2 a_k \xi_k \,.$$

With this result we get from equation (3.32)

$$Y_{\text{lin}} = \sum_{i=1}^{n} a_i \xi_i (2 X_i - \xi_i),$$

and by application of equations (3.34) and (3.35) the remainder

$$R(X_1, \ldots, X_n) = \sum_{i=1}^{n} a_i (X_i - \xi_i)^2 \,.$$

From this equation we conclude that $\xi_i - 0.01/\sqrt{n a_i} \leq X_i \leq \xi_i + 0.01/\sqrt{n a_i}$ can be allowed, so that $R(X_1, \ldots, X_n) \leq 10^{-4}$ is valid.

We still have to handle the case where the model equation is not given explicitly, but rather by the implicit function $f(Y, X_1, \ldots, X_n) = 0$, which depends non-linearly on the input quantities $X_1, \ldots, X_n$ and the output quantity Y. For this purpose we consider the following example.

Example 3.18

A hemisphere above the (X_1, X_2) plane, with its centre at the origin of the co-ordinate system and radius R, can be described by the implicit equation

$$f(Y, X_1, X_2) = Y^2 + X_1^2 + X_2^2 - R^2 = 0, \tag{3.36}$$

with the constraint

$$Y \geq 0. \tag{3.37}$$

We choose the operating point $(X_1 = \xi_1, X_2 = \xi_2)$ with the additional constraints $-R < \xi_1 < +R$ and $-R < \xi_2 < +R$. For this point we obtain from equation (3.36) and with the condition (3.37) the solution

$$y = \sqrt{R^2 - \xi_1^2 - \xi_2^2}. \tag{3.38}$$

In addition, equation (3.36) yields

$$\left[\frac{\mathrm{d}f(Y, X_1, X_2)}{\mathrm{d}Y} \right]_{Y=y} = 2\sqrt{R^2 - \xi_1^2 - \xi_2^2} \neq 0.$$

Thus, according to proposition 3.1, the equation (3.36) is uniquely solvable for Y and is therefore, under the condition (3.37), a valid implicit model equation that is non-linear with respect to the input and output quantities.

We are looking for a linear approximation $Y = c_0 + c_1 X_1 + \cdots + c_n X_n$ of the implicit model equation $f(Y, X_1, \ldots, X_n) = 0$, assuming that this equation is uniquely solvable for the output quantity Y, because if this requirement is not met, the model would be unusable anyway. We restrict the linearization to the neighbourhood of a point given by $X_1 = \xi_1, X_2 = \xi_2, \ldots, X_n = \xi_n$, which we call the operating point as before.

Now we consider the explicit non-linear equation $Z = f(X_1, \ldots, X_{n+1})$, which becomes identical with the original equation $f(Y, X_1, \ldots, X_n) = 0$, if we use the replacements $Z = 0$ and $X_{n+1} = Y$. Analogous to equation (3.26), we start with the approach

$$f(X_1, \ldots, X_{n+1}) = f(\xi_1, \ldots, \xi_{n+1}) + \sum_{i=1}^{n+1} c_i(X_i - \xi_i) + R(X_1, \ldots, X_{n+1}), \tag{3.39}$$

with the coefficients

$$c_k = \left[\frac{\partial f(X_1, \ldots, X_{n+1})}{\partial X_k} \right]_{X_1 = \xi_1, \ldots, X_{n+1} = \xi_{n+1}}, \qquad k = 1, \ldots, n+1, \tag{3.40}$$

where the remainder $R(X_1, \ldots, X_{n+1})$ again describes the deviation of the linear approximation from the original model equation.

We now return to the original equation by setting $Z = 0$ and $X_{n+1} = Y$, i.e. we have $f(X_1, \ldots, X_{n+1}) = 0$ and $\xi_{n+1} = y$, where y is—in accordance with the requirements— the unique solution of the equation $f(y, \xi_1, \ldots, \xi_n) = 0$. In doing so, we obtain from equations (3.39) and (3.40)

$$\sum_{i=1}^{n} c_i(X_i - \xi_i) + c_{n+1}(Y - y) + R(Y, X_1, \ldots, X_n) = 0, \tag{3.41}$$

with the coefficients

$$c_k = \left[\frac{\partial f(Y, X_1, \ldots, X_n)}{\partial X_k}\right]_{Y=y, X_1=\xi_1, \ldots, X_n=\xi_n}, \qquad k = 1, \ldots, n,$$

and

$$c_{n+1} = \left[\frac{\partial f(Y, X_1, \ldots, X_n)}{\partial Y}\right]_{Y=y, X_1=\xi_1, \ldots, X_n=\xi_n}.$$

Since we have presupposed that the equation $f(Y, X_1, \ldots, X_n) = 0$ is uniquely solvable for Y, we have $c_{n+1} \neq 0$ according to proposition 3.1, i. e. we can divide equation (3.41) by c_{n+1} and subsequently solve for Y. Moreover, we can omit the output quantity Y in the expression for the remainder, because Y is a function of the input quantities and consequently the remainder is merely a function of the input quantities. Overall, our considerations lead to the following proposition:

Proposition 3.4 (Linearization of an implicit model function)

An implicit model function $f(Y, X_1, \ldots, X_n) = 0$, which is differentiable at the point $X_1 = \xi_1, \ldots, X_n = \xi_n$, can be approximated at this point by the linear function

$$Y_{\text{lin}} = y - \sum_{i=1}^{n} \frac{c_i}{c_{n+1}}(X_i - \xi_i), \tag{3.42}$$

with the coefficients

$$c_k = \left[\frac{\partial f(Y, X_1, \ldots, X_n)}{\partial X_k}\right]_{Y=y, X_1=\xi_1, \ldots, X_n=\xi_n}, \qquad k = 1, \ldots, n, \tag{3.43}$$

and

$$c_{n+1} = \left[\frac{\partial f(Y, X_1, \ldots, X_n)}{\partial Y}\right]_{Y=y, X_1=\xi_1, \ldots, X_n=\xi_n}, \tag{3.44}$$

where y denotes the solution of the implicit equation $f(y, \xi_1, \ldots, \xi_n) = 0$, provided that the deviations represented by remainder

$$R(X_1, \ldots, X_n) = Y - Y_{\text{lin}} \tag{3.45}$$

are negligibly small in the neighbourhood of $X_1 = \xi_1, \ldots, X_n = \xi_n$.

This proposition requires some additional remarks: In equation (3.42) we have to require $c_{n+1} \neq 0$. However, this requirement is—according to proposition 3.1—already guaranteed by the presupposed unique solvability of the implicit function $f(Y, X_1, \ldots, X_n) = 0$

with respect to Y. But this solvability does not necessarily mean that the output quantity Y and its value y at the operating point, occurring in equations (3.42) and (3.45), can actually be written as explicit functions of the input quantities $X_1, \ldots, X_n$ and their values $\xi_1, \ldots, \xi_n$ at the operating point. Thus, whenever possible, it is strongly recommended to solve the equation $f(Y, X_1, \ldots, X_n) = 0$ explicitly for Y and use proposition 3.3 for the linearization instead of proposition 3.4.

Proposition 3.4 is important whenever an explicit solvability with respect to the quantity Y is either impossible or inappropriate. In this case a linearization requires the application of numerical methods for the calculation of the value y as a solution of the equation $f(y, \xi_1, \ldots, \xi_n) = 0$. This applies also for the calculation of the quantity Y whose values we determine as solution of the equation $f(Y, X_1, \ldots, X_n) = 0$ for different values $X_1 = \xi_1 + h_1, \ldots, X_n = \xi_n + h_n$ in the neighbourhood of the operating point.

We demonstrate the application of proposition 3.4 by continuing example 3.18. In this case we are still able to find an analytical solution.

Example 3.19 (Continuation of example 3.18)

By insertion of equation (3.36) into relation (3.43) and application of equation (3.29) we obtain the coefficients

$$
c_1 = \left[\frac{\partial f(Y, X_1, X_2)}{\partial X_1} \right]_{Y=y, X_1=\xi_1, X_2=\xi_2} =
$$

$$
\left[\lim_{X_1 \to \xi_1} \frac{f(y, X_1, \xi_2) - f(y, \xi_1, \xi_2)}{X_1 - \xi_1} \right]_{X_1=\xi_1} =
$$

$$
\left[\lim_{X_1 \to \xi_1} \frac{X_1^2 - \xi_1^2}{X_1 - \xi_1} \right]_{X_1=\xi_1} = \left[\lim_{X_1 \to \xi_1} \frac{(X_1 + \xi_1)(X_1 - \xi_1)}{X_1 - \xi_1} \right]_{X_1=\xi_1} = 2\xi_1
$$

and

$$
c_2 = \left[\frac{\partial f(Y, X_1, X_2)}{\partial X_2} \right]_{Y=y, X_1=\xi_1, X_2=\xi_2} =
$$

$$
\left[\lim_{X_2 \to \xi_2} \frac{f(y, \xi_1, X_2) - f(y, \xi_1, \xi_2)}{X_2 - \xi_2} \right]_{X_2=\xi_2} =
$$

$$
\left[\lim_{X_2 \to \xi_2} \frac{X_2^2 - \xi_2^2}{X_2 - \xi_2} \right]_{X_2=\xi_2} = \left[\lim_{X_2 \to \xi_2} \frac{(X_2 + \xi_2)(X_2 - \xi_2)}{X_2 - \xi_2} \right]_{X_2=\xi_2} = 2\xi_2 .
$$

Insertion of equation (3.36) into relation (3.44) and application of equations (3.29) and (3.38) yields

$$c_3 = \left[\frac{\partial f(Y, X_1, X_2)}{\partial Y} \right]_{Y=y, X_1=\xi_1, X_2=\xi_2} =$$

$$\left[\lim_{Y \to y} \frac{f(Y, \xi_1, \xi_2) - f(y, \xi_1, \xi_2)}{Y - y} \right]_{Y=y} = \left[\lim_{Y \to y} \frac{Y^2 - y^2}{Y - y} \right]_{Y=y} =$$

$$\left[\lim_{Y \to y} \frac{(Y + y)(Y - y)}{Y - y} \right]_{Y=y} = 2y = 2\sqrt{R^2 - \xi_1^2 - \xi_2^2}.$$

If in turn these results are inserted into equation (3.42) and equation (3.38) is used, the linear approximation

$$Y_{\text{lin}} = \frac{R^2 - \xi_1 X_1 - \xi_2 X_2}{\sqrt{R^2 - \xi_1^2 - \xi_2^2}} \tag{3.46}$$

arises. Solving equation (3.36) for Y yields

$$Y = \sqrt{R^2 - X_1^2 - X_2^2}. \tag{3.47}$$

If the results (3.46) and (3.47) are inserted into equation (3.45), the expression

$$R(X_1, X_2) = \sqrt{R^2 - X_1^2 - X_2^2} - \frac{R^2 - \xi_1 X_1 - \xi_2 X_2}{\sqrt{R^2 - \xi_1^2 - \xi_2^2}} \tag{3.48}$$

for the reminder will be obtained.

The results (3.46) and (3.48) would, of course, have been obtained as well if the explicit function given by equation (3.47) had directly been linearized by application of proposition 3.3.

Finally, we will demonstrate the linearisation of an implicit model equation in a case where an explicit solution with respect to the output quantity is not possible, and thus numerical calculations are necessary.

Example 3.20
We consider the implicit equation

$$f(Y, X_1, X_2) = Y^5 - Y - X_1^2 - X_2^2 = 0. \tag{3.49}$$

It is well-know, that this equation is not explicitly solvable for Y.

We chose the operating point $(X_1 = 1, X_2 = 0)$. For this point we obtain from equation (3.49)

$$y^5 - y - 1 = 0,$$

with the numerical solution $y = 1.1673$ (this is the only real solution).

By insertion of equation (3.49) into relation (3.43) and application of equation (3.29) we obtain the coefficients

$$c_1 = \left[\frac{\partial f(Y, X_1, X_2)}{\partial X_1} \right]_{Y=y, X_1=1_1, X_2=0} =$$

$$\left[\lim_{X_1 \to 1} \frac{f(y, X_1, 0) - f(y, 1, 0)}{X_1 - 1} \right]_{X_1=1} =$$

$$\left[\lim_{X_1 \to 1} \frac{1 - X_1^2}{X_1 - 1} \right]_{X_1=1} = \left[\lim_{X_1 \to 1} \frac{(1 + X_1)(1 - X_1)}{X_1 - 1} \right]_{X_1=1} = -2$$

and

$$c_2 = \left[\frac{\partial f(Y, X_1, X_2)}{\partial X_2} \right]_{Y=y, X_1=1_1, X_2=0} =$$

$$\left[\lim_{X_2 \to 0} \frac{f(y, 1, X_2) - f(y, 1, 0)}{X_2} \right]_{X_2=0} = 0.$$

Insertion of equation (3.49) into relation (3.44) and application of equation (3.29) yields

$$c_3 = \left[\frac{\partial f(Y, X_1, X_2)}{\partial Y} \right]_{Y=y, X_1=1_1, X_2=0} =$$

$$\left[\lim_{Y \to y} \frac{f(Y, 1, 0) - f(y, 1, 0)}{Y - y} \right]_{Y=y} = \left[\lim_{Y \to y} \frac{Y^5 - Y - y^5 + y}{Y - y} \right]_{Y=y} =$$

$$\left[\lim_{Y \to y} (Y^4 + yY^3 + y^2Y^2 + y^3Y + y^4 - 1) \right]_{Y=y} = 5y^4 - 1.$$

If we insert these results into equation (3.42), we obtain the linear approximation

$$Y_{\text{lin}} = y + \frac{2}{5y^4 - 1}(X_1 - 1),$$

and with $y = 1.1673$ finally

$$Y_{\text{lin}} = 1.1673 + 0.2414(X_1 - 1).$$

This approximation does no longer depend on X_2.

In order to get an impression for the magnitude of the deviation of the linear approximation, we numerically calculate, using equation (3.49), for some values of the input quantities X_1 and X_2 the value of the output quantity Y and also, by using equation (3.45), the respective values of the remainder. The calculated results are given in the following table.

X_1	X_2	Y	$R(X_1, X_2)$
0.99	-0.01	1.1649	-0.0024
0.99	0	1.1649	-0.0024
0.99	+0.01	1.1649	-0.0024
1	-0.01	1.1673	$1.2 \cdot 10^{-5}$
1	0	1.1673	0
1	+0.01	1.1673	$1.2 \cdot 10^{-5}$
1.01	-0.01	1.1697	+0.0024
1.01	0	1.1697	+0.0024
1.01	+0.01	1.1697	+0.0024

We note that, for the ranges $0.99 \leq X_1 \leq 1.01$ and $-0.01 \leq X_2 \leq +0.01$, the value of the remainder $R(X_1, X_2)$ is virtually independent of X_2 and that its absolute value remains smaller than $2.4 \cdot 10^{-3}$.

The linearization of the model equations as shown in this section is based on the TAYLOR expansion, but only up to the linear term. This is the method proposed in the *Guide to the Expression of Uncertainty in Measurement (GUM)*. However, we have shown that such a linearization always requires the differentiability of the non-linear function at the operating point. Non-linear functions which do not fulfil this requirement cannot be linearized by an application of the TAYLOR expansion. An example is the function $f(X) = |X|$ which is not differentiable at the operating point $X = 0$. But even if the function under consideration is differentiable everywhere, it is still possible that the derivative at the operating point yields an infinite value and thus a linearization becomes impossible as well. This is true even for such a simple function as $f(X) = \sqrt{R^2 - X^2}$ at the operating point $X = R$, as the reader may verify as an exercise.

3.6   Quadratic approximation

In practice, there are sometimes cases where the sensitivity coefficients, obtained by a linear approximation of the model equation at the operating point, all become zero at the same time. We will see later that this would have the consequence that the respective uncertainty contributors also become zero. But since the uncertainty must generally be non-zero, a linear approximation of the model equation is useless in these cases, i. e. we must at least use a quadratic approximation. This is also necessary if the non-linearities of the model function are so large that the deviations of the linear approximation from the original model equation in the neighbourhood of the operating point cannot be considered as negligibly small.

As an example we consider a case which is of practical importance in length measurement, where the linear approximation of the model equation fails, and we must therefore proceed to a quadratic approximation:

Example 3.21 (Cosine error in length measurement)

The "cosine error" is a well known systematic error in precision length measurement. It occurs when the object whose length is to be measured is not properly aligned with the scale (see Fig. 3.4).

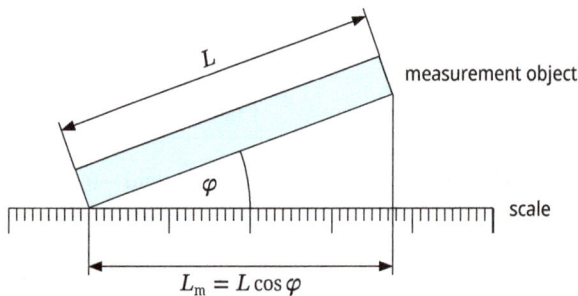

Fig. 3.4: Cosine error in length measurement.

The actual length L of the measurement object results from the measured length L_m according to the equation

$$L = \frac{L_\mathrm{m}}{\cos \varphi}. \tag{3.50}$$

Thus, we obtain the measurement error

$$\Delta L = L_\mathrm{m} - L = L(\cos \varphi - 1). \tag{3.51}$$

The equation (3.50) is our model equation with the output quantity L and the input quantities L_m and φ. We note that this is a non-linear model.

Since we are interested in the change of length in the vicinity of $\varphi = 0$, we decide to use this value as operating point for the linearization of our model equation. Applying proposition 3.3 yields the sensitivity coefficients

$$c_1 = \left(\frac{\partial \Delta L}{\partial L} \right)_{\varphi=0} = 0 \quad \text{and} \quad c_2 = \left(\frac{\partial \Delta L}{\partial \varphi} \right)_{\varphi=0} = 0. \tag{3.52}$$

It turns out that both sensitivity coefficients are zero at the operating point, and we therefore obtain the result $\Delta L = 0$ from the linearization, which obviously leads to an unusable model and is consequently unacceptable.

Whenever a linear approximation turns out to be unusable, it is recommended in the *Guide to the Expression of Uncertainty in Measurement (GUM)* to use a quadratic approximation instead. We will show how this approximation can be obtained for an explicit model equation $Y = f(X_1, \dots, X_n)$.

We choose $X_i = \xi_i$ ($i = 1, \dots, n$) as operating point and require that no deviation between the approximation and the original model equation shall occur at this point. This time we use the quadratic approach

$$Y = f(\xi_1, \dots, \xi_n) + \sum_{i=1}^{n} c_i (X_i - \xi_i)$$

$$+ \sum_{i=1}^{n} \sum_{j=1}^{n} C_{ij}(X_i - \xi_i)(X_j - \xi_j) + R(X_1, \dots, X_n), \tag{3.53}$$

where $R(X_1, \dots, X_n)$ denotes the remainder that describes the expected deviations of the quadratic approximation. It is obvious that $R(\xi_1, \dots, \xi_n) = 0$, i. e. the approach fulfils our requirement.

In the previous section we obtained the coefficients c_k ($k = 1, \dots, n$) by the demand $R(\xi_1, \dots, \xi_n) = 0$ and the additional requirement that the relative deviation between the approximation equation and the original model equation at the operating point must also be zero. This yielded (see equation (3.31))

$$c_i = \left[\frac{\partial f(X_1, \dots, X_n)}{\partial X_i} \right]_{X_1 = \xi_1, \dots, X_n = \xi_n}, \qquad i = 1, \dots, n. \tag{3.54}$$

It is reasonable to keep this condition here as well, because if the model equation is linear, the approach of equation (3.53) should still be valid, i. e. for a linear approximation this equation shall turn into expression (3.31). But this can only be the case if the relations (3.54) are valid.

We now calculate the first partial derivation of equation (3.53) with respect to the particular input quantity X_k according to the rules given in the preceding section (see also the footnote on page 60) and obtain

$$\frac{\partial f(X_1, \dots, X_n)}{\partial X_k} = c_k + \sum_{i=1}^{n} (C_{ik} + C_{ki})(X_i - \xi_i) + \frac{\partial R(X_1, \dots, X_n)}{\partial X_k},$$

$$k = 1, \dots, n. \tag{3.55}$$

Using requirement (3.54) yields at the operating point

$$\left[\frac{\partial R(X_1, \dots, X_n)}{\partial X_k} \right]_{X_1 = \xi_1, \dots, X_n = \xi_n} = 0, \qquad k = 1, \dots, n. \tag{3.56}$$

This result is in agreement with equation (3.30), as it should be.

If we now look somewhat closer at equation (3.55), we realize that the function $(\partial f(X_1, \dots, X_n)/\partial X_k)$ will be approximated linearly by this equation, i. e. we can apply proposition 3.3 to this function. If we require analogous to the relation (3.56)

$$\left[\frac{\partial^2 R(X_1, \dots, X_n)}{\partial X_k \partial X_l}\right]_{X_1=\xi_1, \dots, X_n=\xi_n} = 0, \qquad k, l = 1, \dots, n,$$

we obtain

$$\left[\frac{\partial^2 f(X_1, \dots, X_n)}{\partial X_k \partial X_l}\right]_{X_1=\xi_1, \dots, X_n=\xi_n} = C_{lk} + C_{kl}, \qquad k, l = 1, \dots, n.$$

If we finally set $C_{lk} = C_{kl}$, we achieve that the quadratic approximation of the model equation agrees with the respective TAYLOR polynomial of second degree. This specification is in agreement with the recommendations given in the *Guide to the Expression of Uncertainty in Measurement (GUM)*.

We summarize our considerations about a quadratic approximation of an explicit function by the following proposition:

Proposition 3.5 (Quadratic approximation of an explicit model function)
An explicit model function $Y = f(X_1, \dots, X_n)$ which is twice differentiable at the point $X_1 = \xi_1, \dots, X_n = \xi_n$, can be approximated there by the quadratic function

$$Y_{qu} = f(\xi_1, \dots, \xi_n) + \sum_{i=1}^{n} c_i(X_i - \xi_i) + \sum_{i=1}^{n}\sum_{j=1}^{n} C_{ij}(X_i - \xi_i)(X_j - \xi_j) \qquad (3.57)$$

with the coefficients

$$c_i = \left[\frac{\partial f(X_1, \dots, X_n)}{\partial X_i}\right]_{X_1=\xi_1, \dots, X_n=\xi_n}, \qquad i = 1, \dots, n,$$

and

$$C_{ij} = C_{ji} = \frac{1}{2}\left[\frac{\partial^2 f(X_1, \dots, X_n)}{\partial X_i \partial X_j}\right]_{X_1=\xi_1, \dots, X_n=\xi_n}, \qquad i, j = 1, \dots, n, \qquad (3.58)$$

provided that the deviation represented by the remainder

$$R(X_1, \dots, X_n) = Y - Y_{qu}$$

is negligibly small in the neighbourhood of $X_1 = \xi_1, \dots, X_n = \xi_n$.

We demonstrate the application of this proposition by continuing example 3.21, in order to obtain a reasonable approximation.

Example 3.22 (Continuation of example 3.21)

Now we calculate the quadratic approximation of the model equation (3.51). We have already shown that in doing so the coefficients c_1 and c_2 will not change, and we therefore can continue to use the results (3.52). By application of equation (3.58) we obtain the remaining coefficients

$$C_{11} = \left(\frac{\partial^2 \Delta L}{\partial L^2}\right)_{\varphi=0} = 0, \qquad C_{22} = \left(\frac{\partial^2 \Delta L}{\partial \varphi^2}\right)_{\varphi=0} = -\frac{L}{2}$$

and

$$C_{12} = C_{21} = \left(\frac{\partial^2 \Delta L}{\partial L \partial \varphi}\right)_{\varphi=0} = 0.$$

This yields, according to equation (3.57), the quadratic approximation[8]

$$\Delta L_{qu} = -\frac{L}{2}\varphi^2.$$

This is a reasonable result. The length deviation is proportional to the square of the tilt angle φ and, regardless of the sign of φ, is always negative, i. e. the measured length is always too small, as long as φ is non-zero.

The cosine error is a so-called second order error, because it depends quadratically on the deviation causing the error. Since this deviation is comparatively small, the length deviations to be expected are generally very small. The tilt angle $\varphi = \pm 5°$, for example, yields only a relative deviation $(\Delta L_{qu}/L) \approx -3.8 \cdot 10^{-3}$.

If we restrict ourselves to the range $-5° \leq \varphi \leq +5°$, we find $R \leq 1.2 \cdot 10^{-5}$ for the remainder. The approximation is therefore sufficiently good.

The requirements for the application of proposition 3.5 are considerably more restrictive than those presupposed in the case of a linearization of an explicit function by the application of proposition 3.3, since the function must be twice differentiable instead of only once. This means that in some particular cases neither a linear nor a quadratic approximation of a given model equation may be possible, because the first derivatives are all zero and the second derivatives do not exist at all. However, a quadratic approximation is also not possible if a second derivative becomes infinite at the operating point, as is the case e. g. for the function $f(X) = X^{3/2}$, which at the operating point $X = 0$ yields zero for the first derivative, and all higher derivatives turn out to be infinite at this particular point.

---

8 Note that in mathematics, the value of a plane angle is usually given in radians (rad), which is also the SI unit of plane angle. Using any other unit instead can cause problems. To convert from degrees to radians, or *vice versa*, use the conversion formulas

$1° = (\pi/180)\,\text{rad} \approx 0.017\,453\,\text{rad}$ and

$1\,\text{rad} = (180/\pi)° approx 57.295\,780°$,

respectively.

If neither a linear nor a quadratic approximation can be found for a particular model equation, because all first and second derivatives are zero at the operating point, we must, strictly speaking, proceed to the next higher term of the TAYLOR polynomial. This approach, however, is not mentioned in the *Guide to the Expression of Uncertainty in Measurement (GUM)* and is not recommended due to its disproportionate complexity. It is better to solve such a problem, if possible, by a small shift of the operating point. This practice is permissible because we are generally free to choose the operating point. Another possibility is to use other methods not covered in this book, e. g. Monte Carlo methods.

3.7 Graphical modelling

When modelling large and complex measurement problems, clarity can quickly be lost. This risk is reduced when submodels are used, but even then it can be difficult to make the structure of a model accessible from the model equations alone. In contrast, the human mind is very good at recognizing patterns and structures in graphical representations. It therefore seems reasonable to use graphical methods in modelling. We will show in this section that graph theory can be very helpful in this task.

Graph theory is a branch of mathematics concerned with the properties of graphs. Problems of very different kinds can be modelled using graphs. Graphs are abstract structures composed of simple elements that are used to represent information in terms of relations between objects.

Definition 3.4 (Directed graph)
A directed graph is an ordered pair $(\mathfrak{B}, \mathfrak{E})$ of a finite non-empty set $\mathfrak{B}$ and a relation $\mathfrak{E} \subseteq (\mathfrak{B} \times \mathfrak{B})$.
The elements of $\mathfrak{B}$ are called vertices and the elements of $\mathfrak{E}$ edges.

The edges of a directed graph, i. e. the elements of $\mathfrak{E}$, are ordered pairs of vertices, i. e. of the elements of $\mathfrak{B}$. We denote an edge by the ordered pair $(X, Y) \in \mathfrak{E}$, where $X \in \mathfrak{B}$ and $Y \in \mathfrak{B}$. We say, the edge (X, Y) connects the vertices X and Y and we call X the start vertex and Y the end vertex of the edge. If start vertex and end vertex are identical, the edge is called a loop.

A directed graph can be depicted (hence the term "graph") by representing the vertices by dots, circles, or any other symbols, which are pairwise connected by lines, where one end of each line is terminated by an arrow, in order to specify a direction (see Fig. 3.5). Which symbols are used for the vertices and how the edges are drawn, whether as curves or straight lines, or whether the lines cross each other or not, is at the discretion of the user and depends only on the convenience and aesthetics of the representation. The formal mathematical definition of a graph is independent of its graphical representation, which only serves as a visualization.

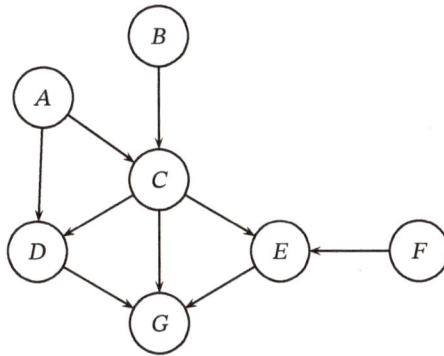

Fig. 3.5: A typical directed graph.

The graph shown in Fig. 3.5 consists of the set of vertices $\mathfrak{B} = \{A, B, C, D, E, F\}$ and the set of edges

$$\mathfrak{E} = \{(A, C), (A, D), (B, C), (C, D), (C, G), (C, E), (D, G), (E, G), (F, E)\}.$$

There are no loops in the graph.

In order to be able to differentiate between the vertices, we have assigned names to them. Hence, it is a named graph.

Definition 3.5 (Named graph)

A named graph $(\mathfrak{B}, \mathfrak{E})$ is a graph, where the set of vertices $\mathfrak{B}$ corresponds to a set of unique names.

The names of the vertices of the graph illustrated in Fig. 3.5 are the letters inscribed in the circles.

We notice that the graph shown in Fig. 3.5 is free of loops. No path returns to a previous vertex. The generalization of a loop is a cycle. In order to define this term, we need the definition of a path in a graph.

Definition 3.6 (Path)

A path in a directed graph is a sequence $(V_1, \dots, V_n)$ of mutually distinct vertices, where successive vertices V_i and V_{i+1} ($i = 1, \dots, n-1$) are connected by edges. The vertex V_1 is named initial vertex of the path and the vertex V_n final vertex.

In the directed graph which is shown in Fig. 3.5 there are a total of eight paths of different lengths, namely (A, D, G), (A, C, G), (A, C, D, G), (A, C, E, G), (B, C, G), (B, C, D, G), (B, C, E, G) and (F, E, G). Note that the order of the vertices in a path cannot be changed, as it is a directed graph and a path must therefore always follow the direction indicated by the arrows.

By means of definition 3.6 we can now define what a cycle in a graph is.

Definition 3.7 (Cycle)

A cycle is a path in a directed graph, where the initial vertex and the final vertex of the path are identical.

A cycle is thus a path in a graph which, starting from any vertex, returns to the same vertex where it started, whereby arbitrarily many vertices are visited on that path through the graph. A loop is a special cycle, where a path immediately returns to a vertex without visiting another vertex.

In the graph shown in Fig. 3.5 no cycle exits, because for none of its vertices there is a closed path returning to it.

By means of definitions 3.4 and 3.7 we can state the definition of a directed acyclic graph:

Definition 3.8 (Directed acyclic graph)

A directed acyclic graph is a directed graph without cycles.

The graph shown in Fig. 3.5 is a directed acyclic graph, because we have already recognized that there is no cycle in this graph.

We now turn to the classification of the vertices in a graph. For that purpose we need some additional definitions.

Definition 3.9 (Predecessor, successor, isolated vertex)

If an edge exists in a graph which connects the vertices X and Y, then X is called direct predecessor of Y and Y direct successor of X. A vertex which has neither a predecessor nor a successor is called isolated vertex.

Definition 3.10 (Connected graph)

A non-empty graph with no isolated vertex is called connected graph.

Thus, in a connected graph each vertex is at least connected with one other vertex by an edge. The graph shown in Fig. 3.5 is a connected graph, because none of its vertices is isolated.

Each vertex in a graph has a certain number of direct predecessors and direct successors, respectively. This number may also be zero.

Definition 3.11 (Input degree, output degree)

The input degree of a vertex is equal to the number of its direct predecessors.
The output degree of a vertex is equal to the number of its direct successors.

By means of this definition, we can strictly distinguish three different types of vertices in a connected graph:

- **input vertex**: vertex with input degree zero.
- **output vertex**: vertex with output degree zero.
- **inner vertex**: vertex which is neither an input vertex nor an output vertex.

In the graph shown in Fig. 3.5 we can classify the vertices as follows:

- input vertices: A, B, F;
- output vertex: G;
- inner vertices: C, D, E.

In order to distinguish between the types of vertices in a graph, we use a named, coloured vertex graph:

Definition 3.12 (Named, coloured vertex graph)

A named, coloured vertex graph $(\mathfrak{B}, \mathfrak{C})$ has a set of vertices $\mathfrak{B} \subseteq \mathfrak{N} \times \mathbb{N}$, where $\mathfrak{N}$ denotes a set of vertex names.

In a named, coloured vertex graph each vertex is an ordered pair of a name and a natural number. When representing a vertex we can also use different colours (hence the name "coloured vertex graph") or different symbols instead of natural numbers.

In order to represent the different types of vertices, we will use the symbolism shown in the following table.

Tab. 3.1: Types of vertices and the numbers assigned to them.

○	(1) input vertices
◎	(2) output vertices
○	(3) inner vertices

We say, that a vertex Y depends on a vertex X, if an edge (X, Y) exists which connects the two vertices. In order to register this dependency, we can mark the end vertices of the graph.

Definition 3.13 (Named, marked vertex graph)

A named, marked vertex graph $(\mathfrak{B}, \mathfrak{C})$ has a set of vertices $\mathfrak{B} \subseteq \mathfrak{N} \times \mathbb{N} \times \mathfrak{M}$, where $\mathfrak{N}$ denotes a set of vertex names and $\mathfrak{M}$ a set of vertex marks.

We will use the kind of marking shown in Fig. 3.6, in order to describe the functional dependency of a vertex on its direct predecessors.

Now we are able to define a dependency graph:

A dependency graph is a named, marked vertex, directed, acyclic graph.

$$X_2$$
$$X_1 \qquad X_3$$
$$Y \;-\;-\; f_Y(X_1, X_2, X_3)$$

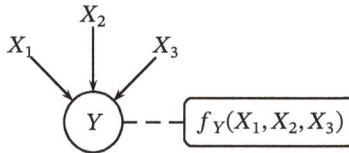

Fig. 3.6: Marking of a vertex.

Due to the definitions given above, a dependency graph is thus an ordered pair $(\mathfrak{B}, \mathfrak{E})$ of a finite non-empty set of vertices $\mathfrak{B} \subseteq \mathfrak{N} \times \mathbb{N} \times \mathfrak{M}$, where $\mathfrak{N}$ denotes a set of vertex names and $\mathfrak{M}$ a set of vertex marks, and a relation $\mathfrak{E} \subseteq (\mathfrak{B} \times \mathfrak{B})$. Moreover, the graph is connected, i. e. it contains no isolated vertices, and the graph is free of cycles, i. e. there is no path which, starting from one of the vertices, returns directly or indirectly to the same vertex.

Each vertex of a dependency graph corresponds to an ordered triple of a vertex name, an identification of the type of vertex and a marking. The vertex name is the name of the quantity assigned to the vertex, which is inscribed in the circle that symbolizes the vertex in the graph. The identification of the type of vertex is given by a natural number— or in its graphical representation by a symbol—according to Tab. 3.1. A function is used to mark an output vertex or an inner vertex, which describes the mathematical dependency of the quantity related to the vertex on the quantities related to its direct predecessors. For the graphical representation, a marking as shown in Fig. 3.6 is used. Input vertices are not marked because the quantities related to them are independent.

Fig. 3.7 shows a typical dependency graph with three input vertices, which are named by X_1, X_2 and X_3, one output vertex, named by Y, and three inner vertices, named by Z_1, Z_2 and Z_3. This graph represents the mathematical model given by the equations

$$Z_1 = f_1(X_1, Z_2),$$

$$Z_2 = f_2(X_1, X_2),$$

$$Z_3 = f_3(Z_2, X_3),$$

$$Y = f_Y(Z_1, Z_2, Z_3).$$

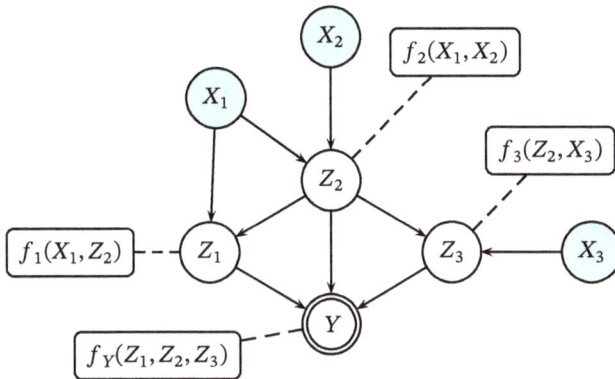

Fig. 3.7: A typical dependency graph.

The input quantities X_1, X_2 and X_3 of the model are assigned to the input vertices of the graph, the output quantity Y to the output vertex and the intermediate quantities Z_1, Z_2 and Z_3 to the inner vertices.

Since there exists at least one path in the graph between each of the three input quantities X_1, X_2 and X_3 and the output quantity Y, this quantity depends on all three input quantities. But this dependency is not a direct one, because none of the direct predecessors of the output vertex is an input vertex. Each path between the input vertices and the output vertex passes through the inner vertices. This creates the complex structure of the model. Without inner vertices a dependency graph has a very simple structure, like the one shown in Fig. 3.6, whose model is given by a single model equation of the form $Y = f(X_1, X_2, X_3)$.

In the literature frequently a statement can be found that a mathematical model represents a causal relation of the quantities involved. Such a statement, however, is not correct, because every model equation is merely a description of a functional relation. In order to understand the difference between a causal and a functional relation, we consider two examples.

Example 3.23 (OHM's law)

If we connect the terminals of a battery with a wire, an electric current flows, whose strength depends on the voltage of the battery and the resistance of the wire.

The functional relation between the current I, the voltage U, and the resistance R is given by OHM's law

$$I = \frac{U}{R}.$$

The dependency graph corresponding to this model is depicted in Fig. 3.8.

We know that, according to the laws of physics, an electric current is caused by an electric potential difference, i. e. by a voltage. Ohm's law thus describes not only a functional relation, but also the causal relation between voltage and current,

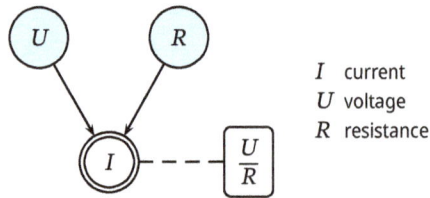

Fig. 3.8: OHM's law.

represented in the dependency graph by the edge pointing from the vertex named *U* to the vertex named *I*. On the other hand, the edge from the vertex named *R* to the vertex named *I* does not correspond to a *causal* relation, because the resistance is *not* the cause of the current. However, there is a *functional* relation between these two quantities, because the current depends on the resistance.

Example 3.24 (Resistance of an incandescent light bulb)

We are interested in the resistance of a light bulb. The information available to calculate this resistance is the nominal power and the nominal voltage of the light bulb.

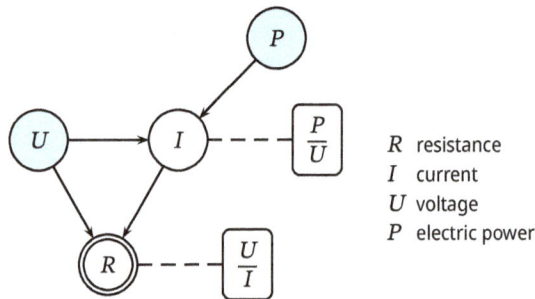

Fig. 3.9: Resistance of an incandescent lamp.

It is known from the principles of electrical engineering that the electric power *P* depends on the voltage *U* and the current *I* and that the relation existing between current, voltage and resistance *R* can be described by OHM's law. By introducing the current as an intermediate quantity, we obtain the dependency graph shown in Fig. 3.9, which corresponds to the model equations

$$ I = \frac{P}{U} \quad \text{and} \quad R = \frac{U}{I} $$

However, by these two equations only a *functional* relation between the respective quantities is described.

The only *causal* relation in the dependency graph, represented by the edge from the vertex named U and the vertex named I, exists between voltage and current. All other edges in the dependency graph do *not* correspond to a causal relation between the respective quantities, because neither the voltage nor the current is the cause of the resistance, and the electric power is not the cause of the electric current, but the opposite is true.

These two examples clearly show that in a dependency graph the connection of two vertices X and Y by an edge (X, Y) does not necessarily mean that the quantity X is the cause of the quantity Y or, *vice versa*, that the quantity Y is the effect of the quantity X. It only means that there is a *functional* relation between the respective quantities. However, it is of course not explicitly excluded that a *causal* relation may exist between them as well.

We will now demonstrate with an example how a dependency graph can be used for graphical modelling. For this purpose, we will construct the graph step by step from the underlying submodels.

Example 3.25 (Electric power loss in a wire)

An electrical load is connected to a power supply via a two-core cable. Each of the two cores of the cable consists of a wire with a circular cross-section.

When an electric current flows through the line to the electrical load, an electric power loss

$$P = UI, \tag{3.59}$$

occurs which is converted into heat, where I denotes the current and U the voltage drop created in the line due to the current. In order to represent the model given by equation (3.59) by a dependency graph, we need three vertices, one output vertex to represent the power loss P and two input vertices to represent the current I and the voltage drop U, respectively. We also need two edges to represent the functional (and causal) dependency of the power loss on the current and the voltage drop. Both edges are directed from the input vertices to the output vertex.

Finally, we place a marking on the output vertex, in order to capture in the graph the functional dependency of the output quantity on the input quantities as expressed by the equation (3.59).

Fig. 3.10 shows the dependency graph obtained after the first modelling step. The two input vertices of this graph emphasize that we have to measure both the current through the line and the voltage drop across the line due to the current in order to determine the resulting power loss.

We decide to measure the current, but not the resulting voltage drop, because if the line is long, this will require considerable effort and, because of the small value to be expected, it is very likely that we will encounter a large measurement uncertainty.

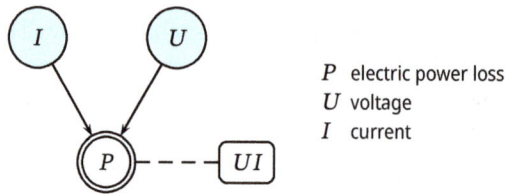

Fig. 3.10: Electric power loss in a wire (dependency graph after the first modelling step).

P electric power loss
U voltage
I current

In addition, we are also interested to know how the resistance of the line affects the power loss. The relation between this resistance and the current through the line is given by OHM's law

$$U = RI,\qquad(3.60)$$

where R denotes the resistance of the line, I the current flowing through the line and U the voltage drop caused by this current. We therefore extend the dependency graph by an additional vertex named R, which symbolizes the resistance of the line. At the same time, we convert the vertex denoted by U, which symbolizes the voltage drop, into an inner vertex and add two more edges to the graph to represent the functional dependency of the voltage drop on the current and the resistance.

The two additional edges are from the existing input vertex named I and from the new input vertex named R to the now new inner vertex named U. On the latter vertex we also place a marking to capture in the graph the functional dependency of the voltage drop U on the input quantities I and R, as expressed by the equation (3.60).

The dependency graph obtained after the second modelling step is shown in Fig. 3.11. The two input vertices of this graph emphasize that we have to measure both the current through the line and the resistance of the line to determine the resulting electrical power loss.

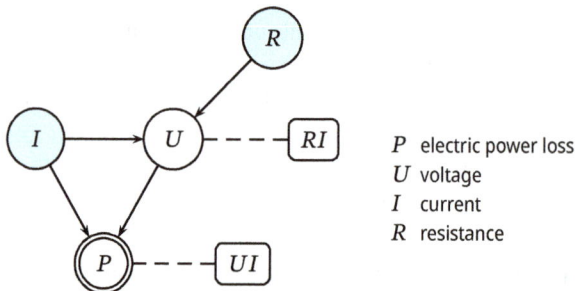

P electric power loss
U voltage
I current
R resistance

Fig. 3.11: Electric power loss in a wire (dependency graph after the second modelling step).

Again we decide to measure the current through the line, however, we do not wish to measure the resistance of the line directly, but rather to determine it from the length of the line L, the cross-section A of the wires used in the line, and its specific resistance ρ. The relation of these quantities with the resistance R of the line is given by the equation

$$R = 2\rho\frac{L}{A} \tag{3.61}$$

where the factor two is necessary, because we assume that the line consists of two wires of the same length, cross-section, and material. We add three new input vertices to the dependency graph, named by the quantities L, ρ and A. At the same time, we convert the vertex named R into an inner vertex and connect it to the three new input vertices by edges directed to it. This symbolizes the functional dependency of the resistance R on the input quantities L, ρ and A. We also place a marking on this vertex in order to capture in the graph the functional dependency described by the equation (3.61).

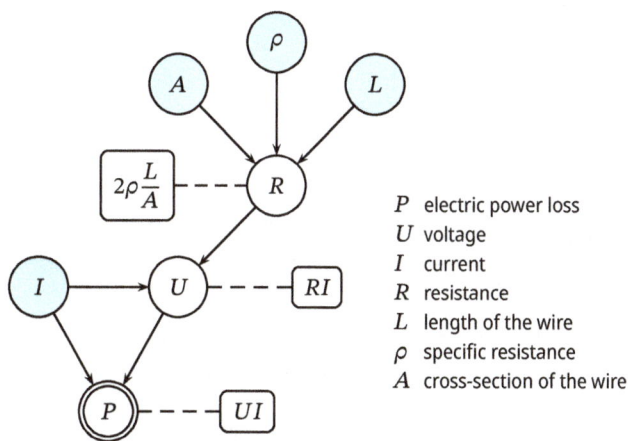

P electric power loss
U voltage
I current
R resistance
L length of the wire
ρ specific resistance
A cross-section of the wire

Fig. 3.12: Electric power loss in a wire (dependency graph after the third modelling step).

The dependency graph obtained after the third modelling step is shown in Fig. 3.12. It is evident from this graph that we can determine the power loss generated in the line by measuring the current I flowing through it, the length L of the line, the cross-section A of the wire contained in the line and the specific resistance ρ of the material used for the wire. However, we will not measure the material constant ρ, but look it up in an appropriate handbook.

Since a direct measurement of the cross-section A of the wire is usually not possible, we decide to determine this quantity from the diameter D of the wire, which is easier to measure, by using the well-known geometrical relation

$$A = \frac{\pi}{4}D^2 \tag{3.62}$$

between the cross-section and the diameter. Therefore, we modify our dependency graph again by introducing a new input vertex named D and converting the vertex named A into an inner vertex. To symbolize the functional dependency of the cross-section on the diameter, we connect these two vertices with an edge directed towards the inner vertex. This dependency, given by the equation (3.62), is additionally captured in the graph by a marking placed on the vertex named A. In this way, we obtain the dependency graph shown in Fig. 3.13. This graph represents the complete model of our problem. It visualizes how the electric power loss P in a line can be determined if we know the current I flowing through the line, the length L of the line, as well as the diameter D and the specific resistance ρ of the wires that make up the line.

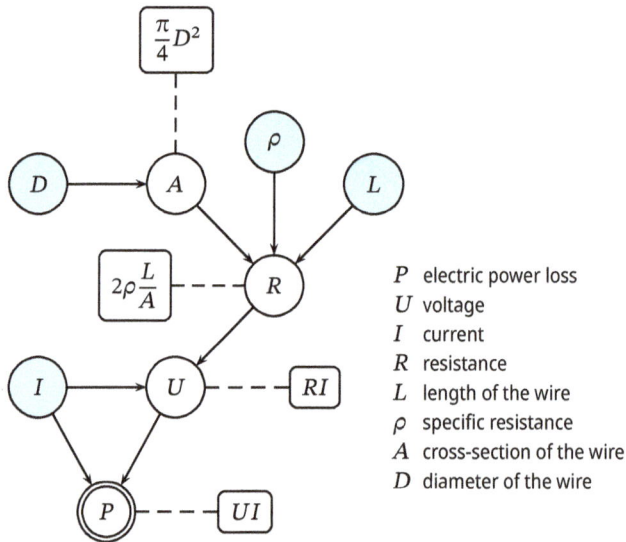

Fig. 3.13: Electric power loss in a wire (dependency graph after the last modelling step).

This example illustrates how the modelling procedure can be facilitated by a stepwise construction of a dependency graph and by using the equations of the underlying submodels. Starting from the output vertex, the dependency graph of the model is developed by gradually adding input vertices, converting some of the vertices into inner vertices whenever they become dependent on newly added input vertices. The functional dependency of the quantities assigned to the vertices is symbolized by the directed edges which, at each step, connect the newly inserted input vertices to the vertices that depend on them. The markings placed on the output vertex and all the inner vertices allow a complete inclusion of the submodel equations in the dependency graph.

The dependency graph of a mathematical model is not only useful for visualizing the structure of the model, but it also represents an algorithm that allows the value of the output quantity to be calculated from the values of the input quantities. To understand this statement, we imagine that the values of the input quantities—obtained for example by measurements—are entered at the respective vertices of the graph assigned to them. From there these values can reach neighbouring inner vertices along the connecting directed edges. All inner vertices carry a marking with a rule (usually a formula) for calculating a value from the values that a particular vertex receives from its direct predecessors. In this way, the corresponding calculation can be performed locally at each of the inner vertices and the result can then be transferred to its direct successors. In this way, the information modified by the inner vertices moves along the directed edges until it finally reaches the output vertex, which also carries a marking with a rule for the calculation of a value from all the values received from its direct predecessors. Thus, the value of the output quantity can also be calculated locally at the output vertex.

We can make the dependency graph even more useful by putting "weights" on its edges. We then obtain an edge-weighted graph:

Definition 3.15 (Edge-weighted graph)

An edge-weighted graph is a graph, where the ordered pair $(\mathfrak{B}, \mathfrak{C})$ is extended to an ordered triple $(\mathfrak{B}, \mathfrak{C}, d)$ by adding a function $d : \mathfrak{C} \to \mathbb{R}$ which assigns a real number to each edge.

Fig. 3.14 shows a section of an edge-weighted graph, where each edge is marked with the name of a function, which has to be used, in order to assign a value to that particular edge. If, as shown in the figure, the vertex of a graph named by Y is marked with the functional relation $f_Y(X_1, X_2, X_3)$, we will use the functions

$$c_{Y, X_i} = \left[\frac{\partial f_Y(X_1, X_2, X_3)}{\partial X_i} \right]_{X_1 = x_1, X_2 = x_2, X_3 = x_3}, \qquad i = 12, \dots , \qquad (3.63)$$

in order to calculate the weights of the directed edges, which connect this particular vertex with its direct predecessors named by X_i ($i = 1, 2, \dots$), where the values of the quantities assigned to these vertices are denoted by x_i ($i = 1, 2, \dots$). The weights c_{Y, X_i} are obviously the sensitivity coefficients arising during a linearization of a submodel equation $Y = f_Y(X_1, X_2, X_3)$.

We demonstrate the application of equation (3.63) by putting weights on the edges of the graph of example 3.25.

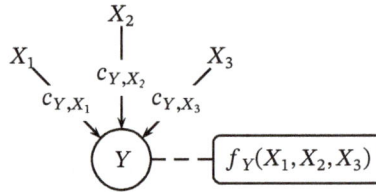

Fig. 3.14: Weighted edges.

Example 3.26 (Continuation of example 3.25)

Applying the relation (3.63) to the equations (3.59) to (3.62) yields

$$c_{P,U} = I_0, \qquad c_{P,I} = U_0, \qquad c_{U,I} = R_0, \qquad c_{U,R} = I_0,$$

$$c_{R,L} = 2\frac{\rho_0}{A_0}, \qquad c_{R,\rho} = 2\frac{L_0}{A_0}, \qquad c_{R,A} = -2\rho_0\frac{L_0}{A_0^2}, \qquad c_{A,D} = \frac{\pi}{2}D_0.$$

The dependency graph of Fig. 3.14 from example 3.25 is shown again in Fig. 3.15, but now extended by the weights placed on the edges.

Since the weights depend only on the information available at the vertex to which the edges are directed, they can be computed locally at each inner vertex and at the output vertex after the quantity values have been computed.

After weighting all edges of a dependency graph, the linearized model can be obtained directly from the graph. For this purpose, we form an equation for each inner vertex as well as for the output vertex by placing on the left side of the equation the difference of the quantity assigned to the vertex under consideration and its quantity value, and on the right side of the equation the sum of all differences of the quantities assigned to the direct predecessors of the vertex under consideration and their values, multiplied by the weight of the respective edge. We demonstrate this procedure by using the graph shown in Fig. 3.15.

Example 3.27 (Continuation of example 3.25)

In order to obtain the equations of the linearized model of example 3.25, we start with the output vertex. The application of the procedure described above yields

$$P - P_0 = c_{P,U}(U - U_0) + c_{P,I}(I - I_0),$$

where the respective value of each quantity is denoted by the quantity name with a subscript 0 attached. If we apply the same procedure to all inner vertices, we obtain the equations

$$U - U_0 = c_{U,R}(R - R_0) + c_{U,I}(I - I_0),$$

$$R - R_0 = c_{R,\rho}(\rho - \rho_0) + c_{R,L}(L - L_0) + c_{R,A}(A - A_0),$$

$$A - A_0 = c_{A,D}(D - D_0).$$

These four equations together give the linearized model, which of course we would also have obtained by linearizing the equations (3.59) to (3.62).

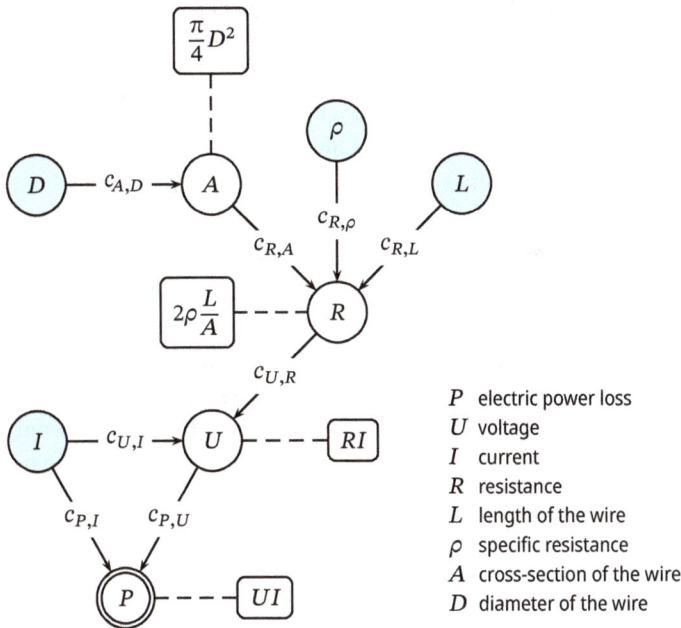

Fig. 3.15: Electric power loss in a wire (dependency graph with weighted edges).

This example shows that a linearized model can easily be read from a dependency graph. Such a graph is therefore very useful not only for modelling but also for supporting the evaluation of measured data.

At the end of this section, we look at some aspects of a software implementation of the graphical modelling approach, without going into all the details involved. Graphical modelling through a step-by-step construction of a dependency graph can be accomplished using a graphical editor with a graphical user interface (GUI), where the user can select the graphical elements necessary to construct the graph, establish the required relations, and add additional information, such as markings on the vertices with appropriate rules or equations, and weights on the edges. During the creation of the dependency graph, the program can build an internal structure containing all the necessary information about the graph. The markings attached to the internal vertices and the output vertex, containing the functional relations between the quantity assigned to each vertex and the quantities assigned to its direct predecessors, could be accumulated in an abstract syntax tree (AST). Abstract syntax trees are structures used in computer science to store the syntactic structure of source code, e. g. mathematical expressions in compilers.

Once the internal structure of the dependency graph has been completed, the program could start calculating the quantity values and edge weights from the measured data available at the input vertices. This can be done by traversing the graph from the input vertices, using the attached abstract syntax tree to control the calculations at each inner vertex and at the output vertex. The derivatives needed to compute the sensitivity coefficients (edge weights) can be obtained by automatic differentiation[9] applied to all abstract syntax trees attached to the inner vertices and the output vertex.

Note that all the computations in the dependency graph can be done locally at the inner vertices and the output vertex, respectively, i. e. by distributed computation, not by sequential computation. This allows parallel processing in general. Each inner vertex can start its assigned computation task as soon as all necessary values are available. This is particularly useful for large and complex models or when the use of matrix algorithms is considered useful, such as when models with more than one output quantity are required.

Since the size and complexity of a dependency graph is limited only by the memory available in the computer (and possibly also by the processing time), modelling using a software implementation of the method presented in this section is limited only by the ability of the user to develop a sufficiently good model that is suitable for the particular measurement task.

---

9 Automatic differentiation is a method in computer science that can be used to compute the derivatives of algorithms (for details see e. g. L. B. RALL [19]). Unlike approximation methods, such as e. g. difference approximation, the results are numerically exact.

4 Basics of probability theory

> If we were not ignorant there would be
> no probability, there could only be
> certainty. But our ignorance cannot be
> absolute …
>
> *(Henri Poincaré, 1902)*

Probability theory is a branch of mathematics that provides tools and methods for describing and evaluating random experimental data. Random here means that it is not possible to predict the outcome of an experiment.

In this chapter we look at the basics of probability theory in order to understand the statistical methods used in measurement data evaluation and uncertainty calculations.

4.1 Probability concepts

Intuitively we all know what probability is, but nobody seems to be able to give a satisfactory explanation of this term. In everyday life we encounter all kinds of probabilities. For example, the questions

- What is the probability of rolling a six with a totally symmetric (ideal) die?
- What is the probability of snow in Munich tomorrow?
- What is the probability that *a particular person* becomes 90 years old?

all have a different quality. The answer to the first question is certainly easy for most people. In the case of the second question, a majority of respondents might argue that the answer depends on the information available. The answer to the third question, however, can only be subjective because everyone will inevitably give a different answer.

There is no simple answer to the question of what probability *actually* is. In fact, there are many different opinions in the literature on how to define this term. We will not, however, enter into the philosophical discussion of which definition of "probability" might be the correct one, or whether there are several reasonable definitions of this term that should coexist on an equal footing, but will limit ourselves to a discussion of three generally accepted definitions, illustrating with examples under what circumstances these definitions are justified.

Classical probability

According to the current view of scientific history, the mathematical methods of probability were first used in the correspondence between PIERRE DE FERMAT and BLAISE PASCAL

https://doi.org/10.1515/9783111453712-004

in 1654. Apart from questions such as the fair distribution of stakes in an interrupted game of chance, one of these letters deals with a complaint by a friend of PASCAL, the CHEVALIER DE MÉRÉ. In his opinion, mathematics was incompatible with the "practicalities of life". For the Chevalier, the "practicalities of life" were games of chance, especially dice. He had changed the rules of a then popular game of dice, thinking that his chances of winning would remain unchanged. But much to his chagrin, he found that after changing the rules, the players were winning more often than the bank, whereas before it had been the other way round.

In the following we will consider the problem of the CHEVALIER DE MÉRÉ because it leads us directly to the definition of classical probability. We begin with a simple observation:

Example 4.1 (Rolling an ideally symmetric die)
A die has six faces, which we label with the numbers 1, 2, 3, 4, 5 and 6. Thus, there are six different possible outcomes for each roll. Since we assume that the die is ideally symmetric, we are confident that in one out of six cases a six should appear, and in five out of six cases we expect to see *no* six.

With these preliminary considerations, let us now turn to the game of dice, which was popular in France at the time of the CHEVALIER DE MÉRÉ.

Example 4.2 (French game of dice in the 17th century)
In the 17th century a game of dice, played with a single die, was popular in France. The die was rolled four times in succession. The player won if *no* six appeared in these four rolls, otherwise the bank won. We are interested in the odds of the bank winning the game, assuming an ideally symmetric die.

We represent the outcomes of the four dice rolls by the sequence $(\mathfrak{E}_1, \mathfrak{E}_2, \mathfrak{E}_3, \mathfrak{E}_4)$, where $\mathfrak{E}_k$ ($k = 1, \ldots, 4$) denotes the outcome of the k-th roll. Since these outcomes can take exactly six different values for each k, there are $6^4 = 1296$ different possible outcomes when the die is rolled four times. But the outcomes $\mathfrak{E}_k$ can only take five of these values for each k that differ from six, i. e. there are only $5^4 = 625$ different outcomes where *no* six occurs during the four throws.

We conclude that if a die is thrown four times, *no* six will appear 625 times out of 1296, i. e. in the long run the players could only win about $(5/6)^4 \approx 48.2\,\%$ of all games. So it was a profitable business for the bank.

In order to make the game more interesting, the CHEVALIER DE MÉRÉ came up with the idea of changing the rules. He suggested playing 24 times simultaneously with two dice instead of four times with just one, with the player winning if no double six occurs. He actually believed that this would not change the odds and argued that the odds were originally 4 : 6 (four rolls and six possible outcomes for one die) and with the changed rules $24 : 36$ (24 rolls and $6 \cdot 6 = 36$ possible outcomes for two dice), i. e. $4 : 6$.[1] However, as we demonstrated in the previous example, the Chevalier was already wrong about the odds of the unchanged game.

In order to reveal the second mistake of the CHEVALIER DE MÉRÉ, we now calculate the odds of the game of dice with the rules changed according to his suggestions.

Example 4.3 (The game of dice of the CHEVALIER DE MÉRÉ)

According to the new rules proposed by the CHEVALIER DE MÉRÉ, the gamblers have to roll two dice simultaneously 24 times in succession. If no double six occurs, the gambler wins, otherwise the bank wins.

Each of the two dice has six faces, which we label with the numbers 1, 2, 3, 4, 5 and 6. Thus, we have six different possible outcomes for each die, i. e. after one roll with both dice $6 \cdot 6 = 36$ different possible outcomes. Since we are assuming that the dice are ideally symmetric, we believe that only in one of these 36 cases a double six will occur, and in 35 cases either no six or at most one six will be seen.

If the dice are thrown 24 times, there are obviously 36^{24} possible outcomes, of which 35^{24} are no six or at most one six. Thus, the chances for the gamblers to win are $(35/36)^{24}$, i. e. the gamblers win about 50.9 % of all games. The business is no longer profitable for the bank.

Instead of 24 rolls, the chevalier should have required 26 rolls in order not to change substantially the chances of the bank, because then the gamblers would have won only about 48.1 % of the games.

The examples have demonstrated how the chances of winning or losing in a game of pure chance can be estimated. For that purpose the quotient of the number of favourable cases to the number of possible cases has to be determined.[2] Today this quotient is called classical probability.

Definition 4.1 (Classical probability)

The probability $P(\mathfrak{E})$ of an event $\mathfrak{E}$ is the quotient of the number of favourable cases of this event $g(\mathfrak{E})$ to the number of all possible cases n, i. e.

$$P(\mathfrak{E}) = \frac{g(\mathfrak{E})}{n} \, .$$

Note that we always have $0 \le g(\mathfrak{E}) \le n$, i. e. $0 \le P(\mathfrak{E}) \le 1$.

---

1 BLAISE PASCAL commented on these naive considerations on July 26th 1654 in his letter to PIERRE DE FERMAT [20] with the words: "... *il a trés bon esprit, mais il n'est pas géomètre; c'est, comme vous savez, un grand défaut;* ..." [... he has a very good mind, but he is not a mathematician; that is, as you know, a great deficiency; ...]

2 This recommendation was made in 1812 by P. S. Laplace in the comprehensive introduction to his influential work on probability theory [21].

We demonstrate the application of this definition by two simple examples taken from well known games of pure chance.

Example 4.4 (Throwing a fair coin)

A fair coin has two different sides, which we call "head" and "tail", i. e. the number of possible outcomes for each flip of the coin is two, if we exclude the unlikely event that the coin could stand still on the edge. If we ask for the probability of the coin landing "head up", the number of favourable cases is one for each flip. Therefore, the probability of this event is $1/2$.

Example 4.5 (Drawing a card from a deck of cards)

A common deck of cards contains 52 different cards, i. e. for each draw the number of possible outcomes is 52. If we ask for the probability of drawing a court card, the number of favourable cases is 12, namely four jacks, four queens and four kings. Therefore, the probability of this event is $12/52 = 3/13$.

In these two examples the number of possible cases is known in advance (a fair coin has two different sides and a common deck contains 52 different cards). In addition, we *implicitly assume* that all possible cases are equally likely. These two assumptions are characteristic of classical probability. But it is by no means self-evident that we can always state the number of possible cases at once, as the following example shows.

Example 4.6 (Sums of the number of dots when rolling two dice)

We simultaneously roll two dice, which we assume to be ideally symmetric, and calculate the sum of the number of dots occurring at a time. We are interested in the probability that this value equals the number 7.

In order to determine how frequently the respective sums occur, we use a tabular compilation of all possible cases:

Value of the sum	Number of cases	Cases
2	1	⚀+⚀
3	2	⚀+⚁ ⚁+⚀
4	3	⚀+⚂ ⚁+⚁ ⚂+⚀
5	4	⚀+⚃ ⚁+⚂ ⚂+⚁ ⚃+⚀
6	5	⚀+⚄ ⚁+⚃ ⚂+⚂ ⚃+⚁ ⚄+⚀
7	6	⚀+⚅ ⚁+⚄ ⚂+⚃ ⚃+⚂ ⚄+⚁ ⚅+⚀
8	5	⚁+⚅ ⚂+⚄ ⚃+⚃ ⚄+⚂ ⚅+⚁
9	4	⚂+⚅ ⚃+⚄ ⚄+⚃ ⚅+⚂
10	3	⚃+⚅ ⚄+⚄ ⚅+⚃
11	2	⚄+⚅ ⚅+⚄
12	1	⚅+⚅
	36	

Obviously there are eleven different sums possible, namely 2, 3, 4, 5, 6, 7, 8, 9, 10, 11, and 12. Thus, it seems that the sum 7 is one favourable case of eleven possible cases, and we could be tempted to state 1/11 as the probability of this particular sum. But this would be wrong. The eleven sums are *not* equally probable, because they do not occur equally frequent.

According to this table the sum 7 occurs in six of a total of 36 cases, i. e. the probability of its occurrence is $6/36 = 1/6$.

This example makes it clear that we *always* have to ensure that the events under consideration are *equally probable* before we can calculate the classical probability of a particular event according to its definition.

Up to now we have always assumed that all possible cases are *equally probable*, without really knowing what this phrase actually means. Not only that it is certainly disappointing to base the definition of classical probability on the undefined phrase "equally probable" (this is obviously a circular definition *idem per idem*[3]), but there also exist problems where it is even not clear which cases have to be considered as equally probable. Such examples were given by J. BERTRAND [22].

Example 4.7 (BERTRAND's paradox)

A chord is randomly drawn inside a circle. What is the probability that the chord is longer than a side of an equilateral triangle inscribed in the circle?

This question cannot be answered, as long as it is not known which cases have to be considered as equally probable. We can, for example, make the following assumptions:

Assumption 1: In this case one endpoint of the chord is fixed at a preassigned point on the circumference of the circle, while the other endpoint of the chord is randomly chosen on the circumference of the circle.

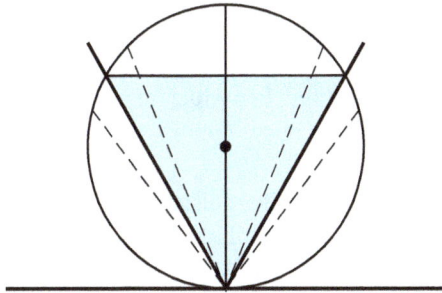

We consider the angle α that the chord forms with the diameter of the circle through the fixed point of the chord. This angle can only take values from the

---

3 lat., the same by the same

interval $(-90°, +90°)$. All chords, where $-30° < \alpha < +30°$ is valid, are longer than a side of an equilateral triangle inscribed in the circle. If we assume that the possible angular values are equally probable, then one third of the values from the interval $(-90°, +90°)$ are favourable, i.e. under the stated assumption the probability has to be 1/3.

Assumption 2: In this case the direction of the chord is preassigned perpendicular to an arbitrarily chosen diameter of the circle, while the centre point of the chord is randomly chosen on the selected diameter of the circle.

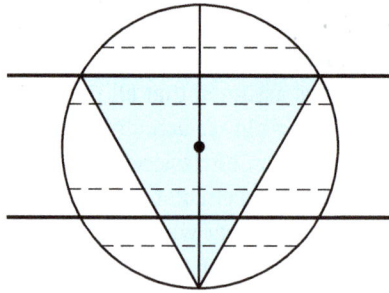

We denote by d the position of the centre point of the chord, measured along the diameter. The quantity d can only take values from the interval $(-R, +R)$, where R denotes the radius of the circle. All chords for which $-R/2 < d < +R/2$ holds, are longer than a side of an equilateral triangle inscribed in the circle. If we assume that the possible values of d are equally probable, then half of the values from the interval $(-R, +R)$ are favourable, i.e. under the stated assumption the probability has to be 1/2.

BERTRAND has given a third example in which the centre of the chord is chosen at random inside the circle. It can be shown that in this case there is a probability of 1/4 that the chord is longer than the side of the triangle. The reader should verify this as an exercise.

It turns out that for different assumptions we obtain different values for the probability we are looking for. But which of the these possibilities is the *correct* answer to the question posed above?

BERTRAND's paradox has been extensively discussed in the literature. It has been shown that there is no unique solution to the problem, i.e. it is not well posed. However, it shows that the probability value we ultimately obtain always depends on the assumptions we make about various possible cases.

Most physicists are aware of this fact. They specify different "statistics" for different elementary particles. Electrons, for example, obey other "statistical laws" than photons. This reveals that it is obviously the user of probability theory who assigns a meaning to the phrase "equally probable" in each particular case.

Besides objective evidence, this fact is a well-known, but often ignored, subjective element of probability theory, which can never get along without hypotheses. Such a statement may certainly be unsatisfactory, but it is necessary to be kept in mind, in order to avoid unpleasant surprises.

Classical probability is sometimes also called combinatorial probability, because in most cases combinatorics can be used to calculate the numbers of the respective possible cases. In very general terms, combinatorics allows us to calculate the number of possible arrangements of distinguishable or indistinguishable objects with or without taking their order into account. This mathematical discipline was independently founded by B. PASCAL [23] and G. W. LEIBNIZ [24] and later brought to a certain perfection by J. BERNOULLI [25], who also used it to calculate probability values.

Frequentist probability

Since classical probability does not seem to be entirely satisfactory, it was natural to ask for an alternative interpretation of probability. It was P. S. LAPLACE who paved the new way when he determined the quotient of the number of male births to the total number of births in Berlin, London, St. Petersburg, and throughout France [26]. In doing so, he used a type of probability as a basis for his considerations that was different from the probability concept customary at his time. Today we call this type of probability *frequentist probability*.

In order to understand the meaning of the term *frequentist probability*, we have at first to deal with the terms *random experiment* and *relative frequency*. To begin with, we consider an example.

Example 4.8 (Multiple rolling of an ideally symmetric die)

We roll an ideally symmetric die 100 times and note in each case which of the six possible faces, which we assume to be marked by the numbers 1 to 6, comes up on each roll. A typical result of such an experiment is shown in Fig. 4.1.

Note that the sequence 5, 2, 1, 3, 4, 3, 5, 5, 3, 4, ... of the faces coming up is apparently completely random, i. e. it is not predictable which face will turn up at the next roll. Moreover, an absolute frequency, equal for each face of the die, which actually should be expected for an ideally symmetric die, can obviously not be observed. The face marked with the number six, for example, does not occur at all during the first ten rolls, while the faces marked with the numbers three and five turn up three times. Even after 100 rolls the expected uniform distribution of the faces turning up is not observable, not even approximately, as the bar chart in Fig. 4.1 clearly shows.

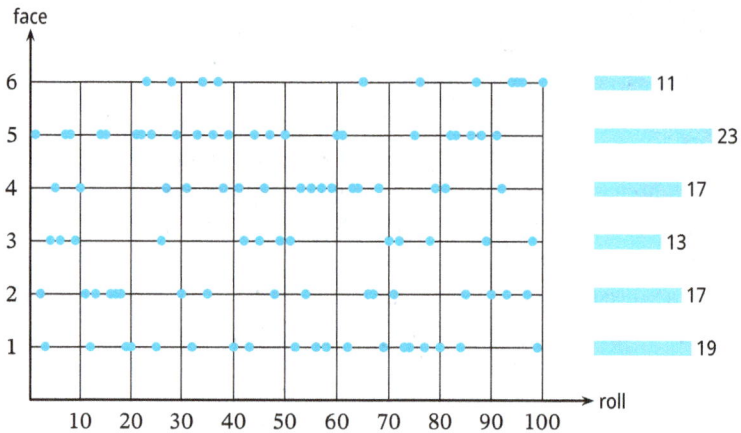

Fig. 4.1: Outcomes after rolling an ideally symmetric die 100 times. The bar chart on the right side of the figure shows the absolute frequency of each face of the die coming up.

This is a typical example of a random experiment. Other examples are: tossing a coin several times, registering births and deaths in a country over a given period of time, counting the number of calls received by a telephone exchange per hour, measuring the average temperature at a fixed location over a given number of days, counting the number of cars crossing a bridge in a day, etc. We can therefore give the following definition:

Definition 4.2 (Random experiment)
A random experiment is a sequence of arbitrarily often repeatable experiments under unchanging conditions, which are independent of each other and whose outcome cannot be predicted.

Note that a random experiment is not necessarily a measurement or a counting process, but has to be understood more generally as a sequence of unpredictable events of any kind under unchanging conditions.

We return to the random experiment described in example 4.8. Instead of the absolute frequency, we are now interested in the relative frequency of a face, denoted by the number k, to come up when we roll an ideally symmetric die a varying number of times. We obtain the relative frequency of the appearance of a particular face by counting the number of times that face appears (absolute frequency) and putting that number in proportion to the total number of rolls. Thus we can define the relative frequency by:

Definition 4.3 (Relative frequency)
The relative frequency $h_n(\mathfrak{E})$ of an event $\mathfrak{E}$ during a random experiment is equal to the ratio of the absolute frequency $H_n(\mathfrak{E})$ of occurrences of that event to the total

number n of experiments, i. e.

$$h_n(\mathfrak{E}) = \frac{H_n(\mathfrak{E})}{n} .$$

Notice that we always have $0 \leq H_n(\mathfrak{E}) \leq n$, i. e. $0 \leq h_n(\mathfrak{E}) \leq 1$.

Now we examine how the relative frequencies considered in the example 4.8 depend on the number of experiments. For an ideally symmetric die we would expect each of its six faces to come up equally often, i. e. the relative frequency should not deviate much from $h_n(\mathfrak{E}_k) = 1/6$ ($k = 1, \ldots, 6$). However, as can be seen from Fig. 4.2, this is by no means the case for a few rolls (i. e. for a small value of n), but only (approximately) for a very large number of rolls. The relative frequencies seem eventually to converge to the values of classical probability, the better, the larger the number of experiments. Accordingly, we assume that if we could perform an infinite number of experiments, we would get a perfect agreement.

This behaviour, illustrated in Fig. 4.2, is characteristic of all random experiments. As the number of experiments increases, the values of the relative frequencies stabilize and appear to approach a certain limit. A mathematical explanation for this observed behaviour was given by J. BERNOULLI [25] for the first time in 1713. He made the following remark:

> However, lest these remarks be misunderstood, it must be noted that we do not wish for this ratio, which we undertake to determine by observations, to be accepted as absolutely accurate (for then the contrary would result from this, and it would be the more unlikely that the true ratio had been found the greater the number of observations made), but rather be taken in some approximation, i. e. bounded by two limits, which though can be assumed to be as close together as one wishes.

Then he proved mathematically rigorously that, with an increasing number of experiments, the mentioned limits get closer and closer, and that it becomes more and more probable that the ratio addressed in his remark will stay within the determined limits, instead of taking a value outside these limits. Today this result is known as the (weak) law of large numbers.

When interpreting BERNOULLI's law of large numbers, attention has to be paid that this law does *not* state that for an increasing number of experiments eventually the values *always* stay within two limits, which get closer and closer, because at any time a value outside these limits can still occur. The law of large numbers only states that the probability of such cases approaches zero for an infinite number of experiments.

The law of large numbers is often misunderstood, especially by gamblers. It does not justify the conclusion that an event, which in the past had occurred less frequently

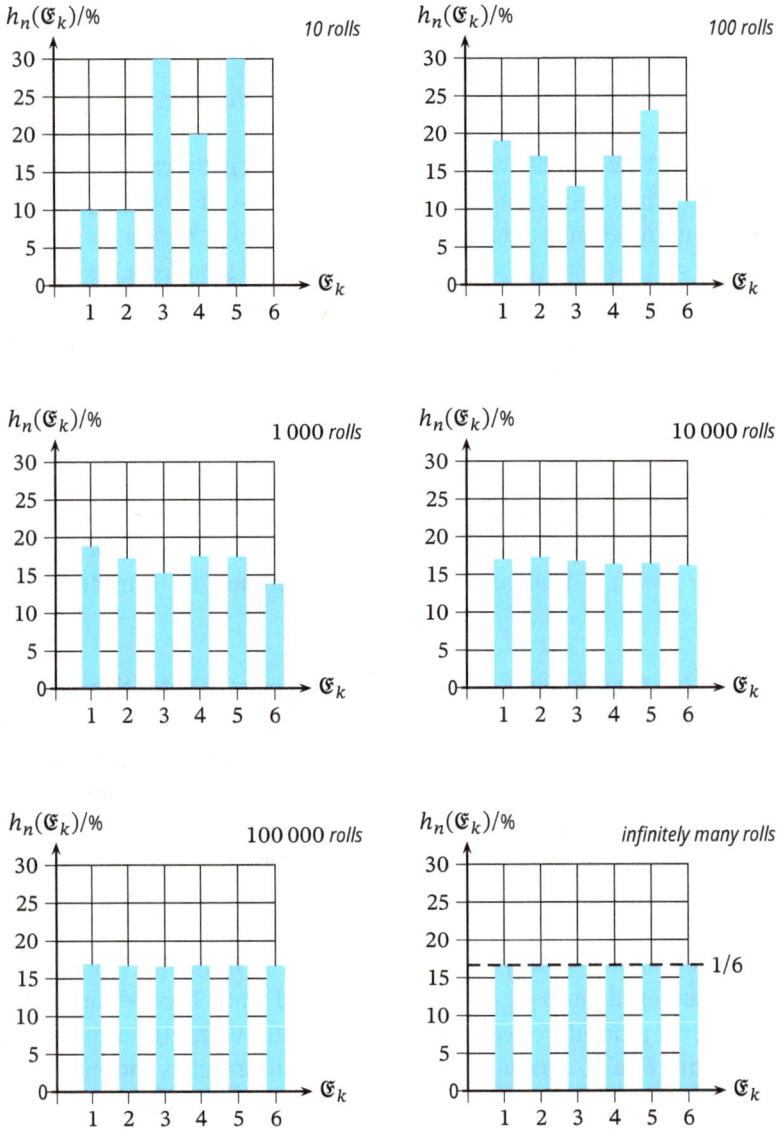

Fig. 4.2: Relative frequency $h_n(\mathfrak{E}_k)$ of the occurrence of the event $\mathfrak{E}_k$ that the face marked by the number k comes up, when an ideally symmetric die is rolled n times as indicated.

than was to be expected by its probability, will catch up its "shortfall" in future and thus occur more frequently.

The definition of the frequentist probability is essentially based on the (weak) law of large numbers as discussed above.

Definition 4.4 (Frequentist probability)

The probability $P(\mathfrak{E})$ of the occurrence of the event $\mathfrak{E}$ is equal to its observed relative frequency $h_n(\mathfrak{E})$ as the number n of experiments approaches infinity, i. e.

$$|P(\mathfrak{E}) - h_n(\mathfrak{E})| \to 0, \quad \text{if} \quad n \to \infty.$$

The limit in this definition does not correspond to the concept of a limit in the sense of mathematical analysis, as it was still believed at the beginning of the last century [27, 28, 29]. Such a limit cannot exist, because then, according to mathematical analysis (see e. g. [30]), a number $n_0(\varepsilon)$ should exist for every arbitrarily small value $\varepsilon > 0$, such that

$$|P(\mathfrak{E}) - h_n(\mathfrak{E})| < \varepsilon, \quad \text{whenever} \quad n > n_0(\varepsilon) > 0.$$

This is not the case, however, because certain variations of the relative frequency $h_n(\mathfrak{E})$, as shown in Fig. 4.3, are always observable, no matter how large the number of experiments.

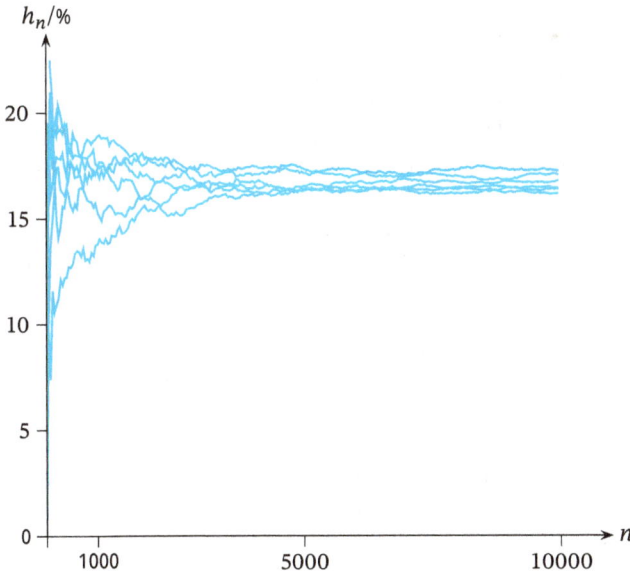

Fig. 4.3: Dependency of the relative frequency h_n on the number of rolls n for six different experiments.

These variations prevent the possibility of giving a number $n_0(\varepsilon)$ for each value ε, such that this value will never be exceeded by $|P(\mathfrak{C}) - h_n(\mathfrak{C})|$ as the number of experiments n increases beyond $n_0(\varepsilon)$. Actually, we have to admit that at any time in the course of a random experiment we do not know how close we actually are to a limit and, moreover, there is even no guarantee in the long run that the observed relative frequency at any time tends to approach a limit. In fact, we can only make a statement about the *probability* of reaching a limit. This probability increases with the number of experiments and seems to approach certainty in the long run. Therefore, the limit used in the definition 4.4 is called "stochastic limit".

Our considerations finally lead to the conclusion that the definition of frequentist probability is ultimately based on the concept of a stochastic limit, which in turn requires the statement of another probability (possibly of a different kind). Here again we have a kind of circular definition, as was the case with the definition of classical probability. But even if we are willing to accept this unsatisfactory situation, the definition of frequentist probability would still be problematic, because, as we have presupposed, the definitions of relative frequency and random experiment are based on the assumption of *unchangeable* conditions. But it is impossible to ensure that there are always constant conditions during a sequence of experiments, or, if we assume that this is indeed the case, to verify that such an assumption is valid. This fact, together with the requirement to be able to repeat an experiment infinitely often, causes a problem in the application of frequentist probability.

Subjective probability

Our everyday view of probability corresponds to neither classical nor frequentist probability. Every day we make many decisions that are inevitably based on more or less incomplete information. Unconsciously, we take into account our knowledge and personal experiences from the past, as far as they are relevant, and estimate the probability of the possible consequences of various possible actions under the given circumstances.

A personal view of probability is always subjective. Different individuals will usually have different opinions about the probability of the same event occurring because they normally do not have the same information, or because they judge the available information differently. We will try to make this point clearer with an example.

Example 4.9 (Probability of rain)

Before we leave the house in the morning, we notice a falling air pressure on the barometer. In addition, when we look at the sky, we observe a heavy accumulation of clouds. We know from experience that our observations are often an indication of rain. From this we conclude that it will almost certainly rain during the day and we should take an umbrella with us.

Our neighbour, on the other hand, might be of the opinion, based on *his* information, that the probability of rain is not particularly high and therefore leaves his umbrella at home.

It is obvious that classical probability methods are not applicable in this case, because it is at least inappropriate, if not impossible, to state a number of equally probable cases and to select from them a number of favourable cases. However, frequentist probability methods, which are solely based on the concept of frequency, are also inappropriate, because here the probability of the occurrence of an event has to be assessed, which cannot be interpreted as a random experiment in the sense of the definition 4.2. In fact, H. REICHENBACH [31, 32] had tried to explain cases such as those given in example 4.9 purely by frequentist probability methods, but his view was not well received by his contemporaries and did not prevail in the sequel. REICHENBACH had argued that the probability of the occurrence of an event could not be assigned to an individual case, but had to be assigned to a class of events, i. e. ultimately it was always possible to speak of frequencies. But such a classification is not always feasible, as illustrated by the following example.

Example 4.10 (Insufficient knowledge)

A container is filled with 1000 balls, which differ only in colour and are otherwise identical. The balls are either black or white. Given this information, what is the probability of drawing a white ball?

In order to answer this question, although we do not have any knowledge of the number of black or white balls, we assume that the probability of drawing a white ball is the same as the probability of drawing a black ball. Furthermore, we assume that the balls are well mixed and evenly distributed, i. e. that not all the black balls or all the white balls are at the top of the container. Under these assumptions it is reasonable to conclude that the probability of drawing a white ball could be 1/2.

The example is less artificial than it seems. In everyday life, decisions like the one in this example almost always have to be made, even if the information available to us is generally inadequate.

In situations like the one in the example, it is obviously impossible to relate the probability to a relative frequency. In this case a probability statement is necessarily based on a different principle. If we have no knowledge at all about the distribution of black and white balls in the container, we are forced to compensate for this lack of information by an assumption if we want to arrive at a probability statement. Since we will either draw a black or a white ball—there is no other possibility—it is reasonable to assume that both events are equally likely. This assumption corresponds to the common sense that different possible events should be considered equally likely if there is no discernible reason to favour one of them.

This way of thinking, now known as the *principle of insufficient reason* (also known as *indifference principle*), was already expressed by P. S. LAPLACE, who stated[4] that probability theory is nothing but common sense subject to calculation. In Bayesian probability theory, this principle is the simplest non-informative prior.[5]

Given the situation in example 4.10, if we were able to get some additional information about the distribution of the black and white balls in the container, even if this information would still be insufficient, we would certainly estimate a different probability of drawing a white ball from the container. We illustrate this with the following example.

Example 4.11 (Additional knowledge obtained by an experiment)

We are in the same situation as in example 4.10. We will now do an experiment. We draw a ball ten times in succession, record its colour and put it back in the container. Suppose the result of this experiment is that we have drawn seven white balls and three black balls. Given this knowledge, what is the probability that the next ball we draw will be white?

Knowing the result of our experiment, we are certainly no longer prepared to maintain that the probability of drawing a white ball next time is 1/2, but we are now confident that the probability should be greater. However, we will also not assume that the probability of drawing a white ball next time is 7/10, because if we had drawn only white balls during the experiment, we would not assume that the container is filled only with white balls, because such a conclusion would contradict our prior knowledge available before the experiment, namely that the container is filled with white *and* black balls. We would probably decide to draw more balls in order to become more confident.

The examples above show that our knowledge at any given time determines the probability value we are willing to assign to a particular event. This is a typical characteristic of subjective probability. Whenever the available information changes, for example due to the results of an experiment, it may be necessary to change the probability value as well, provided that the new information is relevant to the problem under consideration and we have no reasonable doubt about its correctness.

---

4 «*On voit, ... , que la théorie des probabilités n'est, au fond, que le bon sens réduit au calcul; elle fait apprécier avec exactitude ce que les esprits justes sentent par une sorte d'instinct, sans qu'ils puissent souvent s'en rendre compte.*» [21]

"*We see, ... , that the theory of probabilities is, fundamentally, only common sense reduced to calculation; it allows us to appreciate with exactitude what just minds feel by a sort of instinct, without them often being able to realize it.*"

5 The principle is meaningless in the theory of frequency probability, where probabilities are relative frequencies rather than degrees of belief in uncertain statements, conditional on information about the state of events.

The influence of the information available in a specific situation on the value of the probability is illustrated by another example, originally used by J. BERNOULLI [25] for the same purpose.

Example 4.12 (Shipwreck)

Three ships set sail from the harbour. After some time it is reported that one of them was shipwrecked. What do we suspect?

If I only consider the number of ships, I could come to the conclusion that misfortune could equally befall any of them. But as I remember that one of them was more corroded by decay and age than the others, that it was poorly equipped with sails and sail-yards, and that it was also commanded by a new and inexperienced captain, I conclude that it is certainly more likely that this one was affected than the others.

Taking into account the information available at any given time implies that the subjective probability is always a *conditional* probability. This eventually leads to the following definition:

Definition 4.5 (Subjective probability)

The subjective probability $P(\mathfrak{E} \mid I)$ is a measure of the degree of personal belief, based on the available information I, that the event $\mathfrak{E}$ will occur.

In order to illustrate the influence of changing information on the value of probability, we give another example.

Example 4.13 (Influence of changing information on probability)

A common deck contains 52 cards. One of the cards is drawn. According to the principle of insufficient reason, each of the 52 cards has the same chance of being drawn under normal conditions. Therefore, the probability that the card drawn is the King of Clubs, for example, is $P(\clubsuit K \mid I) = 1/52$, based on the information I that there is only one King of Clubs in a common deck.

Now we are told that the card drawn belongs to the club suit. Based on this additional information and the knowledge we already have, what is the probability that the card drawn is the King of Clubs?

This time we use a different piece of information I', namely that there are exactly 13 club cards in a common deck and that only one of these cards is the King of Clubs. Based on the new information I', the conditional probability that the card drawn is the King of Clubs is now $P(\clubsuit K \mid \clubsuit, I') = 1/13$, i. e. the additional information has caused a change in the conditional probability.

This example demonstrates how the probability assigned to a particular event is changed by additional information. Subjective probability is always related to the available information. This is a significant difference from the other two probability concepts we have already discussed.

Another difference is that subjective probability does not allow for an operational definition, unlike the other two definitions of probability, i. e. there is no rule that would make it possible to actually determine a probability value. The reason is that a probability statement in colloquial language is at first only of a comparative kind ("It is more probable that it will rain than that it will not rain.") in the sense of a purely subjective (personal) expectation.

By specifying simple axioms (principles) for the notion of "probability", which seem plausible on the basis of our everyday experience, this notion can be specified more precisely, and it can be shown that for subjective probability—as for the other probability concepts—an objective measure can also be specified [33, 34, 35].

Concluding remarks and comments

Although the term "probability" is used freely in colloquial language as well as in science and engineering, no satisfactory definition has yet been found. Perhaps this is not possible at all.

Historically, the different probability concepts originated due to various questions and are still used today according to their original purpose. In the following, we briefly summarize the three probability concepts introduced in this section, point out their differences and discuss their areas of application.

It always makes sense to use the concept of classical probability when there is a finite and countable set of events, and it is known, or can reasonably be assumed, that all events are equally likely to occur. Under these circumstances, methods of combinatorics can be used to calculate the number of occurrences. Lotteries or games of chance of any kind can be treated in this way, as can many technical problems or, for example, classical statistical mechanics in physics.

The concept of frequentist probability corresponds to the current doctrine and constitutes the basis of mathematical statistics. It should always be noted, however, that there are cases in the natural sciences and engineering which are not really compatible with this interpretation of probability, because many experiments are restricted—mainly for economic reasons—to a few tests. Therefore, the resulting relative frequencies cannot be stable, as we have shown. Moreover, the experiments often do not meet the requirements of a random experiment, because it is not possible to ensure that the conditions remain unchanged, or it is not possible to verify that such conditions actually prevail. There are, however, physical reasons for justifying frequentist probability. The decay of radioactive materials, for example, is largely consistent with the concept of frequentist probability. This is also true of other mass phenomena where a very large number of similar events occur. In all these cases it is reasonable to consider the experimentally determined relative frequencies as good approximations to the underlying probabilities. However, it is not possible to say anything about the quality of these approximations.

The concept of subjective probability is advantageous in all cases where probability cannot be reasonably interpreted by either combinatorial methods or relative frequencies, such as in the case of single events or learning from experience. But even in cases where one of the other two probability concepts may be reasonably applied, subjective probability concepts can be used because on the one hand they are independent of the (often implicit) assumptions of the other two probability concepts, and on the other hand they allow all available knowledge to be used in order to arrive at better justified probability statements.

In principle, we can always use prior knowledge, because it is inconceivable that we would have no information about a quantity to be measured before performing a measurement, otherwise we would not be able to design an experiment. It is therefore appropriate to use such additional knowledge whenever we evaluate measurement data. Note, however, that neither the classical nor the frequentist probability concept allows prior knowledge to be taken into account.

Probability statements are always based on the information available at any given time, or on hypotheses. This fact is obvious in the case of subjective probability, but it also applies to the other probability concepts. In the case of classical probability, for example, it is implicitly assumed that possible events are equally likely; in the case of frequentist probability, it is implicitly assumed that the conditions for a random experiment according to the definition 4.2 are fulfilled.

By definition, classical and frequentist probability values can only be rational numbers, whereas subjective probability values are not restricted to rational numbers, but can in principle also be real numbers. In practice, this difference is not essential, since real numbers can be approximated by rational numbers as accurately as we wish. Furthermore, real numbers cannot be used for numerical calculations, where we always have to use numbers with a predetermined limited precision.

Looking for similarities, we can identify the following common (conventional) facts, regardless of which probability concept is used:

- The probability can be represented by a non-negative number that can only take values between and including zero and one.
- The case of impossibility corresponds to a probability value zero.
- The case of certainty corresponds to a probability value one.

It is easy to prove that, in the case of classical and frequentist probability, these properties follow directly from the definitions of these two probability concepts. However, it can be shown that they are also valid for subjective probability [36, 37].

At the beginning of this section we already stated that we do not really know what probability actually is. From a philosophical point of view this is certainly disappointing, but in practice it does not prevent us from doing calculations with probability values.

However, this requires a sound mathematical foundation, which will be the subject of the following sections.

4.2 Events and outcomes

In the previous section we talked about events several times without defining the term. This was only possible because we intuitively know what an event is, and we use this term frequently in the normal course of life. For a rigorous formulation of probability theory, however, we need an unambiguous definition.

In the conventional probability theory used today, the definition of the term "event" is based on the term "sample space", which is defined in an ISO standard (ISO 3534:2006, 2.1 [38]) as follows:

Definition 4.6 (Sample space)
A sample space is a set of all possible outcomes.
The sample space is usually denoted by Ω.

However, the meaning of the term "outcome" used in this definition is not given in the standard. This was not considered necessary because in conventional probability theory, as currently taught in universities, this term is used only in the sense of an outcome of a random experiment according to the definition 4.2. We will not follow this view, however, because it would imply an unnecessary restriction to frequentist probability.

In order to be able to include subjective probability in our considerations, we will understand an outcome as the realization of a circumstance, which we will specify as a logically meaningful proposition. To make this specification more precise, we add the definition of a (logical) proposition attributed by the historians of mathematics to CHRYSIPPUS OF SOLI.[6]

Definition 4.7 (Proposition)
A proposition is something that can be denied or affirmed as it is. A proposition is true if and only if the circumstances it expresses exist.

Examples of logically meaningful propositions in the sense of this definition are the following:
–   It is night.
–   A six has been thrown with this die.
–   There is life on Mars.

---

[6] Greek Stoic philosopher (* 281/276 BC in Soli, Cilicia, † 208/204 BC probably in Athens). He is considered to be the founder of classical axiomatic propositional logic.

It is not necessary to know whether a proposition is true or false, but it is essential that one and only one of these two possibilities can be true.

Of course, it is possible to make statements about the outcomes of random experiments, as well as about circumstances that are not related to random outcomes. We can assign probabilities to all different kinds of outcomes.

In order to illustrate the term "sample space" according to the definition 4.6, some examples are given:

Example 4.14 (Sample spaces)

a) If we flip a coin, we have the sample space

$$\Omega = \{H, T\},$$

where H denotes the result "the coin landed head up" and T denotes the result "the coin landed tail up".

b) If we roll a die, we have the sample space

$$\Omega = \{\boxdot, \boxdot, \boxdot, \boxdot, \boxdot, \boxdot\},$$

where $\boxdot$ denotes the outcome "the number one turned up", $\boxdot$ denotes the outcome "the number two turned up", ..., and $\boxdot$ denotes the outcome "the number six turned up".

c) If we flip two distinguishable coins, we have the sample space

$$\Omega = \{(H, H), (H, T), (T, H), (T, T)\}.$$

Note that in the case of two distinguishable coins the outcomes (H,T)—"the first coin landed head up and the second coin landed tail up"—and (T,H)—"the first coin landed tail up, and the second coin landed head up"—are not identical.

d) We are interested in the number of atoms in the universe. In this case we choose the sample space[7] $\Omega = \mathbb{N}$, because the number of atoms in the universe is certainly huge, but we are not able to give a reliable upper bound for it.

e) We are interested in the position of the next meteorite impact on the surface of the Earth. If we represent the position by longitude λ and latitude φ, respectively, the sample space is determined by

$$\Omega = \{(\varphi, \lambda) \,|\, -180° \leq \lambda \leq +180°, -90° \leq \varphi \leq +90°\} \subset \mathbb{R}^2,$$

i. e. Ω is a subset of the two-dimensional continuum[8] $\mathbb{R}^2$.

---

7 In mathematics, the symbol $\mathbb{N}$ denotes the set of natural numbers.
8 In mathematics, the symbol $\mathbb{R}$ denotes the set of real numbers. The set $\mathbb{R}$ and with it also the Cartesian product $\mathbb{R}^2 = \mathbb{R} \times \mathbb{R}$ are non-denumerable sets, also referred to as continua.

These examples show that not only finite sample spaces exist but also countably infinite or even continuous sample spaces. In applications of probability theory in metrology the latter are even the most important.

Having clarified the meaning of the terms "outcome" and "sample space", we are now ready to define the term "event". The standardized definition is (ISO 3534:2006, 2.2 [38]):

Definition 4.8 (Event)
An event is a subset of the sample space.

This definition shows that events can be represented by sets. However, this approach is only a convention that is common in probability theory today[9] and has proved its worth for about a century now.

An event $\mathfrak{E}$ is a *subset* of the sample space Ω, i. e. $\mathfrak{E} \subseteq \Omega$ is valid, where the symbol $\subseteq$ has the meaning "is subset of" (note that the relation $\mathfrak{E} \subseteq \Omega$ does not exclude the possibility that $\mathfrak{E}$ and Ω are equal). An outcome ω, on the other hand, is always an *element* of the sample space, i. e. $\omega \in \Omega$ is valid, where the symbol $\in$ has the meaning "is element of". Unfortunately, the two terms "event" and "outcome" are not always consistently distinguished in the literature.

As an example of representing events by sets, consider some of the possible events that can occur when a single die is rolled:

Example 4.15 (Some events when rolling a single die)
We represent the events by sets and quote the respective event:

$\{⚅\}$	"the number six turned up"
$\{⚀, ⚁, ⚂, ⚃, ⚄\}$	"the number six did *not* turn up"
$\{⚁, ⚃\}$	"the number two or the number four turned up"
$\{⚀, ⚂, ⚄\}$	"an odd number turned up"
$\{⚁, ⚃, ⚅\}$	"an even number turned up"
$\{\}$	"no number turned up"
$\{⚀, ⚁, ⚂, ⚃, ⚄, ⚅\}$	"any number turned up"

From this example we can deduce that there are events (represented by sets) which contain only a single outcome (element of the set). These events are called *elementary events* and are always represented by sets of the form $\{\omega\}$, where $\omega \in \Omega$. Therefore, we can give the following definition:

---

9 It should be mentioned that probability theory can also be based on mathematical logic, as proposed by R. CARNAP [39]. This approach, however, did not gain recognition.

Definition 4.9 (Elementary event)

An elementary event is represented by a set containing only a single element (outcome) of the sample space.

In addition, the example 4.15 shows that when using sets, the impossible event is represented by the empty set $\emptyset = \{\}$, while the certain event is represented by the sample space $\Omega = \{\boxdot, \boxdot, \boxdot, \boxdot, \boxdot, \boxdot\}$ itself. This is immediately understandable, since it is impossible for a die to show no number at all, and it is certain that one of the six possible outcomes will always occur. Our considerations lead to two further definitions:

Definition 4.10 (Impossible event)

An impossible event is represented by the empty set.

Definition 4.11 (Certain event)

A certain event is represented by the sample space itself.

Definition 4.8 states that an event, represented by a set, is a subset of the sample space Ω. The subset relation $\subseteq$ can generally be applied to any two events, as is evident from the example 4.15, since we have, for example, $\{\boxdot, \boxdot\} \subseteq \{\boxdot, \boxdot, \boxdot\}$. This particular subset relation can be interpreted as follows: The event "the number two or the number four turned up" implies the event "an even number turned up". This interpretation is obviously meaningful and easy to understand. Given two events, denoted by $\mathfrak{A}$ and $\mathfrak{B}$, we generally interpret the relation $\mathfrak{A} \subseteq \mathfrak{B}$ by a statement like "the event $\mathfrak{A}$ implies the event $\mathfrak{B}$", or "if the event $\mathfrak{A}$ occurs, then the event $\mathfrak{B}$ also occurs". Since the subset relation does not exclude equality, the events $\mathfrak{A}$ and $\mathfrak{B}$ can also be equal, but the event $\mathfrak{B}$ should of course never be the impossible event. If the equality of the events $\mathfrak{A}$ and $\mathfrak{B}$ is to be explicitly excluded, this can be expressed by the relation $\mathfrak{A} \subset \mathfrak{B}$, i. e. $\mathfrak{A}$ is a proper subset of $\mathfrak{B}$.

In addition to the subset relation, there are other set operations that can be applied to events represented by sets. The basic set operations and their corresponding event relations, as used in probability theory, are represented in Tab. 4.1 by EULER-VENN diagrams.[10] This compilation shows that each set operation can be associated with an event relation and *vice versa*. In this way events become mathematically treatable, i. e. we can apply the rules of the algebra of sets to events and thus establish an algebra of events. This is the first step towards a mathematically rigorous treatment of probability theory.

Set operations obey the rules of a Boolean algebra [41, 42] which are introduced and proved in set theory and mathematical logic. For further details we refer the interested reader to the relevant textbooks (e. g. the very good and comprehensible book by P. R. HALMOS [43]).

---

10 Similar diagrams were used by L. EULER and later, in a modified version, by the English logician J. VENN, to visualize logical relations between a finite collection of different sets. [40]

Tab. 4.1: Basic set operations and their corresponding event relations represented by EULER-VENN diagrams.

EULER-VENN diagram	Event relation
$\mathfrak{A} \subseteq \mathfrak{B}$	Subset relation of two sets: Event $\mathfrak{A}$ implies event $\mathfrak{B}$, i. e. if event $\mathfrak{A}$ occurs, event $\mathfrak{B}$ has to occur as well.
$\mathfrak{C}^c$	Complement of a set: $\mathfrak{C}^c$ is the complementary event of the event $\mathfrak{C}$, i. e. if event $\mathfrak{C}^c$ occurs, event $\mathfrak{C}$ does not occur.
$\mathfrak{A} \cup \mathfrak{B}$	Union of two sets: At least one of the two events $\mathfrak{A}$ and $\mathfrak{B}$ occurs, but they can conjointly occur as well.
$\mathfrak{A} \cap \mathfrak{B}$	Intersection of two sets: Events $\mathfrak{A}$ and $\mathfrak{B}$ conjointly occur.
$\mathfrak{A} \setminus \mathfrak{B}$	Difference of two sets: Event $\mathfrak{A}$ occurs without event $\mathfrak{B}$ occurring.

We add two other important definitions. The first follows from the following consideration. If the relation $\mathfrak{A} \cap \mathfrak{B} = \varnothing$ is valid for two arbitrary events $\mathfrak{A}$ and $\mathfrak{B}$, then this obviously means that a conjoint occurrence of these two events is the impossible event, i. e. the events are mutually exclusive. Therefore, we can give the following definition:

Definition 4.12 (Mutually exclusive events)
Two events $\mathfrak{A}$ and $\mathfrak{B}$ are mutually exclusive, if and only if
$$\mathfrak{A} \cap \mathfrak{B} = \varnothing.$$

Mutually exclusive events are also called disjoint events. They do not share any outcome of the sample space.

The second definition establishes the relationship between the set operations set-difference and set-complement.

Definition 4.13 (Set difference)
The difference of two sets $\mathfrak{A}$ and $\mathfrak{B}$ is defined by
$$\mathfrak{A} \setminus \mathfrak{B} = \mathfrak{A} \cap \mathfrak{B}^c.$$

Substituting $\mathfrak{A} = \Omega$ into the equation of this definition yields $\Omega \setminus \mathfrak{B} = \Omega \cap \mathfrak{B}^c$ and finally, by the rules of set theory, $\Omega \setminus \mathfrak{B} = \mathfrak{B}^c$. Thus, the set complement is always formed with respect to the sample space Ω. This justifies the corresponding EULER-VENN diagram of the complementary event in Tab. 4.1.

In certain cases, we are interested in decomposing an event into two or more disjoint events, if possible (note that elementary events cannot be decomposed, since they contain only one outcome). We call this decomposition a disjoint partition.

Definition 4.14 (Disjoint partition)
A disjoint partition of an event $\mathfrak{C}$ into two events $\mathfrak{A}$ and $\mathfrak{B}$ is given by
$$\mathfrak{A} \cap \mathfrak{B} = \varnothing \qquad \text{and} \qquad \mathfrak{A} \cup \mathfrak{B} = C.$$

Now we turn to the question of how many events are possible for a given sample space Ω. It is immediately obvious that this question is meaningful only in the case of a finite sample space, because the number of elementary events in a countably infinite or in a continuous sample space is already unlimited.

To begin with, we will only consider a finite sample space. We already know that events are subsets of the sample space Ω, i. e. all possible events are completely contained in the power set[11] $\mathcal{P}(\Omega)$.

---

[11] In set theory the power set $\mathcal{P}(\mathfrak{M})$ is defined as the set of all subsets of a given basic set $\mathfrak{M}$. According to the definition of a subset—which does not exclude the possibility that the sets are equal—the basic set is always contained in the power set. The empty set is also included in the power set.

If there are N possible outcomes in the sample space Ω, then—according to the rules of combinatorics—the power set $\mathcal{P}(\Omega)$ contains a total of 2^N subsets, i. e. a maximum of 2^N events are possible. We illustrate this fact with some examples:

Example 4.16 (Power sets of the sample space)
a) If there is no possible outcome, the sample space is empty, i. e. $\Omega = \emptyset$. This yields the power set $\mathcal{P}(\Omega) = \{\emptyset\}$, which contains only the impossible event. This case is of no interest in probability theory.

b) If only one outcome is possible, which we denote by ω, the sample space is $\Omega = \{\omega\}$. This yields the power set $\mathcal{P}(\Omega) = \{\emptyset, \Omega\}$, which contains two events, namely the impossible event and the certain event. This case is also not very interesting in probability theory, because an event either occurs or it does not occur.

c) The simplest case which is of interest in probability theory arises if two outcomes, denoted by ω_1 and ω_2, are possible. In this case the sample space is $\Omega = \{\omega_1, \omega_2\}$. This yields the power set $\mathcal{P}(\Omega) = \{\emptyset, \mathfrak{E}_1, \mathfrak{E}_2, \Omega\}$, with $\mathfrak{E}_1 = \{\omega_1\}$ and $\mathfrak{E}_2 = \{\omega_2\}$. The events $\mathfrak{E}_1$ and $\mathfrak{E}_2$ are complementary to each other, i. e. $\mathfrak{E}_1 = \mathfrak{E}_2^c$ and $\mathfrak{E}_2 = \mathfrak{E}_1^c$ are valid, because if one of the events occurs, the other one does not occur. This is called a BERNOULLI experiment. For example, flipping a coin can be considered a BERNOULLI experiment.

d) If three outcomes, denoted by ω_1, ω_2, and ω_3, are possible, the sample space is $\Omega = \{\omega_1, \omega_2, \omega_3\}$, yielding $\mathcal{P}(\Omega) = \{\emptyset, \mathfrak{E}_1, \mathfrak{E}_2, \mathfrak{E}_3, \mathfrak{E}_4, \mathfrak{E}_5, \mathfrak{E}_6, \Omega\}$, with the events $\mathfrak{E}_1 = \{\omega_1\}$, $\mathfrak{E}_2 = \{\omega_2\}$, $\mathfrak{E}_3 = \{\omega_3\}$, $\mathfrak{E}_4 = \{\omega_1, \omega_2\}$, $\mathfrak{E}_5 = \{\omega_1, \omega_3\}$, and $\mathfrak{E}_6 = \{\omega_2, \omega_3\}$.
Note that in each case the events $\mathfrak{E}_1$ and $\mathfrak{E}_6$, $\mathfrak{E}_2$ and $\mathfrak{E}_5$, as well as $\mathfrak{E}_3$ and $\mathfrak{E}_4$, are complementary to each other, i. e. only one of the respective events occurs at a time.

These examples show that in probability theory every power set has the property—besides always containing the impossible and the certain event—that, if it contains an event $\mathfrak{E}$, it always also contains the complementary event $\mathfrak{E}^c$ associated with it. It can be shown that this is always true for the power set of a finite set. Furthermore, it can be shown that any union or intersection of events is an event as well, i. e. it is also an element of the power set. We say that a finite power set is closed with respect to unions and intersections.

Finite power sets are useful in the probabilistic treatment of games of chance, but they are of little importance in metrology. Therefore, we also need to consider cases where the sample space Ω does not contain a finite number of outcomes. For example, in the example 4.14 d) we have the case $\Omega = \mathbb{N}$, i. e. all natural numbers are possible outcomes. Since the set $\mathbb{N}$ has infinitely many elements, the number of possible outcomes is also infinite. In the case of an infinite set it no longer makes sense to ask for the number of elements, but the term cardinality of the set is used instead. Cardinality can be considered as a generalization of the number of elements of a set. Although the set of natural numbers is not finite, it is in principle possible to count its elements, $1, 2, 3, \dots$ *ad infinitum*, i. e. the counting process never stops. However, this is no longer possible for the power set $\mathcal{P}(\mathbb{N})$. In set theory it is shown that the power set $\mathcal{P}(\mathbb{N})$ of the set of natural numbers $\mathbb{N}$ has the same cardinality as the set of real numbers $\mathbb{R}$, i. e. this power set is a continuum. Thus, in principle, we can assign a real number (just as a label) to each event of an uncountably infinite set of events, such as $\mathcal{P}(\mathbb{N})$, which is based on a countably infinite sample space, but in such a case it does not make sense to talk about counting events, because the real numbers are not countable.

The situation becomes even more difficult if we consider the case $\Omega = \mathbb{R}$, i. e. all real numbers are possible outcomes, as it is common in many applications in metrology (measurement results are usually real numbers, unless the measurement is a counting). In this case the cardinality of the power set is even greater than that of the continuum, because the set of real numbers is already the continuum and the power set of a set has always a greater cardinality than the set itself. In this case it is beyond human imagination to consider all subsets of the sample space contained in the power set still as events.

Therefore, we realize that in all cases where the sample space is not finite, the power set can no longer serve as a set of possible events because its cardinality is much too large. Since such cases generally occur in metrology, we need a set of events which, in the case of a finite sample space, coincides with the power set of the sample space, but whenever the sample space is infinite, this set shall have a smaller cardinality than the power set, but shall still have the essential properties of a power set as mentioned above. Such a set of events, called σ-algebra,[12] can be found among the subsets of the respective power set of a sample space that obey certain conditions.

Definition 4.15 (σ-algebra)

A σ-algebra $\mathcal{A}$ on a given sample space Ω is a subset of the power set $\mathcal{P}(\Omega)$ satisfying the following conditions:

a) The particular event Ω is contained in $\mathcal{A}$.

b) If $\mathcal{A}$ contains an event $\mathfrak{E}$, it also contains the complementary event $\mathfrak{E}^c = \Omega \setminus \mathfrak{E}$ with respect to Ω.

c) Any union of an arbitrary (not necessarily finite) number of events contained in $\mathcal{A}$ is also an event contained in $\mathcal{A}$.

Note that every σ-algebra based on this definition contains the impossible event, but it is not necessary to add this requirement explicitly to the conditions of the definition. The

---

12 Here the term "algebra" does *not* mean a branch of mathematics.

reason for this is that the impossible event is complementary to the certain event and therefore, according to condition b) of the definition, is also contained in the σ-algebra.

Definition 4.15 also assures that for a finite sample space Ω the power set $\mathcal{P}(\Omega)$ is *a fortiori* a σ-algebra, because it satisfies all conditions of this definition. Thus, whenever the sample space is finite, its power set can be used as a set of events. However, if the sample space is infinite, we will use the smallest proper subset of its power set, which contains all the events we are interested in, and still fulfils the conditions of a σ-algebra, as stated in the definition 4.15. We illustrate this concept with some examples of σ-algebras:

Example 4.17 (σ-algebras)
a) For every non-empty sample space Ω there exists a smallest σ-algebra, which is given by $\mathcal{A} = \{\emptyset, \Omega\}$.
b) For every finite and non-empty sample space Ω, there exists a largest σ-algebra, which is given by $\mathcal{A} = \mathcal{P}(\Omega)$.
c) If Ω denotes a sample space and $\mathfrak{E}$ any subset of Ω, then $\mathcal{A} = \{\emptyset, \mathfrak{E}, \mathfrak{E}^c, \Omega\}$ is a σ-algebra (see Fig. 4.4).
d) In the case of $\Omega = \mathbb{R}$, the so-called BOREL σ-algebra $\mathcal{B}(\mathbb{R})$ (E. BOREL [44]) is usually used, which contains all real intervals.

   The BOREL σ-algebra does not contain *all* subsets of the real numbers $\mathbb{R}$. However, it can be proved that $\mathcal{B}(\mathbb{R})$ has the same cardinality as $\mathbb{R}$, i. e. is also a continuum. In contrast, the power set $\mathcal{P}(\mathbb{R})$, being the set of all subsets of $\mathbb{R}$, has a greater cardinality than $\mathbb{R}$.

We summarize the essential points of this section. All possible outcomes of an experiment are contained in a—not necessarily finite—sample space Ω. All possible events are subsets of this sample space and form the set of events $\mathcal{A}$—a σ-algebra on Ω which is equal to the power set $\mathcal{P}(\Omega)$, provided that Ω is finite—which contains all events of interest and all complements of these events, as well as the certain and the impossible event. Now, if ω is any outcome in the sample space Ω and $\mathfrak{E}$ any event of the set of events $\mathcal{A}$, we say that the event can occur, if and only if the outcome ω is an element of the set $\mathfrak{E}$. Conversely, we say that the event cannot occur, if and only if the outcome ω is an element of the set $\mathfrak{E}^c$ which is complementary to $\mathfrak{E}$ (see Fig. 4.4).

Finally, we note that the ordered pair $(\Omega, \mathcal{A})$ of the sample space Ω and the σ-algebra $\mathcal{A}$ on it is called "measurable space". This term has its origin in measure theory, which is a branch of mathematics and is today one of the foundations of conventional probability theory.

4.3 Mathematical probability

The previous section showed that events can be described by sets. In this section we will show how events can be assigned probability values. In contrast to the more philosophical discussion at the beginning of this chapter, we will now deal with mathematical probability, which is in principle free of any interpretation. We can, for example, interpret a resulting probability value as a relative frequency, as defined by frequentist probability, or, in the sense of subjective probability, as a degree of personal belief that an event will occur or has occurred. This statement is necessary because of the common misconception that mathematical probability is necessarily frequentist. This is not true, however, because mathematical probability is merely an abstract theory of measures, without any reference to the real world. Such a statement is, of course, by no means new. It can be found in the publication of A. N. KOLMOGOROV [45, 46] where he published his axioms of probability theory. This is hardly surprising, since axioms are generally without real meaning unless an interpretation is given.

First we want to clarify what it means to assign a probability value to an event. We have already noted in the previous section that events can be represented by sets, and we have seen in section 4.1 that probabilities are conventionally represented by nonnegative real numbers not greater than one. Assigning a probability value to an event is thus just a special case of the problem of uniquely assigning a non-negative real number to a set. The restriction to the interval $[0, 1]$ is just a specific characteristic of probability theory, which we ignore for the moment.

If we can assign a non-negative real number $\mu(\mathfrak{M})$ to a set $\mathfrak{M}$, the set is called measurable and the number is called its measure. However, it soon turned out that not every set is measurable, as was first demonstrated in 1905 by G. VITALI [47]. But it can be proved that sets of a σ-algebra, which we defined in the previous section, are always measurable. This explains why the ordered pair $(\Omega, \mathcal{A})$ of the sample space Ω and the σ-algebra $\mathcal{A}$ on it is called a "measurable space". Similarly, the ordered triple $(\Omega, \mathcal{A}, \mu)$ with measure μ is called "measure space".

A measure can be regarded as a generalization of the length, the area or the volume of a geometric object. Suppose that the sets under consideration are, for example, geometric figures in a plane. Then their areas can be treated as measures of the sets, because

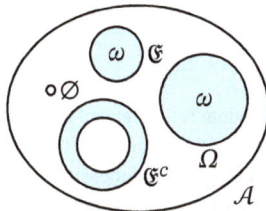

Fig. 4.4: The event $\mathfrak{E}$ can occur, if the outcome ω is contained in $\mathfrak{E}$.

they are non-negative real numbers. Moreover, it is obvious from everyday experience that the total area of two non-overlapping figures in a plane (corresponding to a union of disjoint sets) is equal to the sum of the areas of the two individual figures. These properties are not only characteristic of areas, but of every measure. Thus, we can state the following measure axioms:[13]

Definition 4.16 (Measure axioms)

If $\mathfrak{A}$, $\mathfrak{B}$ and $\mathfrak{M}$ denote measurable sets and $\mu(\mathfrak{M})$ denotes the measure of the set $\mathfrak{M}$, then the following axioms apply:

a) $\mu(\mathfrak{M}) \geq 0$,   (non-negativity)

b) $\mu(\mathfrak{A} \cup \mathfrak{B}) = \mu(\mathfrak{A}) + \mu(\mathfrak{B})$, if $\mathfrak{A} \cap \mathfrak{B} = \varnothing$.   (additivity)

According to this definition, measures are non-negative[14] and additive quantities, which can be interpreted as mappings of arbitrary sets to the set of real numbers, i. e. as functions where the images are non-negative real numbers and the inverse images are sets. In general the inverse functions do not exist, i. e. a given measure does not necessarily give any information about the underlying set. This statement becomes more understandable if we look again at the example of geometric figures in a plane. Two figures of different shapes can certainly have exactly the same area. Similarly, when applied to measurable sets, this means that the same measure can be assigned to different sets. The geometric analogy can also be used to illustrate the concept of null sets in measure theory. These are sets of measure zero. In our geometric model, these sets correspond to points, lines or curves in a plane, to which, obviously, an area of zero has to be assigned.

The non-negativity requirement in the definition 4.16 ensures that a measure is bounded from below, but generally not from above. This makes sense for measures such as length, area, volume, mass, etc. However, under certain circumstances it is also necessary to specify an upper bound. In these cases we speak of a "normalized measure".

Based on these considerations we can now give the definition of mathematical probability, which can be considered as a special measure with the additional restriction that the image is restricted to the real interval $[0, 1]$, i. e. mathematical probability is defined as follows:

Definition 4.17 (Mathematical probability)

Mathematical probability is a measure normalized to unity.

---

13 In mathematics and philosophy, an axiom is a statement that, given a certain interpretation, is immediately evident and is therefore accepted without controversy or proof.

14 In measure theory there are also so-called *signed* measures. This is a generalization of the measure concept and allows measures to take negative values. However, this type of measure is not necessary in probability theory.

The mathematical probability of an arbitrary event, denoted by $\mathfrak{E}$, will be denoted by $P(\mathfrak{E})$ throughout.

In probability theory the term "probability space" is used instead of the more general term "measure space". A probability space will be denoted by the ordered triple $(\Omega, \mathcal{A}, P)$.

The general measure axioms 4.16, along with the additional requirement of a normed measure restricted to the real interval $[0, 1]$, lead to the following axioms of mathematical probability which were first published in 1933 by A. N. KOLMOGOROV [45, 46]:

Definition 4.18 (Axioms of A. N. KOLMOGOROV)
Let $\mathcal{A}$ be a set of events on a sample space Ω, and let $\mathfrak{E}$, $\mathfrak{E}_1$ and $\mathfrak{E}_2$ be events in $\mathcal{A}$, then the measure $P(\mathfrak{E})$ is called the mathematical probability of $\mathfrak{E}$ if and only if $\mathcal{A}$ is a σ-algebra on Ω and the following conditions are satisfied:

K1: $P(\mathfrak{E}) \geq 0$, (non-negativity)
K2: $P(\Omega) = 1$, (normalization)
K3: $P(\mathfrak{E}_1 \cup \mathfrak{E}_2) = P(\mathfrak{E}_1) + P(\mathfrak{E}_2)$, if $\mathfrak{E}_1 \cap \mathfrak{E}_2 = \emptyset$. (additivity)

These axioms state that probability values must be non-negative, that the probability value of the certain event is given by the number one, and that probability is additive provided that the underlying events are disjoint. There is no axiom stating that the number zero has to be assigned to the impossible event. This is not necessary, since this property can be derived as a theorem from the axioms.

Proposition 4.1 (Probability of the impossible event)
The impossible event has probability zero, i. e.

$$P(\emptyset) = 0.$$

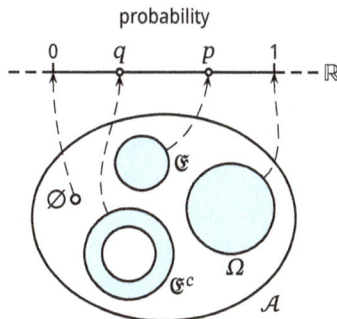

Fig. 4.5: Illustration of an assignment of probabilities to the events contained in the set of events $\mathcal{A} = \{\emptyset, \mathfrak{E}, \mathfrak{E}^c, \Omega\}$ (σ-algebra on Ω). The event $\mathfrak{E}$ occurs with probability $P(\mathfrak{E}) = p$, the complementary event $\mathfrak{E}^c$ with probability $P(\mathfrak{E}^c) = q$.

Note that the impossible event has a zero probability, but it is not true that every event with a zero probability is impossible. We must strictly distinguish between these two cases, because there exists only one impossible event which has a zero probability of occurring, but on the contrary, events with a zero probability are possible but will not happen. In principle, all events are possible (except, of course, the impossible event), but just because something is possible does not necessarily mean that it is probable. Fig. 4.5 on page 117 illustrates by a simple example the assignment of probability values to events of a set of events $\mathcal{A}$ on the sample space Ω.

The axioms of KOLMOGOROV do not completely specify mathematical probability, because the term "conditional probability" is still missing. This term was not introduced axiomatically by A. N. KOLMOGOROV, but specified by a definition [45, 46]. This is a permissible approach, but it has the disadvantage that conditional probability is treated as a special case. We will demonstrate in the next section that axioms of mathematical probability can be stated which already include conditional probability. The axioms of KOLMOGOROV will turn out to be just a special case of these more general axioms.

4.4 Conditional probability

We have already used the term "conditional probability" in section 4.1, where we discussed subjective probability, and also in the example 4.13, but in both cases without explaining its meaning. However, it was intuitively clear that a probability with an additional specified condition was meant.

Although conditional probability is essential to the subjective concept of probability, it is not restricted to it. Conditional probability can be defined for any concept of probability.

In order to establish an understanding of the importance of the concept of conditional probability, we begin with some examples which will lead us step by step to the definition of conditional probability.

Example 4.18 (Drawing cards)
Given a standard deck of cards with 52 cards, we consider the following two cases:
a) What is the probability of drawing another club card, if a club card has already been drawn before?
    There are a total of 13 club cards in the deck of cards. If a club card has already been drawn, there are only 12 club cards in the remaining 51 cards in the deck. Thus, the probability is $P(\clubsuit \mid \clubsuit) = 12/51$.
b) What is the probability of drawing a club card, if a diamond card has already been drawn before?

There are a total of 13 club cards in the deck of cards. If a diamond card (i. e. *no* club card) has been drawn before, there are still 13 club cards among the remaining 51 cards in the deck. Thus, the probability is $P(\clubsuit \mid \diamondsuit) = 13/51$.

We observe that the condition has an effect on the probability value, i. e. different conditions generally give different probability values.

Example 4.19 (Two rolls with an ideally symmetric die)
We roll an ideally symmetric die twice and calculate the sum of the number of dots coming up. A simple consideration shows that only five possibilities exist to obtain the number six for this sum (see the following table).

		2nd roll					
		1	2	3	4	5	6
1st roll	1	2	3	4	5	6	7
	2	3	4	5	6	7	8
	3	4	5	6	7	8	9
	4	5	6	7	8	9	10
	5	6	7	8	9	10	11
	6	7	8	9	10	11	12

We are interested in the probability that the sum of the number of dots coming up is six, given that four dots come up on one of the two rolls.

The table summarizes all 36 possible values of the sum of the number of dots coming up when the die is rolled twice. We consider the following:
- $\mathfrak{A}$ denotes the event that the number of dots coming up adds up to six.
- $\mathfrak{B}$ denotes the event that four dots come up on only one of the two rolls.

We can see from the table that the event $\mathfrak{B}$ (four dots come up in one of the two rolls) can occur in ten of the 36 possible cases (these ten events are underlined in the table), i. e. we have the probability $P(\mathfrak{B}) = 10/36 = 5/18$.

Moreover, we can see from the table that the event $\mathfrak{A} \cap \mathfrak{B}$ (both the number of dots coming up adds up to six and four dots coming up in only one of the two rolls) can occur in two of the 36 possible cases (these two possible events are framed in the table), i. e. we get the probability $P(\mathfrak{A} \cap \mathfrak{B}) = 2/36 = 1/18$.

But under the condition $\mathfrak{B}$ (four dots come up in one of the two rolls) there exist only ten (instead of 36) possible cases for the event $\mathfrak{A}$ (the number of dots that come up adds up to six), because all other cases are out of the question right from the start, since they do not fulfil the condition $\mathfrak{B}$. Of these ten cases, only two yield a sum of the number of dots coming up equal to six, i. e. we obtain the conditional probability $P(\mathfrak{A} \mid \mathfrak{B}) = 2/10 = 1/5$.

This example shows that there is a difference between the probability $P(\mathfrak{A} \cap \mathfrak{B})$, that the events $\mathfrak{A}$ and $\mathfrak{B}$ occur together, and the probability $P(\mathfrak{A} \mid \mathfrak{B})$, that the event $\mathfrak{A}$ occurs under the condition that the event $\mathfrak{B}$ occurs. The reason for this is that in the first case *all* possible events are taken into account, while in the second case only those events $(\mathfrak{A} \cap \mathfrak{B})$ are counted where the number four *definitely* comes up in one of the two rolls.

Example 4.20 (Random experiment)

We perform a random experiment n times, counting both how often the events $\mathfrak{A}$ and $\mathfrak{B}$ occur, and how often the event $\mathfrak{A} \cap \mathfrak{B}$ occurs (i. e. the events $\mathfrak{A}$ and $\mathfrak{B}$ occur together). We get the result:

– the event $\mathfrak{B}$ occurred $n(\mathfrak{B})$ times, and
– the event $\mathfrak{A} \cap \mathfrak{B}$ occurred $n(\mathfrak{A} \cap \mathfrak{B})$ times.

We ask for the conditional probability $P(\mathfrak{A} \mid \mathfrak{B})$ that event $\mathfrak{A}$ will occur, given that event $\mathfrak{B}$ has occurred.

We assume that the number of trials n is large enough that the relative frequencies can be assumed to be equal to the probabilities of the respective events. With this assumption we obtain

$$P(\mathfrak{A} \cap \mathfrak{B}) = \frac{n(\mathfrak{A} \cap \mathfrak{B})}{n} .$$

Here we have related the number of occurrences of the event $(\mathfrak{A} \cap \mathfrak{B})$ to the total number of experiments performed. But this procedure is no longer correct if we want to determine the conditional probability $P(\mathfrak{A} \mid \mathfrak{B})$, because in this case only the number $n(\mathfrak{B})$ is relevant in which the event $\mathfrak{B}$ actually occurred, not the total number n of experiments performed, i. e. we restrict the counting of the occurrences of the event $(\mathfrak{A} \cap \mathfrak{B})$ to these cases only. Therefore the denominator has to be $n(\mathfrak{B})$ instead of n. The numerator, however, remains unchanged, i. e. is still $n(\mathfrak{A} \cap \mathfrak{B})$, because only those cases of the event $\mathfrak{A}$ are counted where the event $\mathfrak{B}$ also occurred. Thus, we have

$$P(\mathfrak{A} \mid \mathfrak{B}) = \frac{n(\mathfrak{A} \cap \mathfrak{B})}{n(\mathfrak{B})} .$$

Note that the probabilities $P(\mathfrak{A} \cap \mathfrak{B})$ and $P(\mathfrak{A} \mid \mathfrak{B})$ differ by their denominator.

We see that for the conditional probability $P(\mathfrak{A} \mid \mathfrak{B})$ it is no longer reasonable to take into account *all* outcomes contained in the original sample space Ω, but we have to restrict them to the outcomes only contained in $\mathfrak{B}$, because our requirement is that the event $\mathfrak{B}$ certainly occurs. This means in the end that we have to consider the probability space $(\mathfrak{B}, \mathcal{A}', P')$ instead of the original probability space $(\Omega, \mathcal{A}, P)$, where $\mathcal{A}'$ contains only those events, which are proper subsets of $\mathfrak{B}$, and P' denotes the probability of the newly created probability space. This probability is just the conditional probability under the condition given by the occurrence of the event $\mathfrak{B}$, i. e. for any event $\mathfrak{E}$ in the set of possible events $\mathcal{A}'$ we have $P'(\mathfrak{E}) = P(\mathfrak{E} \mid \mathfrak{B})$.

Since the event $\mathfrak{B}$ plays the same rôle in the new probability space $(\mathfrak{B}, \mathcal{A}', P')$ as the event Ω plays in the original probability space $(\Omega, \mathcal{A}, P)$—both events are indeed the certain events in their respective probability spaces—and since the probability $P'(\mathfrak{A}) = P(\mathfrak{A} \mid \mathfrak{B})$ shall be proportional to the probability $P(\mathfrak{A} \cap \mathfrak{B})$, because the former is merely obtained from the latter by a new normalization, we can state the following requirements, which have to be fulfilled by the conditional probability

$$P(\mathfrak{B} \mid \mathfrak{B}) = 1, \tag{4.1}$$

and

$$P(\mathfrak{A} \mid \mathfrak{B}) = c(\mathfrak{B})P(\mathfrak{A} \cap \mathfrak{B}), \text{ if } P(\mathfrak{B}) > 0, \tag{4.2}$$

with a real constant $c(\mathfrak{B})$ depending on $\mathfrak{B}$, with the additional condition $P(\mathfrak{B}) > 0$ in relation (4.2), because it makes no sense to consider a condition with a probability of zero. If we set $\mathfrak{A} = \mathfrak{B}$ in condition (4.2), we initially obtain $P(\mathfrak{B} \mid \mathfrak{B}) = c(\mathfrak{B})P(\mathfrak{B})$, provided that $P(\mathfrak{B}) > 0$ is valid, and with condition (4.1) this immediately results in $c(\mathfrak{B}) = 1/P(\mathfrak{B})$, provided that $P(\mathfrak{B}) > 0$ is valid. Finally, inserting this constant into the condition (4.2) leads to the definition:

Definition 4.19 (Conditional probability)
Let $(\Omega, \mathcal{A}, P)$ be a probability space and $\mathfrak{A}$ and $\mathfrak{B}$ be events in $\mathcal{A}$, then the conditional probability of $\mathfrak{A}$, given $\mathfrak{B}$, is

$$P(\mathfrak{A} \mid \mathfrak{B}) = \frac{P(\mathfrak{A} \cap \mathfrak{B})}{P(\mathfrak{B})}, \tag{4.3}$$

provided that $P(\mathfrak{B}) > 0$.

Note that the conditional probability $P(\mathfrak{A} \mid \mathfrak{B})$ is undefined if $P(\mathfrak{B}) = 0$. Thus, we must always require that the condition $\mathfrak{B}$ has a non-zero probability.

If we set $\mathfrak{B} = \Omega$ in equation (4.3) and apply axiom K2, we obtain

$$P(\mathfrak{A} \mid \Omega) = \frac{P(\mathfrak{A} \cap \Omega)}{P(\Omega)} = \frac{P(\mathfrak{A})}{1} = P(\mathfrak{A}), \tag{4.4}$$

where in this case, of course, the condition $P(\Omega) > 0$ is always satisfied. The result (4.4) shows that every absolute probability $P(\mathfrak{A})$ can be considered as a conditional probability $P(\mathfrak{A} \mid \Omega)$, i. e. there are only conditional probabilities, because Ω is just a special case of a given condition.

On the basis of these results, we can conjecture that the axioms of KOLMOGOROV should also hold for any conditional probability. It can be shown that this conjecture is *de facto* the case:

– The non-negativity of the conditional probability $P(\mathfrak{A} \mid \mathfrak{B})$ follows immediately from equation (4.3) and the non-negativity of the absolute probabilities $P(\mathfrak{A} \cap \mathfrak{B})$ and $P(\mathfrak{B})$, which is guaranteed by axiom K1.

– The normalization $P(\mathfrak{B} \mid \mathfrak{B}) = 1$ of the conditional probability also follows from equation (4.3) if we set $\mathfrak{A} = \mathfrak{B}$, because $\mathfrak{B} \cap \mathfrak{B} = \mathfrak{B}$ is valid.
– To verify the additivity of the conditional probability, we set $\mathfrak{A} = \mathfrak{C}_1 \cup \mathfrak{C}_2$ in equation (4.3), assuming $\mathfrak{C}_1 \cap \mathfrak{C}_2 = \varnothing$ to hold. This yields

$$P(\mathfrak{C}_1 \cup \mathfrak{C}_2 \mid \mathfrak{B}) = \frac{P((\mathfrak{C}_1 \cup \mathfrak{C}_2) \cap \mathfrak{B})}{P(\mathfrak{B})}, \text{ if } P(\mathfrak{B}) > 0 \text{ and } \mathfrak{C}_1 \cap \mathfrak{C}_2 = \varnothing.$$

Using the distributive law of set theory, this equation becomes

$$P(\mathfrak{C}_1 \cup \mathfrak{C}_2 \mid \mathfrak{B}) = \frac{P((\mathfrak{C}_1 \cap \mathfrak{B}) \cup (\mathfrak{C}_2 \cap \mathfrak{B}))}{P(\mathfrak{B})}, \text{ if } P(\mathfrak{B}) > 0 \text{ and } \mathfrak{C}_1 \cap \mathfrak{C}_2 = \varnothing.$$

Applying axiom K3 to this result and using equation (4.3) again, we finally obtain

$$P(\mathfrak{C}_1 \cup \mathfrak{C}_2 \mid \mathfrak{B}) = P(\mathfrak{C}_1 \mid \mathfrak{B}) + P(\mathfrak{C}_2 \mid \mathfrak{B}), \text{ if } P(\mathfrak{B}) > 0 \text{ and } \mathfrak{C}_1 \cap \mathfrak{C}_2 = \varnothing.$$

This proves the additivity of conditional probability.

Rearranging the equation (4.3) yields the product rule:

Proposition 4.2 (Product rule)
Let $\mathcal{A}$ be a set of events on a sample space Ω, and let $\mathfrak{A}$ and $\mathfrak{B}$ be events in $\mathcal{A}$, then

$$P(\mathfrak{A} \cap \mathfrak{B}) = P(\mathfrak{A} \mid \mathfrak{B}) P(\mathfrak{B}),$$

provided that $P(\mathfrak{B}) > 0$.

Note that because of its deduction, the product rule is only valid under the condition $P(\mathfrak{B}) > 0$. Sometimes it is argued that this additional condition is not necessary, because $P(\mathfrak{B}) = 0$ implies $P(\mathfrak{A} \cap \mathfrak{B}) = 0$. Admittedly, this statement is true. We can illustrate this with a geometrical example. Suppose that $\mathfrak{B}$ represents an arbitrary curve in a plane, which of course has no area, while $\mathfrak{A}$ represents an extended planar figure of any shape. Then $\mathfrak{A} \cap \mathfrak{B}$ obviously represents the part of the curve inside the planar figure, which still does not have an area. Despite this fact we cannot dispense with the requirement $P(\mathfrak{B}) > 0$, because the conditional probability $P(\mathfrak{A} \mid \mathfrak{B})$ is undefined whenever $P(\mathfrak{B}) = 0$.

We summarize all the results derived in this section by stating the axioms of conditional probability first given by A. RÉNYI [48]:

Definition 4.20 (Axioms of conditional probability)
Let $\mathcal{A}$ be a set of events on a sample space Ω and $\mathcal{B}$ be a proper subset of $\mathcal{A}$. Let $\mathfrak{B}$ be an event in $\mathcal{B}$ with $P(\mathfrak{B}) > 0$ and $\mathfrak{C}$, $\mathfrak{C}_1$ and $\mathfrak{C}_2$ be events in $\mathcal{A}$, then the measure $P(\mathfrak{C} \mid \mathfrak{B})$ is called the conditional probability of $\mathfrak{C}$ given $\mathfrak{B}$ if and only if $\mathcal{A}$ is a σ-algebra on Ω and the following conditions are satisfied:
A1: $P(\mathfrak{C} \mid \mathfrak{B}) \geq 0,$ (non-negativity)

A2: $P(\mathfrak{B} \mid \mathfrak{B}) = 1$, (normalization)

A3: $P(\mathfrak{C}_1 \cup \mathfrak{C}_2 \mid \mathfrak{B}) = P(\mathfrak{C}_1 \mid \mathfrak{B}) + P(\mathfrak{C}_2 \mid \mathfrak{B})$, if $\mathfrak{C}_1 \cap \mathfrak{C}_2 = \varnothing$, (additivity)

A4: $P(\mathfrak{C} \cap \mathfrak{B}) = P(\mathfrak{C} \mid \mathfrak{B})P(\mathfrak{B})$. (multiplicativity)

Note that the additional requirement $P(\mathfrak{B}) > 0$ is a necessary condition for these axioms, because otherwise the conditional probability is undefined and thus all the axioms would become meaningless.

The definition 4.20 of the product rule has now become an axiom and therefore is no longer necessary. This rule, like the definition 4.19 of the conditional probability, is only necessary with the axioms of KOLMOGOROV.

In the case of $\mathfrak{B} = \Omega$ and with the help of the equation (4.4), the axioms A1, A2 and A3 become identical to the axioms of KOLMOGOROV as stated in section 4.3, while using KOLMOGOROV's axiom K2 and the equation (4.4), the axiom A4 becomes an identity. Thus, KOLMOGOROV's axioms are only a special case of the more general axioms of conditional probability given by A. RÉNYI.

4.5 Rules of probability calculations

In this section we will draw some conclusions from the axioms of conditional probability given in the previous section. This will give us the rules that underlie probability calculations.

First, we make a statement about the conditional probability of the impossible event:

Proposition 4.3 (Conditional probability of the impossible event)
The conditional probability of the impossible event is zero, i. e.

$$P(\varnothing \mid \mathfrak{B}) = 0,$$

provided that $P(\mathfrak{B}) > 0$.

This property of the conditional probability already results from its axioms, and thus it is not necessary to postulate it as a separate axiom. In the case of the unconditional probability we have already been able to draw a similar conclusion from KOLMOGOROV's axioms.

The next proposition relates the probability of an event to the probability of its complementary event.

Proposition 4.4 (Conditional probability of the complementary event)
Let $\mathfrak{C}$ be any event and $\mathfrak{C}^c$ be its complementary event, then

$$P(\mathfrak{C}^c \mid \mathfrak{B}) = 1 - P(\mathfrak{C} \mid \mathfrak{B}),$$

provided that $P(\mathfrak{B}) > 0$.

In order to illustrate the application of this proposition, we give some simple examples:

Example 4.21 (Probability of complementary events)
a) When rolling a die, the probability of a six coming up is 1/6. Thus, the probability of *no* six coming up is 5/6.
b) When flipping a fair coin, the probability that "head" will come up is 1/2. The probability that "head" does *not* come up is also 1/2.
c) In figure 4.5 in section 4.3 we have $q = 1 - p$.

If two events $\mathfrak{E}_1$ and $\mathfrak{E}_2$ conjointly occur, i. e. if the event $\mathfrak{E}_1 \cap \mathfrak{E}_2$ occurs, then this conjoint event implies both the event $\mathfrak{E}_1$ and the event $\mathfrak{E}_2$, i. e. $\mathfrak{E}_1 \cap \mathfrak{E}_2 \subseteq \mathfrak{E}_1$ and $\mathfrak{E}_1 \cap \mathfrak{E}_2 \subseteq \mathfrak{E}_2$ is valid (see also table 4.1 in section 4.2). If both events depend on the same condition $\mathfrak{B}$, the following proposition is valid:

Proposition 4.5 (Conditional probability of conjoint events)
Let $\mathfrak{E}_1$ and $\mathfrak{E}_2$ be conjoint events that depend on the same condition $\mathfrak{B}$, then both

$$P(\mathfrak{E}_1 \cap \mathfrak{E}_2 \,|\, \mathfrak{B}) \leq P(\mathfrak{E}_1 \,|\, \mathfrak{B})$$

and

$$P(\mathfrak{E}_1 \cap \mathfrak{E}_2 \,|\, \mathfrak{B}) \leq P(\mathfrak{E}_2 \,|\, \mathfrak{B}),$$

are valid, provided that $P(\mathfrak{B}) > 0$.

We illustrate this proposition by an example:

Example 4.22 (Drawing coloured objects)
A container is filled with coloured geometric objects (balls, cubes, tetrahedra, etc.). We randomly draw one of the objects from the container. Assuming that the condition $\mathfrak{B}$ is known, we ask for the conditional probabilities $P(\mathfrak{E}_1 \,|\, \mathfrak{B})$, $P(\mathfrak{E}_2 \,|\, \mathfrak{B})$ and $P(\mathfrak{E}_1 \cap \mathfrak{E}_2 \,|\, \mathfrak{B})$, where
- $\mathfrak{B}$ represents the information "the container is filled with geometric objects of different colours and shapes",
- $\mathfrak{E}_1$ represents the event "the object is a ball",
- $\mathfrak{E}_2$ represents the event "the colour of the object is red", and
- $\mathfrak{E}_1 \cap \mathfrak{E}_2$ represents the event "the object is a red ball".

Obviously the probability of drawing a red ball is smaller than the probability of drawing a ball of any colour or a red object of any shape. However, this is not true, if $\mathfrak{B}$ represents the information "the container is only filled with red balls" or "there are no red balls in the container", because then the probabilities of the events $\mathfrak{E}_1$, $\mathfrak{E}_2$ and $\mathfrak{E}_1 \cap \mathfrak{E}_2$ are all equal.

The following proposition can be derived from proposition 4.5:

Proposition 4.6 (Monotonicity relation of conditional probability)

If the event $\mathfrak{C}_2$ implies the event $\mathfrak{C}_1$, i. e. if $\mathfrak{C}_2 \subseteq \mathfrak{C}_1$, then the monotonicity relation

$$P(\mathfrak{C}_2 \mid \mathfrak{B}) \leq P(\mathfrak{C}_1 \mid \mathfrak{B})$$

of conditional probability holds, provided that $P(\mathfrak{B}) > 0$.

This proposition means that an event that implies another event (see table 4.1 in section 4.2) cannot have a greater probability than the more general event. We illustrate this statement with the following example:

Example 4.23 (A shipwrecked person on an island)

A shipwrecked person is stranded on an island. Using the notations
– $\mathfrak{B}$ represents the information "the island is inhabited",
– $\mathfrak{C}_1$ represents the event "the shipwrecked person meets a human being",
– $\mathfrak{C}_2$ represents the event "the shipwrecked person meets a woman",

and assuming that $\mathfrak{B}$ is already known, we ask for the conditional probabilities $P(\mathfrak{C}_1 \mid \mathfrak{B})$ and $P(\mathfrak{C}_2 \mid \mathfrak{B})$.

Obviously, the probability that the shipwrecked person will meet a woman is smaller than the probability that this person will meet a human being on the island. However, this will *not* be true, if $\mathfrak{B}$ represents the information "there are only women on the island" or "the island is uninhabited", because then the probabilities of the events $\mathfrak{C}_1$ and $\mathfrak{C}_2$ are equal.

The next proposition is a direct consequence of proposition 4.6.

Proposition 4.7 (Boundedness of the conditional probability)

Conditional probability is bounded, i. e. if $\mathfrak{C}$ is some event and $\mathfrak{B}$ some condition, then

$$0 \leq P(\mathfrak{C} \mid \mathfrak{B}) \leq 1$$

is valid, provided that $P(\mathfrak{B}) > 0$.

The lower bound in this proposition is taken for $\mathfrak{C} = \emptyset$, i. e. in the case of the impossible event. This is a consequence of the proposition 4.3. The upper bound is taken for $\mathfrak{C} = \mathfrak{B}$. This is a consequence of axiom A2.

Proposition 4.7 ensures that the conditional probability can only take values between zero and one (including these values). Hence, this property is also already included in the axioms.

We now turn again to axiom A3 in definition 4.20. This axiom only describes the additivity of probabilities of mutually exclusive events. In practice, however, we need a more general addition theorem for conditional probabilities. This can be derived from the axioms. We state it here without proof.

Proposition 4.8 (Addition theorem of conditional probability)
Let $\mathfrak{E}_1$ and $\mathfrak{E}_2$ be arbitrary events, not necessarily disjoint, then

$$P(\mathfrak{E}_1 \cup \mathfrak{E}_2 \mid \mathfrak{B}) = P(\mathfrak{E}_1 \mid \mathfrak{B}) + P(\mathfrak{E}_2 \mid \mathfrak{B}) - P(\mathfrak{E}_1 \cap \mathfrak{E}_2 \mid \mathfrak{B}),$$

provided that $P(\mathfrak{B}) > 0$.

We demonstrate the application of the general addition theorem for conditional probabilities with a simple example:

Example 4.24 (Flipping a fair coin two times)
A fair coin is flipped two times. What is the probability that "head" will appear on at least one flip?

This random experiment has $\Omega = \{(H, H), (H, T), (T, H), (T, T)\}$ as its sample space, where the letters H and T represent the outcomes "head" and "tail", respectively, and the expression (ω_1, ω_2) means that the outcome is ω_1 on the first flip and ω_2 on the second flip. If we specify that

– $\mathfrak{B}$ represents the information "the coin is fair",
– $\mathfrak{E}_1 = \{(T, H), (T, T)\}$ represents the event "tail appears on the first flip",
– $\mathfrak{E}_2 = \{(H, T), (T, T)\}$ represents the event "tail appears on the second flip",
– $\mathfrak{E}_1 \cap \mathfrak{E}_2 = \{(T, T)\}$ represents the event "tail appears on both flips", and
– $\mathfrak{E}_1 \cup \mathfrak{E}_2 = \{(T, H), (H, T), (T, T)\}$ represents the event "tail appears on at least one flip",

the addition theorem 4.8 yields

$$P(\mathfrak{E}_1 \cup \mathfrak{E}_2 \mid \mathfrak{B}) = P(\mathfrak{E}_1 \mid \mathfrak{B}) + P(\mathfrak{E}_2 \mid \mathfrak{B}) - P(\mathfrak{E}_1 \cap \mathfrak{E}_2 \mid \mathfrak{B}) = \frac{1}{2} + \frac{1}{2} - \frac{1}{4} = \frac{3}{4}.$$

A direct calculation of the conditional probability $P(\mathfrak{E}_1 \cup \mathfrak{E}_2 \mid \mathfrak{B})$—e. g. by using classical probability—of course yields the same result.

The condition $\mathfrak{B}$, i. e. the information "the coin is fair", is essential here, because otherwise it is not possible to use the methods of combinatorics—in the sense of classical probability—to calculate the particular probabilities.

Axiom A4 establishes a link between conditional and unconditional probabilities. However, it can be generalized in a way that only conditional probabilities are used. This leads to the following theorem:

Proposition 4.9 (Multiplication theorem of conditional probability)
Let $\mathfrak{E}_1$ and $\mathfrak{E}_2$ be arbitrary events, then

$$P(\mathfrak{E}_1 \cap \mathfrak{E}_2 \mid \mathfrak{B}) = P(\mathfrak{E}_1 \mid \mathfrak{E}_2 \cap \mathfrak{B})P(\mathfrak{E}_2 \mid \mathfrak{B}),$$

provided that $P(\mathfrak{E}_2 \cap \mathfrak{B}) > 0$ and $P(\mathfrak{B}) > 0$.

The next proposition is based on the following generalization of the definition 4.14 regarding disjoint partitioning:

Definition 4.21 (Disjoint partition)

A disjoint partition of an event $\mathfrak{E}$ into n mutually disjoint events $\mathfrak{E}_i$ $(i = 1, \dots, n)$ is given by

$$\mathfrak{E} = \mathfrak{E}_1 \cup \mathfrak{E}_2 \cup \cdots \cup \mathfrak{E}_{n-1} \cup \mathfrak{E}_n = \bigcup_{i=1}^{n} \mathfrak{E}_i \quad \text{and} \quad \mathfrak{E}_i \cap \mathfrak{E}_j = \varnothing, \quad \text{if} \quad i \neq j.$$

A typical application of this definition is the frequently used disjoint partition of the certain event Ω (sample space):

$$\Omega = \bigcup_{i=1}^{n} \mathfrak{E}_i, \quad \text{with} \quad \mathfrak{E}_i \cap \mathfrak{E}_j = \varnothing, \quad \text{if} \quad i \neq j.$$

Note that the events $\mathfrak{E}_i$ $(i = 1, \dots, n)$ in this relation are not necessarily elementary events in Ω. We give some simple examples:

Example 4.25 (Disjoint partition applied to a game of dice)

Consider a game of dice with only one die. In this case the certain event is given by the sample space $\Omega = \{\boxdot, \boxdot, \boxdot, \boxdot, \boxdot, \boxdot\}$. The following disjoint partitions of Ω are, for example, possible:

a) $\Omega = \{\boxdot\} \cup \{\boxdot\} \cup \{\boxdot\} \cup \{\boxdot\} \cup \{\boxdot\} \cup \{\boxdot\}$ (elementary events),
b) $\Omega = \{\boxdot, \boxdot, \boxdot, \boxdot, \boxdot\} \cup \{\boxdot\}$ (complementary events),
c) $\Omega = \{\boxdot, \boxdot, \boxdot\} \cup \{\boxdot, \boxdot, \boxdot\}$ (even and odd numbers).

We now turn to a proposition that is frequently used, in order to calculate conditional probabilities:

Proposition 4.10 (Law of total probability)

If a disjoint partition of the event $\mathfrak{E}$ by the events $\mathfrak{E}_i$ $(i = 1, \dots, n)$ is given, then

$$P(\mathfrak{E} \mid \mathfrak{B}) = \sum_{i=1}^{n} P(\mathfrak{E} \mid \mathfrak{E}_i \cap \mathfrak{B}) P(\mathfrak{E}_i \mid \mathfrak{B}), \tag{4.5}$$

provided that $P(\mathfrak{E}_i \cap \mathfrak{B}) > 0$ $(i = 1, \dots, n)$ and $P(\mathfrak{B}) > 0$.

We apply this proposition to an example:

Example 4.26 (Reliability of products)

A company purchases similar components needed to produce one of its products from three different suppliers. The following information is known about the quantities delivered and the reliability of the components:

a) Supplier 1 delivers 5000 components,
   of which 99.9 % meet the specifications,
b) Supplier 2 delivers 3000 components,
   of which 99.5 % meet the specifications,
c) Supplier 3 delivers 2000 components,
   of which 99 % meet the specifications.

The reliability of all delivered components is tested by an incoming inspection. We are interested in the conditional probability that a randomly selected component will meet the specifications, given that it has been tested by the incoming inspection.

We consider the following events:

- $\mathfrak{B}$ represents the event "the component has been tested by the inspection",
- $\mathfrak{C}_i$ represents the event "the component was delivered by supplier i", and
- $\mathfrak{C}$ represents the event "the component meets the specifications".

The conditional probabilities that the component was delivered by supplier i, under the condition that it was tested by the incoming inspection, are given by

$$P(\mathfrak{C}_1 \mid \mathfrak{B}) = \frac{5000}{5000 + 3000 + 2000} = 0.5,$$

$$P(\mathfrak{C}_2 \mid \mathfrak{B}) = \frac{3000}{5000 + 3000 + 2000} = 0.3,$$

$$P(\mathfrak{C}_3 \mid \mathfrak{B}) = \frac{2000}{5000 + 3000 + 2000} = 0.2.$$

The conditional probabilities that a component meets the specifications, given that it was delivered by supplier i and tested by the incoming inspection, are given by

$$P(\mathfrak{C} \mid \mathfrak{C}_1 \cap \mathfrak{B}) = 0.999,$$

$$P(\mathfrak{C} \mid \mathfrak{C}_2 \cap \mathfrak{B}) = 0.995,$$

$$P(\mathfrak{C} \mid \mathfrak{C}_3 \cap \mathfrak{B}) = 0.990.$$

Using these results, we compute the conditional probability that a component meets the specifications, given that it was tested by the incoming inspection, according to proposition 4.10.:

$$P(\mathfrak{C} \mid \mathfrak{B}) = 0.5 \cdot 0.999 + 0.3 \cdot 0.995 + 0.2 \cdot 0.99 = 0.996.$$

The condition $\mathfrak{B}$, that the component has been tested by the incoming inspection, is essential for the final result, because the relative frequencies of delivered components meeting the specifications can only be determined by statistical methods.

4.6 The theorem of BAYES and LAPLACE

When dealing with conditional events, it often happens that we are actually interested in the probability of an event $\mathfrak{E}_2$ occurring under the condition that the event $\mathfrak{E}_1$ has occurred, but we only know the probability of the event $\mathfrak{E}_1$ occurring under the condition that the event $\mathfrak{E}_2$ has occurred, i. e. we are faced with the problem of determining the conditional probability $P(\mathfrak{E}_2 \mid \mathfrak{E}_1 \cap \mathfrak{B})$ from the conditional probability $P(\mathfrak{E}_1 \mid \mathfrak{E}_2 \cap \mathfrak{B})$, where $\mathfrak{B}$ is any additional condition or information (e. g. in metrology the evaluation model). The solution to this problem was first given by T. BAYES [49] (published posthumously) and independently derived again a few years later by P. S. LAPLACE [50], who published the mathematical form we know today as the theorem of BAYES-LAPLACE (or simply BAYES' theorem). There exist several versions of this theorem. The first version is a direct consequence of the multiplication theorem 4.9.

Proposition 4.11 (Theorem of BAYES-LAPLACE, version 1)
Let $\mathfrak{E}_1$ and $\mathfrak{E}_2$ be two arbitrary events and $\mathfrak{B}$ a given condition, then

$$P(\mathfrak{E}_2 \mid \mathfrak{E}_1 \cap \mathfrak{B}) = P(\mathfrak{E}_1 \mid \mathfrak{E}_2 \cap \mathfrak{B})\frac{P(\mathfrak{E}_2 \mid \mathfrak{B})}{P(\mathfrak{E}_1 \mid \mathfrak{B})}, \tag{4.6}$$

provided that $P(\mathfrak{B}) > 0$, $P(\mathfrak{E}_1 \cap \mathfrak{B}) > 0$, $P(\mathfrak{E}_2 \cap \mathfrak{B}) > 0$, and $P(\mathfrak{E}_2 \mid \mathfrak{B}) > 0$.

Looking at equation (4.6), it is immediately obvious that the two conditional probabilities $P(\mathfrak{E}_2 \mid \mathfrak{E}_1 \cap \mathfrak{B})$ and $P(\mathfrak{E}_1 \mid \mathfrak{E}_2 \cap \mathfrak{B})$ can generally not be equal, because this is only possible, if the probabilities $P(\mathfrak{E}_1 \mid \mathfrak{B})$ and $P(\mathfrak{E}_2 \mid \mathfrak{B})$ are also equal. Unfortunately, this important fact is sometimes overlooked.

We can also give another version of the theorem of BAYES-LAPLACE. If we set $\mathfrak{E}_1 = \mathfrak{E}$ and $\mathfrak{E}_2 = \mathfrak{E}_i$ in equation (4.6) we obtain

$$P(\mathfrak{E}_i \mid \mathfrak{E} \cap \mathfrak{B}) = P(\mathfrak{E} \mid \mathfrak{E}_i \cap \mathfrak{B})\frac{P(\mathfrak{E}_i \mid \mathfrak{B})}{P(\mathfrak{E} \mid \mathfrak{B})}. \tag{4.7}$$

If we now assume that, according to equation (4.5), the events $\mathfrak{E}_i$ $(i = 1, \dots, n)$ form a disjoint partition of the event $\mathfrak{E}$, then the following theorem can immediately be derived from equation (4.7):

Proposition 4.12 (Theorem of BAYES-LAPLACE, version 2)
If the events $\mathfrak{E}_i$ $(i = 1, \dots, n)$ form a disjoint partition of the event $\mathfrak{E}$, then

$$P(\mathfrak{E}_i \mid \mathfrak{E} \cap \mathfrak{B}) = \frac{P(\mathfrak{E} \mid \mathfrak{E}_i \cap \mathfrak{B})P(\mathfrak{E}_i \mid \mathfrak{B})}{\displaystyle\sum_{i=1}^{n} P(\mathfrak{E} \mid \mathfrak{E}_i \cap \mathfrak{B})P(\mathfrak{E}_i \mid \mathfrak{B})}, \qquad i = 1, \dots, n, \tag{4.8}$$

provided that $P(\mathfrak{B}) > 0$ and $P(\mathfrak{E} \cap \mathfrak{B}) > 0$, as well as $P(\mathfrak{E}_i \cap \mathfrak{B}) > 0$ and $P(\mathfrak{E}_i \mid \mathfrak{B}) > 0$ $(i = 1, \dots, n)$.

A typical application of this version of the theorem of BAYES-LAPLACE is medical diagnosis. Most people believe that a positive[15] physical examination result means that they are indeed ill. More critical patients ask the doctor about the sensitivity[16] and the specificity[17] of the test used, because they believe or have heard that these values give evidence about the reliability of the test method used. Unfortunately, this is wrong. Even when using a test with relatively high sensitivity and specificity, the detection efficiency of a particular disease may still be poor if its prevalence[18] is low. We will illustrate this with an example.

Example 4.27 (Surveys of tuberculosis)

Until 1997 surveys of preventive diagnostics of tuberculosis have been performed in Germany by using the tuberculin test according to F. MENDEL and C. MANTOUX. We will justify why such surveys today are no longer reasonable by calculating the probability that a person tested has the disease and the test result is positive.

The following information is available:

a) The sensitivity of the Mendel-Mantoux test is reported to be 80–97 % (see S. Pottumarthy, V. C. Wells, A. J. Morris, A Comparison of Seven Tests for Serological Diagnosis of Tuberculosis, Journ. Clinical Microbiology **38** (2000), pp. 2227–2231).

b) The specificity of the Mendel-Mantoux test is reported to be about 92 % (data from *Deutsches Zentralkomitee zur Bekämpfung der Tuberkulose*).

c) In 2009, 4444 cases of tuberculosis were diagnosed in Germany, of which 683 could not be successfully treated, including 154 cases of death (data from *Robert-Koch-Institut, Bericht zur Epidemiologie der Tuberkulose in Deutschland für 2009*, Berlin 2011); the duration of treatment was about 9–12 months.

d) In 2009, the population of Germany was 81 802 257 (data from *GENESIS-Datenbank der Statistischen Ämter des Bundes und der Länder*).

We consider the following events:

– $\mathfrak{B}$ represents the background information (e. g. medical expert knowledge about the disease),

– $\mathfrak{E}$ represents the event "the result of the examination is positive",

– $\mathfrak{E}_1$ represents the event "the examined person is healthy",

– $\mathfrak{E}_2$ represents the event "the examined person is ill with tuberculosis".

We assign the following probabilities to these events:

---

15 A positive examination result in clinical diagnosis means that a patient is probably ill.

16 The sensitivity in clinical diagnostics corresponds to the proportion of patients who are actually ill and whose disease has been identified by the test.

17 The specificity in clinical diagnostics corresponds to the proportion of patients who are actually healthy and have been identified as such by the test.

18 The prevalence of a disease is the proportion of people in a particular group (population) who have the disease.

- $P(\mathfrak{E} \mid \mathfrak{E}_2 \cap \mathfrak{B})$ denotes the probability that, given the background information, the result of the examination is positive if the person examined is ill. We assume that this probability is equal to the maximum sensitivity. Using the data given in a), we obtain

$$P(\mathfrak{E} \mid \mathfrak{E}_2 \cap \mathfrak{B}) = 0.97 \, .$$

- $P(\mathfrak{E} \mid \mathfrak{E}_1 \cap \mathfrak{B})$ denotes the probability that, given the background information, the result of the examination is positive if the person examined is healthy. We calculate this probability from the specificity, which is given by the probability $P(\mathfrak{E}^c \mid \mathfrak{E}_1 \cap \mathfrak{B})$. Using the data given in b), we obtain

$$P(\mathfrak{E} \mid \mathfrak{E}_1 \cap \mathfrak{B}) = 1 - P(\mathfrak{E}^c \mid \mathfrak{E}_1 \cap \mathfrak{B}) = 1 - 0.92 = 0.08 \, .$$

- $P(\mathfrak{E}_2 \mid \mathfrak{B})$ denotes the probability that the examined person is ill, given the background information. We assume that this probability is equal to the prevalence that we can estimate from the data given in c) and d) above. This yields

$$P(\mathfrak{E}_2 \mid \mathfrak{B}) = 6.4 \cdot 10^{-5} \, .$$

- $P(\mathfrak{E}_1 \mid \mathfrak{B})$ denotes the probability that the examined person is healthy, given the background information. We calculate this probability from the prevalence, which yields

$$P(\mathfrak{E}_1 \mid \mathfrak{B}) = 1 - P(\mathfrak{E}_2 \mid \mathfrak{B}) = 1 - 6.4 \cdot 10^{-5} \approx 1 \, .$$

- $P(\mathfrak{E}_2 \mid \mathfrak{E} \cap \mathfrak{B})$ denotes the probability that, given the background information, an examined person is ill if the result of the examination is positive. This is the probability we are looking for. If we set $n = 2$ and $i = 2$ in equation (4.8) we obtain

$$P(\mathfrak{E}_2 \mid \mathfrak{E} \cap \mathfrak{B}) = \frac{P(\mathfrak{E} \mid \mathfrak{E}_2 \cap \mathfrak{B})P(\mathfrak{E}_2 \mid \mathfrak{B})}{P(\mathfrak{E} \mid \mathfrak{E}_1 \cap \mathfrak{B})P(\mathfrak{E}_1 \mid \mathfrak{B}) + P(\mathfrak{E} \mid \mathfrak{E}_2 \cap \mathfrak{B})P(\mathfrak{E}_2 \mid \mathfrak{B})} \, .$$

Finally, inserting the values already determined into this equation gives

$$P(\mathfrak{E}_2 \mid \mathfrak{E} \cap \mathfrak{B}) = \frac{0.97 \cdot 6.4 \cdot 10^{-5}}{0.08 \cdot 1 + 0.97 \cdot 6.4 \cdot 10^{-5}} = 7.8 \cdot 10^{-4} \, .$$

The probability that a person with a positive test result actually has tuberculosis is therefore very low in the light of the background information. This result justifies why tuberculosis surveys are no longer useful in Germany today.

Of course, this result cannot be generalized, because in many other regions of the world it is absolutely reasonable to conduct surveys for preventive diagnosis of tuberculosis. The difference with Germany is the much higher prevalence in such regions. In the states of the former Soviet Union, for example, the prevalence of multidrug-resistant tuberculosis[19] is up to 28 % (see the report *Global Tuberculosis Control of the World Health Organization (WHO). A short update to the 2009 report,* WHO/HTM/TB/2009.426). Given this prevalence the reader should calculate, as an exercise, the probability that a person examined, given the relevant background information, has the disease if the result of the examination is positive.

Finally, we will give another representation of version 2 of the theorem of BAYES-LAPLACE. By summing over all indices $i = 1, \ldots, n$ in the equation (4.8) we immediately obtain

$$\sum_{i=1}^{n} P(\mathfrak{E}_i \mid \mathfrak{E} \cap \mathfrak{B}) = 1 \,,$$

i. e. the denominator in the equation (4.8) is merely a normalization constant. Therefore, we can also write this equation in the form of the proportionality relation

$$P(\mathfrak{E}_i \mid \mathfrak{E} \cap \mathfrak{B}) \propto P(\mathfrak{E} \mid \mathfrak{E}_i \cap \mathfrak{B}) P(\mathfrak{E}_i \mid \mathfrak{B}) \,, \qquad i = 1, \ldots, n \,.$$

In Bayesian statistics, the probability $P(\mathfrak{E}_i \mid \mathfrak{B})$ of the event $\mathfrak{E}_i$ before the event $\mathfrak{E}$ occurs is usually called *a priori* probability and the probability $P(\mathfrak{E}_i \mid \mathfrak{E} \cap \mathfrak{B})$ after the event $\mathfrak{E}$ has occurred is called *a posteriori* probability.

4.7 Stochastic independence

In everyday life we are familiar with the possibility that there is a relationship between events. For example, we are convinced that lightning and thunder are related. How do we arrive at this conclusion? We observe that thunder often occurs when there has been lightning just before. Therefore, we conclude that lightning is probably the cause of thunder. In principle, the natural sciences use similar empirical methods to detect cause-effect interrelations, using stochastic[20] methods.

Example 4.28 (Lung cancer caused by smoking)
Since the sixties of the last century it can be considered certain that smoking cigarettes most probably contributes to the development of lung cancer [51]. This statement is supported by both retrospective[21] and prospective[22] epidemiological studies. According to the retrospective studies, the conditional probability of a given person being a smoker is greater than 85 % if that person develops lung cancer. Furthermore, according to prospective studies, the conditional probability that a given

---

[19] In the case of multidrug-resistant tuberculosis, at least the two most important drugs (Isoniazid and Rifampicin) are already ineffective.

[20] The term "stochastic" is derived from the ancient Greek στοχαστικὴ τέχνη (art of guessing).

[21] Considering the relationship between a disease and previous circumstances.

[22] Considering the relationship between certain circumstances and a subsequent disease.

person will die of lung cancer is 20 to 64 times greater if that person is a smoker than if that person is a non-smoker, and this probability increases continuously with the amount of tobacco consumed.

This example illustrates that we are usually willing to consider a cause-effect interrelation between two events $\mathfrak{E}_1$ and $\mathfrak{E}_2$, given the same background information $\mathfrak{B}$, as highly probable if the conditional probability $P(\mathfrak{E}_1 \mid \mathfrak{E}_2 \cap \mathfrak{B})$ that event $\mathfrak{E}_1$ will occur, and we already know that event $\mathfrak{E}_2$ has occurred, is especially high compared to the conditional probability $P(\mathfrak{E}_1 \mid \mathfrak{E}_2^c \cap \mathfrak{B})$, that event $\mathfrak{E}_1$ will occur and we already know that event $\mathfrak{E}_2$ has *not* occurred. In this case we consider event $\mathfrak{E}_1$ to be the effect and event $\mathfrak{E}_2$ to be the cause. However, this inductive inference from effect to cause is not necessarily a causal relation between two events, because another (still unknown) cause may have led to the same effect, or even more than one cause may have been active, but we are not able to decide which of them was actually effective.

Unlike deductive reasoning, inductive reasoning is inherently probabilistic in the sense that it always generalizes, often on the basis of a limited number of observations. This may, at best, provide evidence for a *stochastic dependency*, which does not necessarily imply a *causal relation*. There are cases in which there is no causal relationship, although the data suggest this. Even a temporal sequence—the cause comes before the effect—is not a reliable indicator.

If there is no *causal* relation between two events $\mathfrak{E}_1$ and $\mathfrak{E}_2$, we usually expect the conditional probabilities $P(\mathfrak{E}_1 \mid \mathfrak{E}_2 \cap \mathfrak{B})$ and $P(\mathfrak{E}_1 \mid \mathfrak{E}_2^c \cap \mathfrak{B})$ to be approximately equal. In example 4.28, we would therefore expect as many smokers as non-smokers to develop lung cancer if smoking does *not* cause the disease. In such a case we say that the two events are stochastically independent. These considerations lead to the definition of stochastic independence.

Definition 4.22 (Stochastic independence)

Two events $\mathfrak{E}_1$ and $\mathfrak{E}_2$ are stochastically independent, given the condition $\mathfrak{B}$, if

$$P(\mathfrak{E}_1 \cap \mathfrak{E}_2 \mid \mathfrak{B}) = P(\mathfrak{E}_1 \mid \mathfrak{B})P(\mathfrak{E}_2 \mid \mathfrak{B}). \tag{4.9}$$

Textbooks of probability theory and mathematical statistics usually give other definitions of stochastic independence. These definitions are equivalent to the one given here. This statement is supported by the following proposition:

Proposition 4.13 (Stochastic independence)

If two events $\mathfrak{E}_1$ and $\mathfrak{E}_2$ are stochastically independent, given condition $\mathfrak{B}$, then the following applies:

a) $P(\mathfrak{E}_1 \mid \mathfrak{E}_2 \cap \mathfrak{B}) = P(\mathfrak{E}_1 \mid \mathfrak{B})$,

b) $P(\mathfrak{E}_1 \mid \mathfrak{E}_2^c \cap \mathfrak{B}) = P(\mathfrak{E}_1 \mid \mathfrak{B})$, provided that $P(\mathfrak{E}_2 \cap \mathfrak{B}) > 0$ and $P(\mathfrak{E}_2^c \cap \mathfrak{B}) > 0$, i. e. for two stochastically independent events $\mathfrak{E}_1$ and $\mathfrak{E}_2$ $P(\mathfrak{E}_1 \mid \mathfrak{E}_2 \cap \mathfrak{B}) = P(\mathfrak{E}_1 \mid \mathfrak{E}_2^c \cap \mathfrak{B})$ is true.

Still other relations concerning stochastic independence can be derived, such as e. g. $P(\mathfrak{E}_2 \mid \mathfrak{E}_1 \cap \mathfrak{B}) = P(\mathfrak{E}_2 \mid \mathfrak{E}_1^c \cap \mathfrak{B})$, i. e. stochastic independence of two events exists mutually.

We state another important proposition:

Proposition 4.14 (Stochastic independence of complementary events)

If two events $\mathfrak{E}_1$ and $\mathfrak{E}_2$ are stochastically independent, then the events $\mathfrak{E}_1^c$ and $\mathfrak{E}_2^c$, $\mathfrak{E}_1$ and $\mathfrak{E}_2^c$, as well as $\mathfrak{E}_1^c$ and $\mathfrak{E}_2$ are also stochastically independent.

However, this can already be seen from the equation (4.9).

Sometimes people—especially those who are not well-trained in probability theory—are of the opinion that the stochastic independence of two events means that these events are also mutually exclusive. But this is a misconception. Actually, the following is true:

Proposition 4.15 (Stochastic dependence of disjoint events)

Mutually exclusive (disjoint) events are stochastically dependent.

To illustrate why mutually exclusive events are always stochastically dependent, we consider an example.

Example 4.29 (Stochastic dependence in roulette)

In classic French roulette, a total of 37 elementary events can occur, as the ball can land on any number from 1 to 36, or on the special number zero. Thus, under the usual assumption of a fair game, each of the elementary events can occur with the same probability, 1/37.

If we denote by $\mathfrak{E}_1$ the event "the ball lands on zero" and by $\mathfrak{E}_2$ the event "the ball lands on any of the other possible numbers", then $\mathfrak{E}_1 \cap \mathfrak{E}_2 = \varnothing$ is valid, because the two events are mutually exclusive (disjoint). Consequently, we have $P(\mathfrak{E}_1 \cap \mathfrak{E}_2) = 0$. But we also have $P(\mathfrak{E}_1) = P(\mathfrak{E}_2) = 1/37$. This implies $P(\mathfrak{E}_1 \cap \mathfrak{E}_2) \neq P(\mathfrak{E}_1)P(\mathfrak{E}_2)$. Therefore, the two events are stochastically dependent but disjoint.

In this case, the reason for the stochastic dependence is obviously that for two disjoint events $\mathfrak{E}_1$ and $\mathfrak{E}_2$, $P(\mathfrak{E}_1 \cap \mathfrak{E}_2) = 0$ is always true.

The stochastic independence of *two* events is consistent with our daily experience. However, if we extend our considerations to the stochastic independence of *more than two* events, we may encounter a situation that seems paradoxical. That is to say, given more than two events, there may be a case where we actually have stochastic independence of all possible pairs $\mathfrak{E}_i$ and $\mathfrak{E}_j$ ($i \neq j$) of events, such that the equations

$P(\mathfrak{E}_i \cap \mathfrak{E}_j) = P(\mathfrak{E}_i)P(\mathfrak{E}_j)$ $(i \neq j)$ are valid, but that nevertheless more than two events are not necessarily stochastically independent. This fact was first pointed out by S. N. BERNSTEIN [52]. We give an example.

Example 4.30 (Flipping two distinguishable fair coins)

We flip two distinguishable coins that are assumed to be fair. In this case the sample space is $\Omega = \{(H, H), (H, T), (T, H), (T, T)\}$, where H and T denote the outcomes "head up" and "tail up" respectively, such that the ordered pair (T,H) denotes, for example, the outcome "the first coin came up tails and the second coin came up heads". Therefore, there are in total four different possible outcomes. Now consider the following three events:

a) $\mathfrak{E}_1$ denotes the event $\{(H, H), (H, T)\}$, i. e. "the *first* coin came up head", which has probability $P(\mathfrak{E}_1) = 1/2$.

b) $\mathfrak{E}_2$ denotes the event $\{(H, H), (T, H)\}$, i. e. "the *second* coin came up head", which has probability $P(\mathfrak{E}_2) = 1/2$.

c) $\mathfrak{E}_3$ denotes the event $\{(T, H), (H, T)\}$, i. e. "*one* of the coins came up head", which has probability $P(\mathfrak{E}_3) = 1/2$.

We therefore obtain $\mathfrak{E}_1 \cap \mathfrak{E}_2 = \{(H, H)\}$, $\mathfrak{E}_1 \cap \mathfrak{E}_3 = \{(H, T)\}$, and $\mathfrak{E}_2 \cap \mathfrak{E}_3 = \{(T, H)\}$, i. e. the probabilities $P(\mathfrak{E}_1 \cap \mathfrak{E}_2) = P(\mathfrak{E}_1)P(\mathfrak{E}_2) = 1/4$, $P(\mathfrak{E}_1 \cap \mathfrak{E}_3) = P(\mathfrak{E}_1)P(\mathfrak{E}_3) = 1/4$, and $P(\mathfrak{E}_2 \cap \mathfrak{E}_3) = P(\mathfrak{E}_2)P(\mathfrak{E}_3) = 1/4$, respectively. Thus, the events are clearly pairwise stochastically independent. However, each of the events $\mathfrak{E}_1$, $\mathfrak{E}_2$ and $\mathfrak{E}_3$ is determined by the respective other two events, because it only occurs, if exactly one of the remaining other two events occurs as well. Thus, the three events are *not* totally stochastically independent. This is also verified by calculation, because on the one hand we have $\mathfrak{E}_1 \cap \mathfrak{E}_2 \cap \mathfrak{E}_3 = \emptyset$ and accordingly $P(\mathfrak{E}_1 \cap \mathfrak{E}_2 \cap \mathfrak{E}_3) = 0$, while on the other hand we have $P(\mathfrak{E}_1)P(\mathfrak{E}_2)P(\mathfrak{E}_3) = 1/8$. Consequently, all three events are not stochastically independent, because $P(\mathfrak{E}_1 \cap \mathfrak{E}_2 \cap \mathfrak{E}_3) \neq P(\mathfrak{E}_1)P(\mathfrak{E}_2)P(\mathfrak{E}_3)$.

The example demonstrates that in the case of more than two events a *total* stochastic independence of all events is not necessarily guaranteed, even if the events are *pairwise* stochastically independent.

Total stochastic independence is defined as follows:

Definition 4.23 (Total stochastic independence)

Two or more events $\mathfrak{E}_i$ $(i = 1, \dots, n)$ are totally stochastically independent if the probability of each of them does not depend on any of the other events.

It can be shown that the proof of the total stochastic independence of n events requires the examination of a total of $2^n - n - 1$ equations.

4.8 Random quantities

In section 4.3, we introduced the abstract probability space $(\Omega, \mathcal{A}, P)$, where Ω denotes the sample space, i. e. the elements of this set are the possible outcomes. In order to be able to use probability theory to evaluate measurement results, we have to establish a relation between the elements of Ω and the data obtained by the measurements. For this purpose we introduce a function, which uniquely assigns a possible measurement result to each element ω of Ω. Since measurement results are usually real numbers—or natural numbers when counting—the values of this function are real or natural numbers.

The function introduced in this way is called "random quantity" (Zufallsgröße), because this term was originally introduced by A. N. KOLMOGOROV, who published most of his fundamental ideas in probability theory in German. Later, in Anglo-Saxon literature, the term "random variable" was used instead. However, the latter term should be avoided because a random quantity is not a variable but rather a function. Furthermore, this function is by no means random, even if its values appear random due to the unpredictable possible outcomes.

Definition 4.24 (Random quantity)

A function $X : \Omega \to \mathbb{R}$ which assigns a real number x to each element ω of the sample space Ω is called a random quantity if $\{\omega \in \Omega \,|\, X(\omega) \leq x, \, x \in \mathbb{R}\}$ is an event in the set of events $\mathcal{A}$.

It should be noted that random quantities can also be defined which assign a *complex* number to each element of the sample space. However, such a random quantity can be replaced by two random quantities for the real and imaginary parts of the complex number.

Throughout the book, we will denote random quantities by upper case letters and their values by the corresponding lower case letters. For example, we will write $X(\omega) \leq x$ or $Y(\omega) = y$.

Every function has a domain and a range. This is also true for a random quantity. The domain in this particular case is the sample space Ω. The range of a random quantity X is

$$\mathfrak{W}_X = \{x \in \mathbb{R} \,|\, X(\omega) = x, \, \omega \in \Omega\},$$

i. e. a set of real numbers x which are associated with the elements ω of the sample space Ω by the function X.

In order to illustrate the usage of the term "random quantity" we consider some simple examples.

Example 4.31 (Rolling an ideally symmetric die)

Consider rolling an ideally symmetric die. In this case, the sample space is given by $\Omega = \{\omega_1, \omega_2, \omega_3, \omega_4, \omega_5, \omega_6\}$, where ω_i ($i = 1, \dots, 6$) denotes the outcome of the number i coming up on a particular roll.

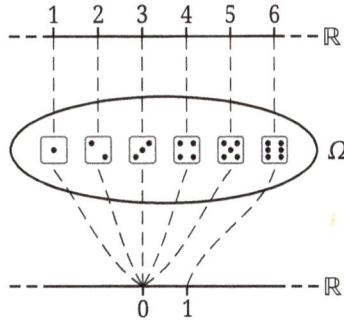

Fig. 4.6: Two different random quantities when rolling a die.

There are several possibilities for specifying the function $X(\omega_i)$. For example, we can use the definition $X(\omega_i) = i$, where the range is the finite set $\mathfrak{W}_X = \{1, 2, 3, 4, 5, 6\}$.

Another possibility would be the definition

$$X(\omega_i) = \begin{cases} 1 & \text{if } i = 6 \\ 0 & \text{otherwise} \end{cases}.$$

We would use this function if we were only interested in the events of whether a six appears or not. In this case the range would be $\mathfrak{W}_X = \{0, 1\}$.

Example 4.32 (Flipping a fair coin several times)

A fair coin is flipped several times. The probability that "head" will come up on any given flip is equal to the probability that "tail" will come up. Nevertheless, if a coin is flipped several times, it is possible that "head" will come up many times and "tail" relatively seldom.[23]

The Hungarian mathematicians P. ERDŐS and A. RÉNYI have shown that it is definitely possible, when flipping a fair coin n times, for "tail" (or "head") to occur $(\log_2 n)$ times in succession without any interruption [53], i. e. if a coin were flipped 100 times, it would not be at all exceptional that, for example, "tail" would occur seven times in succession.

---

23 This seems paradoxical, but can be explained by the fact that a coin does not have a memory. For any fair flip of a coin, the probability of "head" coming up is always $1/2$, regardless of how many times this event has previously occurred.

If we ask—without specifying the number of flips beforehand—for the number of cases in which "tail" occurs consecutively without any interruption, starting to count when "tail" occurs for the first time, the sample space is given by $\Omega = \{T, TT, TTT, ... \}$, where T denotes the outcome "tail".

As the random quantity we choose the number of occurrences of the outcome "tail", i. e. in this case the value of the random quantity is a natural number.[24] Since it is hypothetically possible that it takes an infinitely long time until the outcome "tail" no longer occurs, the range of the random quantity in this case is the set of natural numbers.

Example 4.33 (Rolling a ball)

A ball is set in motion by a kick. It comes to rest after a certain time due to friction. If we ask for the time it takes for the ball to come to rest, then this time is a random quantity and its range is given by the set of non-negative real numbers. In this case, sample space and range are identical.

We could also ask for the distance travelled by the ball during its movement. Then we would have another random quantity with a range, again given by the set of non-negative real numbers.

These examples illustrate that the number of elements in the range of a random quantity can be finite (example 4.31), countably infinite (example 4.32), or uncountably infinite (example 4.33).

Random quantities are classified according to the cardinality[25] of their range into two different classes:

Definition 4.25 (Discrete random quantity)

A random quantity is called discrete if its range is finite or countably infinite.

Definition 4.26 (Continuous random quantity)

A random quantity is called continuous if its range is uncountably infinite.

The classification of random quantities into "discrete" and "continuous" does *not* depend on whether the sample space is a continuum or not. The following example demonstrates that it may be reasonable to consider a discrete random quantity even though the sample space is a continuum.

---

24 As is well known, the natural numbers are considered to be a proper subset of the real numbers.
25 The cardinality card $\mathfrak{M}$ of a set $\mathfrak{M}$ is a measure of its extension (generalized "number of elements"). A finite set $\mathfrak{M}$, containing n elements, has cardinality card $\mathfrak{M} = n$, a countably infinite set $\mathfrak{M}$ has cardinality card $\mathfrak{M} = $ card $\mathbb{N} = \aleph_0$, and an uncountably infinite set $\mathfrak{M}$ has cardinality card $\mathfrak{M} > \aleph_0$.

Example 4.34 (Life expectancy of incandescent lamps)

As part of the quality control, a manufacturer checks the life expectancy of the incandescent lamps it produces. For this purpose a sample is taken from the production and for each lamp of this sample the time until failure (i. e. its life expectancy) is measured. The result of the measurement is then presented in the form of a histogram.

In this case, the sample space is a continuum because the life expectancy of each lamp takes real values. In order to represent the measured values by a histogram, the interval between the smallest and the largest value of the life expectancy is partitioned into equal sub-intervals and for each of these sub-intervals the number of lamps that have reached a life expectancy falling within the respective sub-interval is counted. The number of lamps associated with a given sub-interval is therefore a discrete random quantity.

It is not at all accidental that we use the same denotation—namely capital letters—for quantities (see section 2.1) and random quantities. The reason is that we can always consider quantities as random quantities, because the measured values of a quantity are by no means more predictable than the values of a random quantity. Therefore, in the following, we will not distinguish between a quantity and a random quantity, but consider the measured values of a quantity to be the values of a random quantity.

It is often not well understood in what way chance is involved in a random quantity. Therefore, we explicitly point out that the assignment rule (function) that assigns a real number to each element of the sample space is never random, but always well-defined (deterministic) and fixed once and for all. Only the results of a given experiment are random, and so are the values (in statistics called realizations) taken by the function. For example, suppose we flip a coin several times. If we assign the numbers $+1$ and -1 to the outcomes "head" and "tail" respectively, then this is clearly an invariable assignment rule. However, the corresponding random quantity can randomly take the values $+1$ or -1, depending on which side of the coin comes up, because flipping a coin is a random experiment, i. e. the outcome of each flip is unpredictable.

The special condition

$$\mathfrak{C}_X = \{\omega \in \Omega \,|\, X(\omega) \leq x,\, x \in \mathbb{R}\}, \qquad \mathfrak{C}_X \in \mathcal{A},$$

stated in the definition 4.24 of the term "random quantity" is called regularity condition or measurability condition. The meaning of this condition is that a function $X(\omega)$ is a random quantity if and only if $\mathfrak{C}_X$ is an event in the set of events $\mathcal{A}$. The event $\mathfrak{C}_X$ is the assertion that the value of the random quantity is not greater than a given real number x. Often—especially in practical applications—this event is abbreviated to $(X \leq x)$, and thus the reference to the underlying probability space $(\Omega, \mathcal{A}, \mathsf{P})$ is omitted.

At the end of this section we will give an example of how the regularity condition—which is a precondition for an assignment rule (function) to be regarded as a random quantity—can be verified.

Example 4.35 (Sums of the number of dots when rolling two dice)

We return to the example 4.6 in section 4.1. We roll two ideally symmetric dice at the same time—one white and one black—and calculate the sum of the number of dots that come up. In this case, the sample space Ω is a set of ordered pairs (ω_i, ω_j) $(i, j = 1, \ldots, 6)$, where ω_i denotes the outcome "the white die shows a face with i dots" and ω_j the outcome "the black die shows a face with j dots", respectively.

Since the sample space Ω contains only a finite number of elements (exactly 36), we can use the power set of the sample space (see the considerations in section 4.2) as the set of events, i. e. we have $\mathcal{A} = \mathcal{P}(\Omega)$. This set of events contains a total of $2^{36} = 68\,719\,476\,736$ events. But we are only interested in those eleven events $\mathfrak{E}_k$ $(k = 2, \ldots, 12)$ which yield a sum of the number of dots equal to the number k.

Now, by the assignment rule

$$X[(\omega_i, \omega_j)] = i + j, \qquad i, j = 1, \ldots, 6,$$

we introduce the sum of the number of dots of the two dice as random quantity X. The random quantity introduced in this way satisfies the regularity condition. In order to verify this assertion, we consider the sets

$$\mathfrak{S}_k = \{(\omega_i, \omega_j) \in \Omega \mid X[(\omega_i, \omega_j)] \leq k, k \in \mathbb{R}\}, \qquad k = 2, \ldots, 12,$$

which can be obtained as a union of the events $\mathfrak{E}_k$ $(k = 2, \ldots, 12)$, because

$$\mathfrak{S}_k = \bigcup_{i=2}^{k} \mathfrak{E}_k, \qquad k = 2, \ldots, 12, \tag{4.10}$$

is valid. But we already know that the events $\mathfrak{E}_k$ $(k = 2, \ldots, 12)$ belong to the set of events $\mathcal{A}$. Since $\mathcal{A}$ is a σ-algebra, it finally follows from the definition 4.15 of the σ-algebra that $\mathfrak{S}_k \in \mathcal{A}$ is valid.

It can be shown that in case of discrete random quantities the regularity condition is always satisfied if $\mathcal{A} = \mathcal{P}(\Omega)$, i. e. if the set of events is equal to the power set of the sample space. Thus, the regularity condition is primarily relevant for continuous random quantities.

4.9   Probability distribution functions

In section 4.3, we considered the probability measure P, which assigns probabilities to the possible events contained in the set of events (σ-algebra) $\mathcal{A}$. The elements of $\mathcal{A}$ are subsets of the sample space Ω, which contains all possible outcomes of an experiment. In the previous section we introduced a random quantity $X(\omega)$ as a function which assigns real numbers $x \in \mathbb{R}$ to the elements ω of the sample space Ω. We also stated that a random quantity $X(\omega)$ must satisfy the regularity condition

$$\mathfrak{E}_X = \{\omega \in \Omega \mid X(\omega) \leq x, x \in \mathbb{R}\}, \qquad \mathfrak{E}_X \in \mathcal{A}, \tag{4.11}$$

i. e. the event $\mathfrak{E}_X$ that the relation $X(\omega) \leq x$ is true must always be a possible event of the set of events $\mathcal{A}$.

We can now ask for the probability $P(\mathfrak{E}_X)$ of the particular event $\mathfrak{E}_X$ given by equation (4.11). This probability, often denoted by $P(X \leq x)$, is identical to the probability that the value of the random quantity $X(\omega)$ is less than or equal to a given value x, i. e. it is a function of x. This particular function is called probability distribution function (also abbreviated distribution function or probability distribution) and is denoted by[26] $G_X(x)$, where the index X refers to the random quantity.

Definition 4.27 (Probability distribution function)

If Ω is a sample space, then the real function

$$G_X(x) = P(X \leq x) = P(\{\omega \in \Omega \,|\, X(\omega) \leq x\}), \qquad x \in \mathbb{R},$$

is called probability distribution function of the random quantity X.

This definition applies to both discrete and continuous random quantities. In order to facilitate the understanding of the concept of a probability distribution function, we will demonstrate its application by means of simple examples. We start with an example of a discrete random quantity.

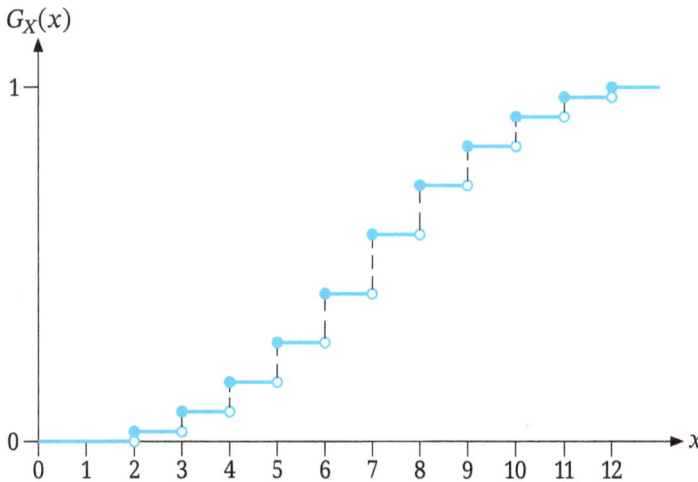

Fig. 4.7: Graph of the probability distribution function of the sum of the numbers of dots when rolling two ideally symmetric dice.

---

26 In deviation from the usual denotation $F_X(x)$ in probability theory, we generally use the denotation $G_X(x)$ in this book, as it is also used in Suppl. 1 of the GUM [54, 55], to avoid any confusion with model functions.

Example 4.36 (Sums of the number of dots when rolling two dice)

We continue the example 4.35. If we denote the random quantity "sum of the number of dots" again by X, we obtain the probability distribution function

$$
G_X(x) = \begin{cases}
0 & \text{if } x < 2 \\
\dfrac{1}{36} & \text{if } 2 \leq x < 3 \\
\dfrac{3}{36} & \text{if } 3 \leq x < 4 \\
\dfrac{6}{36} & \text{if } 4 \leq x < 5 \\
\dfrac{10}{36} & \text{if } 5 \leq x < 6 \\
\dfrac{15}{36} & \text{if } 6 \leq x < 7 \\
\dfrac{21}{36} & \text{if } 7 \leq x < 8 \\
\dfrac{26}{36} & \text{if } 8 \leq x < 9 \\
\dfrac{30}{36} & \text{if } 9 \leq x < 10 \\
\dfrac{33}{36} & \text{if } 10 \leq x < 11 \\
\dfrac{35}{36} & \text{if } 11 \leq x < 12 \\
1 & \text{if } 12 \leq x
\end{cases}
\tag{4.12}
$$

which is shown in Fig. 4.7 on page 141.

In order to derive this probability distribution function, we have made the following observations: When rolling two ideally symmetrical dice, the sample space Ω contains a total of $6 \cdot 6 = 36$ outcomes—six for each die—all of which can be assumed to be equally probable. We have subsumed these outcomes into eleven groups, yielding the events $\mathfrak{E}_k$ ($k = 2, \dots, 12$) where the sum of the number of dots on both dice is equal to k. From these events we can form by the operation of set union (see equation (4.10)) the events $\mathfrak{S}_k$ ($k = 2, \dots, 12$) that the sum of the number of dots is less than or equal to k. In each case, the number of events contained in the sets $\mathfrak{S}_k$ ($k = 2, \dots, 12$) gives the factor by which the probability $1/36$ of the elementary event must be multiplied to obtain the probability of the respective event $\mathfrak{S}_k$ ($k = 2, \dots, 12$). These probabilities are just the values of the probability distribution function for the different cases, as given in equation (4.12).

Another way of obtaining the probability distribution function is as follows. First we determine the probabilities

$$
P(X = k) = P(\mathfrak{E}_k), \qquad k = 2, \dots, 12,
$$

of the events $\mathfrak{E}_k$ ($k = 2, \ldots, 12$) that we are interested in. These eleven probabilities are summarized in the Tab. 4.2.

Tab. 4.2: Probabilities of the events $\mathfrak{E}_k$.

$\mathfrak{E}_2$	$\mathfrak{E}_3$	$\mathfrak{E}_4$	$\mathfrak{E}_5$	$\mathfrak{E}_6$	$\mathfrak{E}_7$	$\mathfrak{E}_8$	$\mathfrak{E}_9$	$\mathfrak{E}_{10}$	$\mathfrak{E}_{11}$	$\mathfrak{E}_{12}$
$\dfrac{1}{36}$	$\dfrac{2}{36}$	$\dfrac{3}{36}$	$\dfrac{4}{36}$	$\dfrac{5}{36}$	$\dfrac{6}{36}$	$\dfrac{5}{36}$	$\dfrac{4}{36}$	$\dfrac{3}{36}$	$\dfrac{2}{36}$	$\dfrac{1}{36}$

From the values given in Tab. 4.2, we calculate by summation the cumulative probabilities

$$p_i = \sum_{k=2}^{i} P(X = k), \quad i = 2, \ldots, 11.$$

The probability distribution function is then given by the equation

$$G_X(x) = \begin{cases} 0 & \text{if } x < 2 \\ p_k & \text{if } k \le x < k+1, \quad k = 2, \ldots, 11, \\ 1 & \text{if } 12 \le x \end{cases}$$

which describes a step function with discontinuities at $x = k$ ($k = 2, \ldots, 12$), as shown in Fig. 4.7 on page 141.

This example illustrates why the very descriptive name "probability distribution function" was introduced in probability theory for this function. The total probability of the 36 equally probable events is distributed over the eleven events we are interested in according to the number of elementary events they contain.

Fig. 4.7 is a typical graph of the probability distribution function of a discrete random variable, because these functions are always step functions which have a limited number of jump discontinuities, exactly one for each of the possible events, where the step height is equal to the probability of the underlying event. Since the probability is non-negative, the value of the probability distribution function can never decrease at the jump discontinuities[27] and remains constant between them, i. e. the probability distribution function is monotonically increasing.

The procedure for the construction of a discrete probability distribution function, as it is presented in the example 4.36, can be generalized. If it is known that a discrete random quantity X can take the values x_k ($k = 1, \ldots, n$) with the probabilities $P(X = x_k)$

---

27 This cumulative behaviour of the probability distribution function motivates the term "cumulative distribution function", abbreviated CDF, which is also often used.

$(k = 1, \dots, n)$, we can calculate the cumulative probabilities using the equations

$$p_i = \sum_{k=1}^{i} P(X = x_k), \qquad i = 1, \dots, n - 1.$$

From these equations the probability distribution function

$$G_X(x) = \begin{cases} 0 & \text{if } x < x_1 \\ p_k & \text{if } x_k \leq x < x_{k+1}, \quad k = 1, \dots, n - 1, \\ 1 & \text{if } x_n \leq x \end{cases} \tag{4.13}$$

can be constructed. Hence, a monotonously increasing step function with step heights 0, p_k ($k = 1, \dots, n - 1$) and 1 is obtained.

We can write this step function in a more elegant form. For this purpose we first introduce the HEAVISIDE function.[28] This function is in a sense the most elementary step function. It is defined by the relation

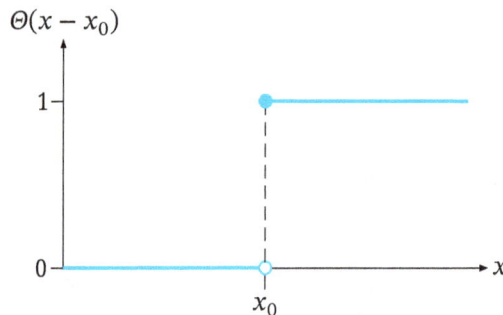

Fig. 4.8: HEAVISIDE function.

$$\Theta(x - x_0) = \begin{cases} 0 & \text{if } x < x_0 \\ 1 & \text{if } x_0 \leq x \end{cases}, \tag{4.14}$$

which is shown in Fig. 4.8.

Using the HEAVISIDE function we can write the equation (4.13) as

$$G_X(x) = \sum_{k=1}^{n} P(X = x_k)\Theta(x - x_k), \tag{4.15}$$

---

[28] The unit step function is named after the engineer, mathematician and physicist O. HEAVISIDE. He was the first to analyse the mathematical properties of such functions in the context of studying the transmission behaviour of overseas cables [56].

as can easily be shown. This representation makes it more obvious that the discrete probability distribution function $G_X(x)$ has a total of n steps of height $P(X = x_k)$ $(k = 1, \ldots, n)$. In the case of $x > x_n$ the equation (4.15) gives the normalization condition

$$\sum_{k=1}^{n} P(X = x_k) = 1$$

of discrete probability distribution functions.

We mention in passing that a similar procedure to the one shown here is used in statistics to determine the so-called "empirical probability distribution function". In this case, however, not the (unknown) probabilities $P(X = x_k)$ $(k = 1, \ldots, n)$ are used, but the relative frequencies of the respective values of the random quantity under consideration. Under certain conditions, the empirical probability distribution function obtained in this way can be regarded as a good approximation of the true probability distribution function of the random quantity.

Next, we give an example of a continuous probability distribution function.

Example 4.37 (Breakage of a thin wire under strain)

Consider a thin wire suspended vertically and loaded by a weight at its lower end. The mass of the weight is assumed to be large enough to cause the wire to break under the force of gravity. The break will occur randomly somewhere along the entire length L of the wire. We can therefore consider the distance of the break point from the attachment point as a random quantity, which we denote by X.

In order to keep the model as simple as possible, we assume that the wire is made of a homogeneous material and has a constant cross-section. We also assume that the wire is very thin so that we can neglect its mass compared to the mass of the weight attached to its end. Under these assumptions we can assume that any point in the interval $[0, L]$ can be considered with equal probability as the breaking point. The probability that the break will occur in the interval $[0, x]$ should therefore be proportional to the ratio x/L, i. e. we can use the probability distribution function

$$G_X(x) = \begin{cases} 0 & \text{if } x < 0 \\ \dfrac{x}{L} & \text{if } 0 \le x < L \\ 1 & \text{if } L < x \end{cases} \tag{4.16}$$

This function is shown in Fig. 4.9.

In this example we have introduced a continuous random quantity and its probability distribution function without mentioning the underlying sample space Ω. It goes without saying, however, that this set must be a continuum, i. e. its power set is too large to serve as a set of possible events $\mathcal{A}$ (see the considerations about the cardinality of sets in section 4.2). By introducing the probability distribution function here without referring to its

$G_X(x)$

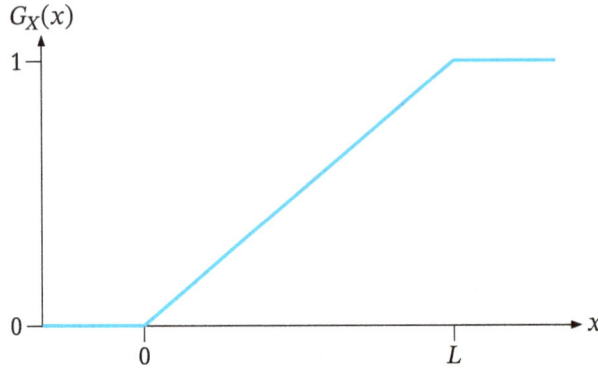

Fig. 4.9: Probability distribution function $G_X(x)$ of the continuous random quantity X (distance of the breaking point of a thin, homogeneous wire under strain from its fixing point).

associated probability space $(\Omega, \mathcal{A}, P)$, we have tacitly avoided the so-called "problem of measure" (for details see e. g. P. BILLINGSLEY [57]). This practice is permissible whenever the existence of the respective probability space can be assured.

The following proposition summarizes the essential mathematical properties of probability distribution functions:

Proposition 4.16 (Properties of a probability distribution function)
A probability distribution function $G_X(x)$ of the random quantity X has the following properties:

a) $G_X(x)$ is monotonically increasing,

b) $G_X(x)$ is continuous from above,

c) $\lim\limits_{x \to -\infty} G_X(x) = 0$,

d) $\lim\limits_{x \to +\infty} G_X(x) = 1$,

e) $\lim\limits_{\varepsilon \to 0} [G_X(x) - G_X(x - \varepsilon)] = P(X = x), \qquad \varepsilon > 0$.

The properties stated in this proposition are valid for both discrete and continuous random quantities. In order to make these properties more understandable, we give some additional explanations. Property a) means that for increasing values of x, a probability distribution function can only increase or remain constant, i. e. if x_1 is less than x_2, then $G_X(x_1)$ is less than or at most equal to $G_X(x_2)$. In order to understand property b), we imagine that we draw the graph of the probability distribution function—contrary to common practice—from right to left. If, while drawing, we have to lift the pen at a jump discontinuity, the last point drawn (in Fig. 4.7 represented by a filled circle) still belongs to the function, i. e. the value of the function at a jump discontinuity is always the upper value of a step. Such functions are called *continuous from above* or *right-continuous*.

Property c) states that a random quantity can never take the value $-\infty$ ($X = -\infty$ is impossible), while property d) states that a random quantity can take any real value ($X < +\infty$ is the certain event[29]). Finally, property e) means that at a jump discontinuity a probability distribution function instantly changes its value by $P(X = x)$, such that the value immediately to the left of the jump discontinuity is smaller than the value at the jump discontinuity itself. In addition, this property also means that the probability $P(X = x)$ that the random quantity X takes the value x has to be equal to zero if there is no jump discontinuity at that particular point.

Property e) is especially significant for continuous probability distribution functions, because *by definition* these functions have no discontinuities. It follows immediately that $P(X = x) = 0$ holds for any value x of a continuous random quantity X, i. e. the probability that a certain value occurs is always zero. On the other hand, this also means that if a random quantity takes a particular value at a given point with non-zero probability, then its probability distribution function necessarily has a jump discontinuity at that point.

The fact that the probability $P(X = x)$ for any value x of a continuous random quantity X is zero is often misunderstood because it is wrongly assumed that a zero probability always implies the impossible event. But this assumption is by no means valid, because there is only *one* impossible event, and it is exclusively represented by the *empty* set (see definition 4.10). In the case considered here, however, we have to assign the probability zero to the event $(X = x)$, similar to assigning the length zero to a point on a line (or the area zero to a curve in a plane), i. e. we are dealing with a null set (see also section 4.3). Given the information that an event has the probability zero, to conclude that it is the impossible event is as absurd as given the information that a point has no length, to conclude that it cannot possibly exist.

It is important to understand that assigning probability zero to an event does not necessarily mean that the event is impossible (this is only true for the impossible event), but rather that it is unlikely, which is a different modality. In probability theory, an event is said to occur almost never if it occurs with probability zero. Similarly, assigning probability one to an event does not necessarily mean that the event is certain (this is only true for the certain event). In probability theory, one says that an event occurs almost surely if it occurs with probability one.

The reader may wonder whether it is not paradoxical that, for instance in the example 4.37, we can be sure—i. e. the probability is one—of finding the break point within the interval $[0, L]$, although the probability of finding the break point exactly at a position x is zero, for any x in $[0, L]$. Actually, this is similar to the fact (which also seems paradoxical) that the length of the interval $[0, L]$ is the unique number L, although the interval contains infinitely many points of the length zero. This is a typical case where

---

29 We do not use the expression $X \leq +\infty$ here, because *by definition* neither $+\infty$ nor $-\infty$ belongs to the set of real numbers.

common sense fails, and we can no longer trust our intuition, as is often the case when infinity is involved. Only a strict application of mathematical rules can prevent us from drawing wrong conclusions.

In order to use probability distribution functions in probability calculus, we need some rules, which are summarized in the following proposition:

Proposition 4.17 (Rules of probability distribution functions)
The following rules apply for a probability distribution function $G_X(x)$:

$$P(X \leq x) = G_X(x), \tag{4.17}$$

$$P(X < x) = G_X(x) - P(X = x), \tag{4.18}$$

$$P(X > x) = 1 - G_X(x), \tag{4.19}$$

$$P(X \geq x) = 1 - G_X(x) + P(X = x), \tag{4.20}$$

$$P(a < X \leq b) = G_X(b) - G_X(a), \tag{4.21}$$

$$P(a \leq X \leq b) = G_X(b) - G_X(a) + P(X = a), \tag{4.22}$$

$$P(a \leq X < b) = G_X(b) - G_X(a) + P(X = a) - P(X = b), \tag{4.23}$$

$$P(a < X < b) = G_X(b) - G_X(a) - P(X = b). \tag{4.24}$$

These rules can be obtained by disjoint partitioning the certain event, applying the two axioms K2 and K3 of KOLMOGOROV (see definition 4.18 on page 117) to the result, and subsequently solving for the desired probability.

We give two examples of the application of proposition 4.17.

Example 4.38 (Continuation of example 4.36)
We are interested in the probabilities that
a)   the sum of the number of dots is greater than 6,
b)   the sum of the number of dots is between 3 and 7.

From equations (4.19) and (4.12) we obtain

$$P(X > 6) = 1 - G_X(6) = 1 - \frac{15}{36} = \frac{7}{12}.$$

Equations (4.22) and (4.12) yield

$$P(2 \leq X \leq 7) = G_X(7) - G_X(3) + P(X = 3) = \frac{21}{36} - \frac{3}{36} + \frac{2}{36} = \frac{5}{9}.$$

In order to determine $P(X = 3)$, the fact that the event "the sum of the number of dots is 3" occurs in two out of 36 cases was used.

Example 4.39 (Continuation of example 4.37)

We are interested in the probability that the breakpoint is in the interval $[a, b]$, where $0 \leq a < b \leq L$ is valid.

From equations (4.22) and (4.16) we obtain

$$P(a \leq X \leq b) = G_X(b) - G_X(a) + P(X = a) = \frac{b}{L} - \frac{a}{L} + 0 = \frac{b - a}{L}.$$

In order to determine $P(X = a)$, the fact was used that the probability distribution function is continuous, i. e. it does not have any jump discontinuities.

In some special cases we are interested in the probability that the value of a random quantity $X(\omega)$ is less than or equal to a certain value x, given that an additional specified condition $\mathfrak{B}$ is satisfied, i. e. in the probability $P(X \leq x \mid \mathfrak{B})$. Using the proposition 4.2 (product rule) leads to the definition of the conditional probability distribution function.

Definition 4.28 (Conditional probability distribution function)

The real function

$$G_X(x \mid \mathfrak{B}) = P(X \leq x \mid \mathfrak{B}) = \frac{P(\{X \leq x\} \cap \mathfrak{B})}{P(\mathfrak{B})}, \text{ provided that } P(\mathfrak{B}) > 0,$$

is called conditional probability distribution function of the random quantity X given the condition $\mathfrak{B}$.

The following example demonstrates a typical application of the conditional probability distribution function.

Example 4.40 (Probability of failure of a system)

We assume that the probability of failure of a system can be described by the probability distribution function $G_T(t)$, where the random quantity denoted by T is the elapsed time to failure. We ask for the probability that the system will fail within the time interval $(t, t + \Delta t)$ if it was still functioning at the time denoted by t. This probability is given by the conditional probability distribution function

$$G_T(T \leq t + \Delta t \mid T > t) = \frac{P(\{t < T \leq t + \Delta t\} \cap \{T > t\})}{P(T > t)}. \tag{4.25}$$

Since

$$\{t < T \leq t + \Delta t\} \cap \{T > t\} = \{t < T \leq t + \Delta t\}$$

is valid, the equation (4.25) becomes

$$G_T(T \leq t + \Delta t \mid T > t) = \frac{P(t < T \leq t + \Delta t)}{P(T > t)}$$

and finally, using equations (4.19) and (4.21),

$$G_T(T \leq t + \Delta t \mid T > t) = \frac{G_T(t + \Delta t) - G_T(t)}{1 - G_T(t)}.$$

This is a well-known relation in reliability theory.

Reliability theory is a branch of probability theory with a strong metrological connection, in which the reliability of technical systems over a given period of time under given conditions is of central interest.

At the end of this section we return to the HEAVISIDE function. This function has all the properties listed in proposition 4.16. This suggests that the HEAVISIDE function can be considered as a probability distribution function. In fact, the HEAVISIDE function is the simplest discrete probability distribution function because it is the simplest step function. It has only one step where the probability jumps from zero to one. In probability theory, this probability distribution function is called "one-point probability distribution". It describes the case where there is only one event that is certain to occur. The relation $G_X(x) = \Theta(x - x_0)$ thus expresses the fact that a random quantity X is certain to take the value x_0 and all other values are excluded. One-point probability distribution functions have applications in astrophysics, for example, where they have been proposed as appropriate statistical measures for dynamical variables [58].

4.10 Probability density functions

In the previous section we showed in detail that the probability $P(X = x)$ that a random quantity X takes exactly the value x can only differ from zero in the case of discrete probability distribution functions, while in the case of continuous probability distribution functions this probability is always zero. For this reason, we obviously cannot obtain a continuous probability distribution function by summing the probabilities of all possible events, as we did for discrete probability distribution functions in the previous section. However, if we can assume that a probability distribution function is continuous, we can use a different procedure.

We consider the random quantity X and ask for the probability that its value can be found in a given interval[30] $[a, b]$. According to the equation (4.21) this probability is given by

$$P(a \leq X \leq b) = G_X(b) - G_X(a), \tag{4.26}$$

where we have already used the fact that $P(X = a) = 0$ is valid due to the assumed continuity of the probability distribution function. If we additionally assume that $G_X(x)$ is differentiable within the considered interval $[a, b]$, we can use the mean value theorem

---

30 We choose the closed interval $[a, b]$ here, but it is possible to choose the open interval (a, b) or one of the half-open intervals $(a, b]$ or $[a, b)$ without changing the result, since, due to the assumed continuity of the probability distribution function, the probability that the random quantity takes the values a or b is zero anyway.

of differential calculus to rewrite equation (4.26) as

$$P(a \leq X \leq b) = G'_X(\xi)(b - a), \tag{4.27}$$

where ξ is a certain value within the interval $[a, b]$ and $G'_X(x)$ is the derivative of the probability distribution function $G_X(x)$. The derivative introduced in this way, if it actually exists, is called probability density function (also abbreviated to density function or probability density) and will be denoted[31] by $g_X(x)$ throughout this book.

Instead of introducing the probability density function as a derivative of the probability distribution function, we can use another approach that leads to the following equivalent definition:

Definition 4.29 (Probability density function)
The real-valued function $g_X(x)$ is called probability density function of the random quantity X, if

$$G_X(x) = \int_{-\infty}^{x} g_X(\xi) \, d\xi \tag{4.28}$$

is the probability distribution function of this random quantity.

This definition of the probability density function is justified by the well-known NEWTON-LEIBNIZ fundamental theorem of calculus.

We give a simple example of the probability density function of a continuous random quantity:

Example 4.41 (Continuation of example 4.37)
We calculate the probability density function by differentiating the probability distribution function given by equation (4.16) with respect to x. This yields

$$g_X(x) = \begin{cases} 0 & \text{if } x < 0 \\ \dfrac{1}{L} & \text{if } 0 \leq x < L \\ 0 & \text{if } L \leq x \end{cases} \cdot \tag{4.29}$$

This probability density function is shown in Fig. 4.10.

---

[31] In deviation from the usual denotation $f_X(x)$ in probability theory, we generally use the denotation $g_X(x)$ in this book, as it is also used in Suppl. 1 of the GUM [54, 55], to avoid any confusion with model functions. This is consistent with the corresponding denotation of the probability distribution function in the previous section.

It is a so-called uniform distribution (rectangular distribution) defined on the interval $[0, L)$, i. e. every point within this interval has the same probability density[32] and at every other point the probability density is zero.

If we insert equation (4.29) into equation (4.28) we get, as expected, the probability distribution function given by equation (4.16).

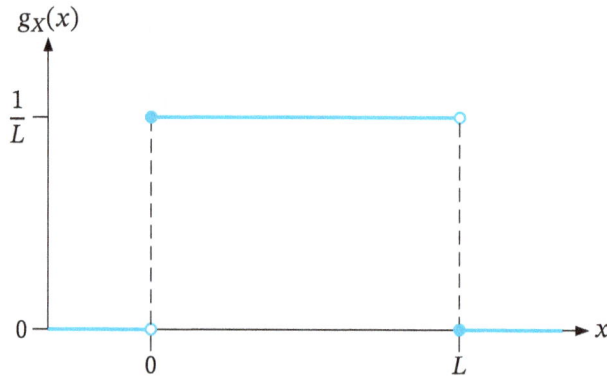

Fig. 4.10: Probability density function $g_X(x)$ of the continuous random quantity X (distance of the breaking point of a thin, homogeneous wire under strain from its fixing point).

By definition 4.29, the probability distribution function is traced back to its associated probability density function, if such a density function actually exists. However, this is guaranteed by the differentiation theorem of H. LEBESGUE [59], which states that every continuous monotone function has a finite derivative at each of its points, except for a countable set of points. Since the probability distribution function $G_X(x)$ satisfies the conditions of this theorem, it is almost everywhere differentiable and thus its probability density function $g_X(x)$ exists. At the countable many points where $G_X(x)$ is continuous but not differentiable, we can choose arbitrary values of the probability density function without changing the probability distribution function, because the latter is only specified up to a null set (see section 4.3). For instance, in the example 4.41, if we arbitrarily changed the values of the probability density function at the endpoints of the interval $[0, L]$, we would still get the same probability distribution function. We could just as well use the open interval $(0, L)$, which again justifies the footnote 30 on page 150.

---

32 Often a statement like "Each point …has the same probability" can be heard. However, such a statement is not correct, because the values of a probability density function are *not* probabilities! Furthermore, in general, the probability density function can take values greater than one. In fact, it should be called a "uniform density function" or "rectangular density function" because it is a density function and not a distribution function.

Not every real-valued function is suitable as a probability density function. The conditions for a probability density function are:

Proposition 4.18 (Conditions of a probability density function)
The probability density function $g_X(x)$ of a random quantity X must satisfy the following conditions:

a) $g_X(x)$ is a non-negative function,

b) $g_X(x)$ fulfils the normalization condition

$$\int\limits_{-\infty}^{+\infty} g_X(\xi)\, d\xi = 1.\qquad(4.30)$$

Condition a) follows from the property of $G_X(x)$ to be a monotonously increasing function. From this property, using equation (4.26), it immediately follows for $b > a$ that $P(a \leq X \leq b) \geq 0$ is valid—as it should be—and thus from equation (4.27) we finally obtain $g_X(x) \geq 0$, i.e. every probability density function must be non-negative. The normalization condition follows from equation (4.28) and property d) in proposition 4.16. Another condition not stated in proposition 4.18, namely that a probability density function must be zero at infinity—i.e. $g_X(-\infty) = 0$ and $g_X(+\infty) = 0$—is already implicit in the normalization condition and therefore does not need to be stated explicitly.

The definition 4.29 of the probability density function is initially only valid under the condition that the probability distribution function is continuous. However, we will show that this condition can be dropped by proving that a probability density function can also be specified for a discrete probability distribution function. For this purpose we use the relation

$$\Theta(x - x_0) = \int\limits_{-\infty}^{x} \delta(\xi - x_0)d\xi,\qquad(4.31)$$

where $\delta(\xi - x_0)$ denotes DIRAC's δ-function delta-function[33] and $\Theta(x - x_0)$ again the HEAVISIDE function given by the equation (4.14). However, due to the definition 4.29, the relation (4.31) only makes sense for our purposes if we consider the HEAVISIDE function as a probability distribution function and DIRAC's δ-function as a probability density function. The first requirement is fulfilled, as we have already shown at the end of the previous section. The confirmation that the second requirement is also fulfilled is done by means of the theory of distributions. Instead of a rigorous proof, we give an informal justification.

---

33 This "function"—actually it is not a function in the strict sense—is named after P. DIRAC, because he was the first to use it in connection with his mathematical approach to quantum mechanics [60], but without giving a mathematically sound definition of it. This was only made possible by the theory of distributions later developed by L. SCHWARTZ [61].

The representation of the unit step function by a density function, as given by equation (4.31), is only a special case of the more general relation

$$f(x_0)\Theta(x - x_0) = \int\limits_{-\infty}^{x} f(\xi)\delta(\xi - x_0)\mathrm{d}\xi, \qquad (4.32)$$

which can be interpreted as a definition of the δ-function proposed by DIRAC.

According to the rules of calculus, it follows from the equation (4.31)—merely from a formal point of view—that DIRAC's δ-function must be the derivative of HEAVISIDE's function. Using the definition of the HEAVISIDE function as given by the equation (4.14), we find by formal differentiation that DIRAC's δ-function must be zero everywhere except at the jump discontinuity of the HEAVISIDE function, where DIRAC's δ-function takes the "value" infinity.[34] Of course, no "reasonable" function behaves in this way. Therefore, the use of DIRAC's δ-function only makes sense in conjunction with its definition given by equation (4.32). In practice, this equation serves as a kind of abbreviated notation for the underlying, more complicated formalism of distribution theory.

We add a note about the graphical representation. DIRAC's δ-function cannot have a graph, since it is not a function in the strict sense. Nevertheless, in some cases we want to have a graphical representation of it. The usual convention is to draw a vertical line at the point where DIRAC's δ-function has its singularity, with an arrow at the top of the line. The length of the line is proportional to the value of the factor by which DIRAC's δ-function must be multiplied (see e. g. Fig. 4.11).

After these brief explanations of the properties of DIRAC's δ-function, as far as they are of interest here, we return to the problem of finding a probability density function of a discrete random quantity. Substituting into the equation (4.32) $x_0 = x_k$, as well as $f(x_0) = P(X = x_k)$ and $f(\xi) = P(X = \xi)$, which is subsequently summed with respect to the index k, and using the equation (4.15), we obtain the representation

$$G_X(x) = \int\limits_{-\infty}^{x} \sum_{k=1}^{n} P(X = \xi)\delta(\xi - x_k)\,\mathrm{d}\xi$$

of the probability distribution function of a discrete random quantity denoted by X. A comparison of this equation with equation (4.28) shows that a discrete random quantity also has a probability density function in the broader sense, which can be represented by

$$g_X(x) = \sum_{k=1}^{n} P(X = x)\delta(x - x_k). \qquad (4.33)$$

The definition 4.29 of the probability density function is thus valid for any real random quantity without any restrictions. This is the desired result.

---

34 This is also called a "singularity".

We give a simple example of the probability density function of a discrete random quantity:

Example 4.42 (Continuation of example 4.36)

If we apply equation (4.33) to the sum of the number of dots obtained by rolling two ideally symmetric dice, and use the probability values given in table 4.2, we obtain the probability density function shown in Fig. 4.11.

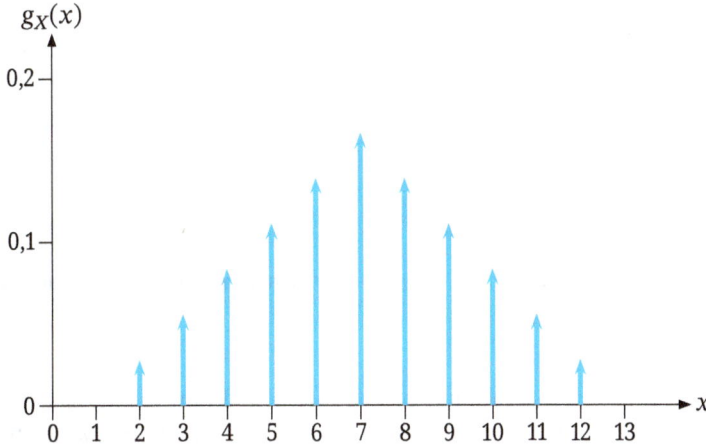

Fig. 4.11: Graph of the probability density function of the sum of the number of dots when rolling two ideally symmetric dice.

Analogous to the probability density function $g_X(x)$, we can define a conditional probability density function, using definitions 4.28 and 4.29 accordingly.

Definition 4.30 (Conditional probability density function)

The real-valued function $g_X(x \mid \mathcal{B})$ is called the conditional probability density function of the random quantity X, if

$$G_X(x \mid \mathcal{B}) = \int_{-\infty}^{x} g_X(\xi \mid \mathcal{B}) \, d\xi$$

is the conditional probability distribution function of this random quantity.

The conditional probability density function, of course, satisfies the requirements of proposition 4.18, i. e. it is also a non-negative, real-valued function normalized to one.

4.11 Transformations of random quantities

In section 4.8, we already mentioned that we will consider measured quantities as random quantities. When we do this, we have to consider every quantity equation—and therefore every model equation—as a functional relation between random quantities.

Technically, we can always transform a given random quantity X by a function f into another random quantity Y. The meaning of such a transformation is illustrated by the commutative diagram shown in Fig. 4.12.

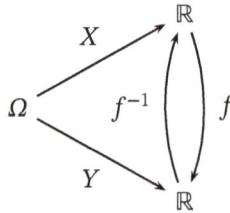

Fig. 4.12: Commutative diagram of the transformation of a random quantity X into a random quantity Y.

The diagram shows that the random quantities X and Y are functions that assign real numbers to each outcome ω of the sample space Ω. The function f, which transforms the image of the function $X(\omega)$ into the image of the function $Y(\omega)$, acts between the two real numbers generated by X and Y. On the other hand, the function f^{-1} transforms the image of the function $Y(\omega)$ into the image of the function $X(\omega)$. Since we do not require f to be uniquely invertible, f^{-1} will generally have several branches[35] (we will see the consequences later).

In order to be a genuine transformation between random quantities, the function f must satisfy certain conditions. This becomes immediately obvious if we consider that Y—similarly to X—must satisfy the definition 4.24, i. e. in particular the regularity condition (measurability condition), which ensures that every subset of the real numbers generated by Y corresponds to the event $\mathfrak{E}_Y$, i. e. to an event of the set of events $\mathcal{A}$. But because of the assumed relation $Y = f(X)$,[36] i. e. $Y(\omega) = f[X(\omega)]$, this regularity condition is equivalent to the condition

$$\{\omega \in \Omega \mid f[X(\omega)] \leq y, \, y \in \mathbb{R}\} \in \mathcal{A},$$

which has to be fulfilled by the function f describing the transformation of the random quantity X into the random quantity Y, i. e. the function f has to be measurable. This

---

[35] For example, if we have $f(X) = X^2$, then the inverse function has the two branches $f^{-1}(X) = +\sqrt{X}$ and $f^{-1}(X) = -\sqrt{X}$.

[36] For the sake of simplicity, we will use the denotation $Y = f(X)$ in the following and throughout the book instead of the correct denotation $Y = f \circ X$.

means nothing else than the requirement that it must be possible to assign a probability to each event $\{f\,[X(\omega)] \leq y\}$.

Another requirement is that the domain of the random quantity X must be equal to or a subset of the domain of the transformation function f, to ensure that each outcome ω of the sample space Ω is actually mapped by the composition of the functions $X(\omega)$ and $f(X)$ onto the same real number as obtained by the mapping with the function $Y(\omega)$ (see Fig. 4.12).

If the random quantity X is transformed into the random quantity Y, the probability density function of X is also transformed. We state the corresponding proposition here without proof.

Proposition 4.19 (Transformation of a probability density function)

Let X be a real random quantity with a given probability density function $g_X(x)$ and let f be a real, continuous and differentiable function which transforms the random quantity X into the random quantity Y. Then the probability density function of the random quantity Y is given by

$$g_Y(y) = \sum_{i=1}^{n+1} \left(\frac{g_X(x)}{|f'(x)|} \right)_{x=f_i^{-1}(y)}, \tag{4.34}$$

where x_i $(i = 1, \ldots, n)$ denotes the roots of the derivative $f'(x)$ of the transformation function $f(x)$ and $f_i^{-1}(y)$ denotes the branch of the inverse of the transformation function restricted to the half-open interval $(x_{i-1}, x_i]$.

We will demonstrate the application of this proposition by looking at three examples of transformations of practical importance.

Example 4.43 (Linear transformation of a continuous quantity)

We consider a continuous random quantity X in the interval $(-\infty, +\infty)$, with a given probability density function $g_X(x)$. This random quantity is transformed into the random quantity Y by the linear transformation $f(X) = aX + b$, with $a \neq 0$. We are interested in the probability density function $g_Y(y)$ of the transformed random quantity.

The derivative of the transformation function is given by $f'(x) = a$, i.e. it is a constant function and so there is no root of the equation $f'(x) = 0$. Therefore, there is a unique inverse function which can be obtained by solving the equation $f(x) = y$. This gives

$$f_1^{-1}(y) = \frac{y - b}{a}, \qquad -\infty < y < +\infty.$$

Substituting this result into equation (4.34), we obtain the probability density function

$$g_Y(y) = \frac{1}{|a|} g_X\left(\frac{y - b}{a}\right), \qquad -\infty < y < +\infty, \tag{4.35}$$

of the transformed random quantity Y. This is a well known result.

Example 4.44 (Quadratic transformation of a continuous quantity)
We consider a continuous random quantity X in the interval $(-\infty, +\infty)$ with a given
probability density function $g_X(x)$. This random quantity is transformed into the
random quantity Y by the transformation $f(X) = aX^2 + bX + c$, with $a \neq 0$. We
are interested in the probability density function $g_Y(y)$ of the transformed random
quantity.

The derivative of the transformation function is given by $f'(x) = 2ax + b$.
Since $f'(x) = 0$ is a linear equation, there is only one root, given by

$$x_1 = -\frac{b}{2a},$$

i. e. the interval $(-\infty, +\infty)$ consists of the two disjoint subintervals $(-\infty, x_1]$ and
$(x_1, +\infty)$. For each of these two intervals there is an inverse function which can be
obtained by solving the equation $f(x) = y$. This yields

$$f_1^{-1}(y) = -\frac{b + \sqrt{4a(y-c) + b^2}}{2a}, \qquad y \geq c - \frac{b^2}{4a},$$

and

$$f_2^{-1}(y) = -\frac{b - \sqrt{4a(y-c) + b^2}}{2a}, \qquad y \geq c - \frac{b^2}{4a}.$$

We therefore obtain

$$|f'(x)|_{f_1^{-1}(y)} = |f'(x)|_{f_2^{-1}(y)} = \sqrt{4a(y-c) + b^2}, \qquad y > c - \frac{b^2}{4a}.$$

Substituting this result into equation (4.34), we obtain the probability density function

$$g_Y(y) = \begin{cases} \dfrac{1}{2aw}\left[g_X\left(-\dfrac{b}{2a} - w\right) + g_X\left(-\dfrac{b}{2a} + w\right)\right] & \text{if } y > c - \dfrac{b^2}{4a} \\ 0 & \text{if } y \leq c - \dfrac{b^2}{4a} \end{cases}$$

of the transformed random quantity Y with the abbreviation

$$w = \sqrt{\frac{y-c}{a} + \frac{b^2}{4a^2}}.$$

In the case of $a = 1$, $b = 0$ and $c = 0$ this gives

$$g_Y(y) = \begin{cases} \dfrac{1}{2\sqrt{y}}\left[g_X\left(-\sqrt{y}\right) + g_X\left(+\sqrt{y}\right)\right] & \text{if } y > 0 \\ 0 & \text{if } y \leq 0 \end{cases}. \tag{4.36}$$

Example 4.45 (Logarithmic transformation of a continuous quantity)

We consider a continuous random quantity X, defined in the interval $(0, \infty)$, with a given probability density function $g_X(x)$. This random quantity is transformed into the random quantity Y by the logarithmic transformation $f(X) = \log(a) + \log \circ X$, with $a > 0$. We are interested in the probability density function $g_Y(y)$ of the transformed random quantity.

The derivative of the transformation function is given by $f'(x) = 1/x$, i. e. $f(x)$ is a positive function in the interval $(0, \infty)$ and thus there is no root of the equation $f'(x) = 0$. Therefore, there is a unique inverse function which can be obtained by solving the equation $f(x) = y$. This results in

$$f_1^{-1}(y) = \frac{e^y}{a}, \qquad 0 < y < \infty.$$

Substituting this result into equation (4.34), we obtain the probability density function

$$g_Y(y) = \frac{e^y}{a} g_X\left(\frac{e^y}{a}\right), \qquad 0 < y < \infty,$$

of the transformed random quantity Y.

It can be shown that the proposition 4.19 is also valid for discrete random quantities. We will show this only for the most important case of a linear transformation. Applying equation (4.35) to the probability density function (4.33) of a discrete random quantity we get

$$g_Y(y) = \frac{1}{|a|} \sum_{k=1}^n P\left(X = \frac{y-b}{a}\right) \delta\left(\frac{y-b}{a} - x_k\right).$$

For DIRAC's δ-function the equality $|c|\delta(cx) = \delta(x)$ is valid (see e. g. [62]), i. e. in our case

$$\frac{1}{|a|}\delta\left(\frac{y-b}{a} - x_k\right) = \delta(y - ax_k - b).$$

Using the sifting property $f(y)\delta(y - c) = f(c)$ of DIRAC's δ-function (see e. g. [63]), this finally yields the probability density function

$$g_Y(y) = \sum_{k=1}^n P(X = x_k)\delta(y - ax_k - b)$$

of the transformed random quantity Y.

4.12 Expectations

The consideration of expectations dates back to the early days of probability theory, when expectations were essentially used to calculate the expected gain from games of chance of all kinds. One of the first instructions to do this was published by C. HUYGENS

[64]. This was commented on and extended by J. BERNOULLI [25], who already used the term "expectation" (lat. *expectatio*) in its present sense and used the following example to illustrate this term.

Example 4.46 (Expectation, J. BERNOULLI [25])

Someone hides 3 (or generally a) balls in one hand and 7 (or generally b) balls in the other. He allows me to take the balls he has in one hand and another person to take the balls he has in the other hand. Thus, both of us together will definitely acquire, and should therefore expect to receive the balls he has in both hands, namely 10 (or in general $a + b$) balls. But each of us has an equal right to what we expect. Therefore, the total expectation should be divided into two equal parts and each of us should get half of the expectation, i. e. 5 (or in general $(a + b)/2$) balls.

This example shows that the expectation is identical to the arithmetic mean if only events with equal probabilities occur. The more general case where the events do not have the same probability was also considered by J. BERNOULLI and leads directly to the definition of the expectation of a discrete random quantity.

Definition 4.31 (Expectation of a discrete random quantity)

Let X be a discrete random quantity taking the values x_k ($k = 1, \ldots, n$) with probabilities $P(X = x_k)$. The expectation of X is given by

$$E[X] = \sum_{k=1}^{n} x_k P(X = x_k). \tag{4.37}$$

It immediately follows from this definition that if all its values occur with equal probability $P(X = x_k) = 1/n$ ($k = 1, \ldots, n$), the expectation of a discrete random quantity is given by

$$E[X] = \overline{x} = \frac{1}{n} \sum_{k=1}^{n} x_k, \tag{4.38}$$

i. e. in this case the expectation is equal to the arithmetic mean, denoted by $\overline{x}$, of the values of the random quantity. We can therefore consider the expectation as a generalization of the arithmetic mean.

We demonstrate the calculation of expectations of discrete random quantities with two simple examples.

Example 4.47 (Rolling an ideally symmetric die)

When rolling an ideally symmetric die, the probability of each of the six possible outcomes is equal, i. e. we can set the expectation equal to the arithmetic mean and thus use the equation (4.38). If we denote the random quantity "number of dots" by X, we get $E[X] = 3.5$.

Example 4.48 (Continuation of example 4.36)

The probabilities of the sums of the dots when two ideally symmetric dice are rolled are given in table 4.2. Since the probabilities are not equal in this case, we cannot equate the expectation with the arithmetic mean, but have to use the equation (4.37). If we denote the random quantity "number of dots" by X, we obtain the expectation

$$E[X] = 2 \cdot \frac{1}{36} + 3 \cdot \frac{2}{36} + 4 \cdot \frac{3}{36} + 5 \cdot \frac{4}{36} + 6 \cdot \frac{5}{36} + 7 \cdot \frac{6}{36}$$

$$+ 8 \cdot \frac{5}{36} + 9 \cdot \frac{4}{36} + 10 \cdot \frac{3}{36} + 11 \cdot \frac{2}{36} + 12 \cdot \frac{1}{36} = 7.$$

These two examples show that the expectation of a discrete random quantity can be equal to one of its values, but that this is not necessarily the case. For instance, in the example 4.47, the expectation 3.5 is not a possible outcome of rolling one die, while in the example 4.47, the expectation 7 is equal to one of the possible sums of the dots when rolling two dice.

People unfamiliar with probability theory often mistakenly believe that the expectation is the arithmetic mean[37] and *vice versa*. These two quantities may be equal (for example, in the case of a symmetric distribution such as the one used in the example 4.36), but they clearly have different meanings.

In order to illustrate the difference between the expectation and the arithmetic mean, we give an example with a discrete random quantity that can take a countably infinite number of values.

Example 4.49 (A six turns up at the k-th roll of a die)

We roll an ideally symmetric die and ask for the probability of the event "a six comes up on the k-th roll". In this case we can consider the number of rolls until a six comes up as a discrete random quantity, which we denote by X.

The probability of a six coming up is $1/6$, i. e. the probability of no six coming up is $1 - 1/6 = 5/6$. All rolls are stochastically independent. Therefore, the probability that no six will come up on $(k - 1)$ rolls and that a six will come up on the k-th roll is equal to

$$P(X = k) = \left(\frac{5}{6}\right)^{k-1} \cdot \frac{1}{6} = \frac{5^{k-1}}{6^k}, \qquad k \geq 1.$$

In principle, it can take an infinitely long time to roll a six, since this probability is non-zero for any value of $k \geq 1$. However, as the number of rolls increases, this event becomes less and less likely, because the probability $P(X = k)$ decreases monotonically and eventually tends to zero for an infinite number of rolls. It becomes

---

[37] Unfortunately, the terms "expectation" and "arithmetic mean" or "average" are not always clearly distinguished in many publications or even textbooks.

apparent that the random quantity X obeys a geometric probability distribution function.[38]

The probabilities of a geometric probability distribution function depending on a parameter p are given by

$$P(X = k) = p(1 - p)^{k-1}, \qquad 0 \le p \le 1, \qquad k \ge 1, \qquad (4.39)$$

where in this example we have $p = 1/6$.

The expectation of a geometric probability distribution function can be obtained by inserting equation (4.39) into equations (4.37). For the limit $n \to \infty$ this gives

$$E[X] = \sum_{k=1}^{\infty} kp(1 - p)^{k-1} = -p\frac{d}{dp}\sum_{k=1}^{\infty}(1 - p)^k = -p\frac{d}{dp}\frac{1}{p} = \frac{1}{p},$$

i. e. in this example $E[X] = 6$.

This example demonstrates that it is possible to obtain a finite result of the expectation even though the random quantity can take a countably infinite number of values. This sounds strange, but the reason is that the probabilities $P(X = k)$ of the underlying probability distribution function decrease monotonically with increasing value of k. If, on the other hand, the underlying probability distribution function were a uniform distribution—notwithstanding the assumption that the random quantity still takes a countably infinite number of values—the expectation would be given by the arithmetic mean according to equation (4.38), which, however, does not exist, because in the case of $n \to \infty$ the sum becomes a divergent series (see e. g. K. KNOPP [65] for more details). In summary, we can state that the arithmetic mean can only be calculated if a discrete random quantity takes a finite number of values, while the expectation can even exist if a discrete random quantity takes a countably infinite number of values.

Our considerations have shown that a finite expectation of a discrete random quantity taking a countably infinite number of values can only exist if (for $n \to \infty$) the equation (4.37) yields an absolutely convergent series (see e. g. K. KNOPP [65] for more details). This requirement is essential because even if the underlying probability distribution function is *not* a uniform distribution, a divergent series may still occur in certain cases.

The definition 4.31 of the expectation of a discrete random quantity can be generalized for a continuous random quantity:

---

[38] Its continuous counterpart is the exponential distribution function.

Definition 4.32 (Expectation of a continuous random quantity)

The expectation of a continuous random quantity X with probability density function $g_X(x)$ is given by

$$E[X] = \int\limits_{-\infty}^{+\infty} x g_X(x) dx \tag{4.40}$$

provided that the integral is absolutely convergent.

The requirement of absolute convergence of the integral in this definition is again essential for the existence of an expectation, because there are probability density functions such that the equation (4.40) does not give a finite expectation, i. e. the expectation is undefined. Well-known examples are the so-called CAUCHY distribution and the LORENTZ function, familiar to physicists.

It is easy to verify that the definition 4.32 of the expectation is valid not only for continuous random quantities, but also for discrete random quantities. In fact, if we substitute the probability density function $g_X(x)$ of a discrete random quantity according to the equation (4.33) into the defining equation (4.40) of the expectation and use the relation (4.32) of DIRAC's δ-function, we again obtain the defining equation (4.37) of the expectation of a discrete random quantity.

At the same time, the requirement of absolute convergence of the integral turns out to be equivalent to absolute convergence of the resulting series. We can therefore consider the definition 4.32 to be generally valid.

We illustrate the calculation of the expectation of a continuous random quantity with a simple example:

Example 4.50 (Continuation of example 4.37)

If we insert the probability density function $g_X(x)$ given by the equation (4.29) into the defining equation (4.40) of expectation, we obtain

$$E[X] = \frac{1}{L} \int\limits_{0}^{L} x dx = \frac{L}{2}, \tag{4.41}$$

i. e. the breaking point is expected to be exactly in the middle of the wire. This result is plausible for a homogeneous wire with a negligible weight.

We now turn to the question of how to compute the expectation $E[Y]$ of a random quantity Y, if this random quantity can be obtained from another random quantity X by applying a transformation function f. At the same time, it is also of interest whether the integrals

$$E[f(X)] = \int\limits_{-\infty}^{+\infty} f(x) g_X(x) dx \tag{4.42}$$

and

$$E[Y] = \int_{-\infty}^{+\infty} y g_Y(y) dy \tag{4.43}$$

give the same results, i. e. whether we can be sure that always $E[Y] = E[f(X)]$ is valid. In fact, if this is the case, the calculation of the expectation of a transformed random quantity will generally be greatly simplified, since the direct calculation of $E[f(X)]$ usually requires less effort than the calculation of $E[Y]$. We will demonstrate this with a simple example.

Example 4.51 (Expectation of a quadratic function)

Let X be a continuous random quantity, defined in the interval $(-\infty, +\infty)$, with a probability density function given by

$$g_X(x) = \begin{cases} \dfrac{1}{3} & \text{if } -1 \le x \le 2 \\ 0 & \text{otherwise} \end{cases}$$

The random quantity X is transformed into the random quantity Y by the quadratic transformation function $f(X) = X^2$. We are interested in a comparison of the two expectations $E[Y]$ and $E[f(X)]$.

Inserting the transformation function $f(x) = x^2$ and the probability density function $g_X(x)$ into equation (4.42) yields

$$E[f(X)] = \frac{1}{3} \int_{-1}^{2} x^2 dx = 1 .$$

On the other hand, using the equation (4.36) we obtain the transformed probability density function

$$g_Y(y) = \begin{cases} 0 & \text{if } -\infty < y \le 0 \\ \dfrac{1}{3\sqrt{y}} & \text{if } 0 < y \le 1 \\ \dfrac{1}{6\sqrt{y}} & \text{if } 1 < y \le 4 \\ 0 & \text{if } 4 < y < +\infty \end{cases} .$$

Finally, inserting this probability density function into equation (4.43) yields

$$E[Y] = \frac{1}{3} \int_{0}^{1} \sqrt{y} dy + \frac{1}{6} \int_{1}^{4} \sqrt{y} dy = 1 .$$

Thus, $E[Y]$ and $E[f(X)]$ are equal.

The result $E[Y] = E[f(X)]$ obtained in this example turns out to be generally valid. This fact allows us to calculate the expectation of a transformed random quantity directly, without the necessity to determine its probability density function beforehand according to the transformation theorem 4.19. Therefore, it is useful—in addition to the definition 4.32—to introduce the expectation of a function of a random quantity.

Definition 4.33 (Expectation of a transformed random quantity)

The expectation of the random quantity X with the probability density function $g_X(x)$, which is transformed by a function f, is given by

$$E[f(X)] = \int_{-\infty}^{+\infty} f(x)g_X(x)\mathrm{d}x,$$

provided that the integral is absolutely convergent.

We can consider this definition as the proper definition of the expectation, as it is common in many textbooks of probability theory today. Consequently, the definition 4.32 is only a special case of the definition 4.33.

Since the expectation is represented by an integral, it inherits to some extent the properties of integration. One of these properties is linearity. For two arbitrary real functions f_1 and f_2 and two real constants a and b,

$$E[af_1(X) + bf_2(X)] = \int_{-\infty}^{+\infty}(af_1(x) + bf_2(x))g_X(x)\mathrm{d}x =$$

$$a\int_{-\infty}^{+\infty} f_1(x)g_X(x)\mathrm{d}x + b\int_{-\infty}^{+\infty} f_2(x)g_X(x)\mathrm{d}x = a\,E[f_1(X)] + b\,E[f_2(X)]$$

is valid. Therefore, because of the linearity of integration, the formation of the expectation of a random quantity is a linear operation.

Proposition 4.20 (Linearity of expectations)

If X is a random quantity, then given the real functions f_1 and f_2 and the real constants a and b, the equation

$$E[af_1(X) + bf_2(X)] = a\,E[f_1(X)] + b\,E[f_2(X)]$$

is valid, provided that the expectations $E[f_1(X)]$ and $E[f_2(X)]$ exist.

Special consequences of this proposition are the relations

$$E[a] = a,$$

i. e. the expectation of a constant is the constant itself, and

$$E[aX + b] = a\,E[X] + b,$$

i. e. shifting and scaling of a random quantity corresponds to shifting and scaling of its expectation. A generalization of proposition 4.20 is

$$E\left[\sum_{i=1}^{n} c_i f_i(X)\right] = \sum_{i=1}^{n} c_i E\left[f_i(X)\right],$$

as can easily be verified. The application of all these rules often facilitates the calculation of expectations.

Another property of the expectation is its monotonicity, which results from the mean value theorem of integral calculus. Accordingly, using the normalization condition (4.30),

$$E\left[f_2(X) - f_1(X)\right] = \int_{-\infty}^{+\infty} \left[f_2(x) - f_1(x)\right] g_X(x)\mathrm{d}x =$$

$$\left[f_2(\xi) - f_1(\xi)\right] \int_{-\infty}^{+\infty} g_X(x)\mathrm{d}x = f_2(\xi) - f_1(\xi)$$

is valid, where ξ denotes a real number, such that this equation is fulfilled. If we now assume $f_1(x) \le f_2(x)$ to be valid, we obtain the expression $f_2(\xi) - f_1(\xi) \ge 0$ on the right side of this equation and on the left side, because of the linearity of the expectation according to proposition 4.20, the result $E[f_2(X)] - E[f_1(X)]$. Hence, we can state the following proposition:

Proposition 4.21 (Monotonicity of expectations)
If X is a random quantity, then given the real functions f_1 and f_2 the relation

$$E[f_1(X)] \le E[f_2(X)], \quad \text{if} \quad f_1(x) \le f_2(x),$$

is valid, provided that the expectations $E[f_1(X)]$ and $E[f_2(X)]$ both exist.

Linearity and monotonicity are essential properties, if we consider the expectation of a random quantity—which we already decided to be equal to a measured quantity—to be the value of a measured quantity.

4.13   Variances and standard deviations

In order to assess the dispersion of the values of a random quantity, various measures of dispersion can be used, such as e. g. the range of the interval (i. e. the difference between the largest and the smallest value of a random quantity), the mean absolute deviation from a reference value (i. e. the expectation of the absolute value of the difference between the value of a random quantity and a reference value), the median of the absolute deviation from a reference value (i. e. the mean of the largest and the smallest absolute value of the deviation of the value of the random quantity from a reference value), the mean square deviation from a reference value (i. e. the expectation of the square of the difference between the value of a random quantity and a reference value), and so on. This list is not exhaustive. Various other measures can also be used.

The various measures of dispersion have been, and still are, favoured or rejected for a variety of reasons. We will only consider the dispersion based on the mean square deviation from a reference value. This measure was introduced by C. F. GAUSS [66][39] and, according to generally accepted conventions, its inverse serves as the measurement precision.

The majority of all measures of dispersion proposed in the past describe the dispersion according to C. F. GAUSS with respect to an arbitrarily chosen reference value, i. e. the dispersion of the values of the random quantity X about the reference value ξ is given by

$$D(X; \xi) = E\left[(X - \xi)^2\right] . \tag{4.44}$$

Since ξ is still arbitrary in this expression, the question arises whether there is a criterion for an optimal choice. In order to derive such a criterion, we transform the equation (4.44) into the equivalent form

$$D(X; \xi) = E\left[(X - E[X])^2 + 2(X - E[X])(E[X] - \xi) + (E[X] - \xi)^2\right] .$$

Using the linearity of expectation according to proposition 4.20 we finally get the relation[40]

$$D(X; \xi) = E\left[(X - E[X])^2\right] + (E[X] - \xi)^2 . \tag{4.45}$$

The dispersion therefore consists of two terms. The first term is a measure of the mean square deviation from the expectation and is called the variance. We can specify the variance using the following definition:

---

39 In contrast to C. F. GAUSS, his contemporary P. S. LAPLACE initially preferred to use a measure of dispersion based on the mean absolute deviation from a reference value [50]. Later, however, he also used the measure of dispersion favoured by C. F. GAUSS.

40 Some authors call this relation STEINER's theorem. This naming seems to be based on an analogy with a theorem in mechanics from which STEINER's theorem actually originates (also known as parallel axis theorem). Indeed, if the values of a random quantity are considered as co-ordinates of the mass points of a rigid body, then the expectation corresponds to the centre of mass, the first part of the dispersion corresponds to the moment of inertia with respect to the axis of rotation through the centre of mass, and the second part of the dispersion corresponds to the moment of inertia with respect to an axis of rotation parallel to the axis of rotation through the centre of mass. Although the use of analogies may be desirable in principle, the naming of mathematical theorems should be done in the correct context in order to avoid confusion.

Definition 4.34 (Variance of a random quantity)

Let X be a random quantity, then its variance is given by the expression

$$\text{Var}(X) = \text{E}\left[(X - \text{E}[X])^2\right].$$

This definition of the variance is given in a general way, so that it is valid for a discrete as well as a continuous random quantity.

In earlier textbooks of probability theory, the variance was often called dispersion. However, we use the latter term here only in the more general sense given above, i. e. as a measure of the dispersion of the values of a random quantity about an *arbitrary* reference value, which is not necessarily equal to the expectation. The distinction between "dispersion" as a more general term and "variance" as a narrower term will prove useful when dealing with measurement uncertainty.

Having recognized that the first term of the dispersion given by equation (4.45) is the variance, we now examine the second term in this equation. This term is the square of the deviation from the expectation[41] and therefore always takes a non-negative value, i. e. we have the relation

$$\text{D}(X; \xi) \geq \text{Var}(X),$$

where the equality sign applies only if $\xi = \text{E}[X]$. Thus, if we use the expectation as the reference value for calculating the dispersion, we obtain the smallest possible value[42] and this value is equal to the variance. The variance is therefore uniquely associated with the expectation. This highlights the importance of expectation and variance in metrology.

Since the variance as the expectation of a squared expression is always non-negative, the positive square root of the variance is well-defined. This leads to the definition of the standard deviation, which provides another measure of the dispersion of the values of a random quantity around its expectation. Moreover, this measure has the same dimension as the expectation.

Definition 4.35 (Standard deviation of a random quantity)

Let X be a random quantity, then the expression

$$\sigma(X) = \sqrt{\text{Var}(X)}$$

is called standard deviation and the expression

$$\sigma_{\text{r}}(X) = \frac{\sigma(X)}{\text{E}[X]}, \qquad \text{E}[X] \neq 0, \tag{4.46}$$

---

[41] In statistics the *deviation of the expectation of the estimator from the true value* is called bias, while in metrology the *estimate of a systematic error*, which is not quite the same, is called bias (VIM, 2.18 [3]). Since "bias" is not a clearly defined term, we will avoid using this term as much as possible.

[42] This fact, which was already known to C. F. GAUSS [66], is used whenever a best estimate is obtained by the method of least squares.

relative standard deviation (coefficient of variation) of this random quantity.

The relative standard deviation is a quotient of two quantities of the same kind, and is therefore always a pure number. However, it can only be applied if the expectation is not equal to zero.

Note that often only the symbols σ and σ_r—instead of the more detailed notations $\sigma(X)$ and $\sigma_r(X)$—are used for the standard deviation and the relative standard deviation, especially when it is clear from the context which random quantity they correspond to.

Like the variance, the standard deviation—being by definition the positive square root of the variance—always takes non-negative values and is similarly associated with the expectation of the corresponding random quantity. We will return to this fact when we discuss measurement uncertainty.

Now we will look at some properties of the variance, but we will concentrate mainly on rules that facilitate the calculation of the variance.

From the definition 4.34 of the variance and the properties of the expectation we obtain

$$\mathrm{Var}(X) = \mathrm{E}\left[(X - \mathrm{E}[X])^2\right] = \mathrm{E}\left[X^2 - 2X\mathrm{E}[X] + (\mathrm{E}[X])^2\right] =$$

$$\mathrm{E}\left[X^2\right] - 2\mathrm{E}[X]\mathrm{E}[X] + (\mathrm{E}[X])^2 = \mathrm{E}\left[X^2\right] - (\mathrm{E}[X])^2,$$

provided that the expectations exist. From this relation[43] in turn, using the property $\mathrm{Var}(X) \geq 0$, we can immediately deduce

$$|\mathrm{E}[X]| \leq \sqrt{\mathrm{E}[X^2]},$$

where the equality sign is only valid if we have $\mathrm{Var}(X) = 0$.

In the previous section we demonstrated how a linear transformation changes the expectation. Now we consider the change of the variance under a linear transformation. Using the definition of the variance 4.34 and the properties of the expectation, we obtain

$$\mathrm{Var}(aX + b) = \mathrm{E}\left[(aX + b - \mathrm{E}[aX + b])^2\right] = \mathrm{E}\left[(aX + b - a\mathrm{E}[X] - b)^2\right] =$$

$$\mathrm{E}\left[a^2(X - \mathrm{E}[X])^2\right] = a^2\mathrm{E}\left[(X - \mathrm{E}[X])^2\right] = a^2\mathrm{Var}(X).$$

This result subsequently yields the standard deviation of a linearly transformed random quantity

$$\sigma(aX + b) = a\,\sigma(X).$$

---

43 This relation is called *displacement law* in some statistics textbooks.

Thus, the variance and the standard deviation are only scaled by a linear transformation, but are invariant with respect to a displacement. The consequence of this property in metrology is that an arbitrarily large systematic measurement error does not cause any change in the variance or standard deviation, i. e. these measures can only be used to assess random measurement errors.

The following proposition summarizes some basic properties of the variance and the standard deviation:

Proposition 4.22 (Properties of the variance and the standard deviation)
Let X be a random quantity, and a and b two arbitrary real numbers, then the variance has the properties

a) $$\text{Var}(X) = \text{E}\left[X^2\right] - \left(\text{E}[X]\right)^2, \tag{4.47}$$

b) $$\text{Var}(aX + b) = a^2\text{Var}(X),$$

and the standard deviation the property

$$\sigma(aX + b) = a\,\sigma(X).$$

In order to demonstrate how the variance and the standard deviation can be calculated using the equation (4.47), we give an example of a discrete and a continuous random quantity, respectively.

Example 4.52 (Continuation of example 4.48)
We use the probabilities of the sums of the dots when two ideally symmetric dice are rolled to calculate the expectation $\text{E}\left[X^2\right]$. This yields

$$\text{E}\left[X^2\right] = 2^2 \cdot \frac{1}{36} + 3^2 \cdot \frac{2}{36} + 4^2 \cdot \frac{3}{36} + 5^2 \cdot \frac{4}{36} + 6^2 \cdot \frac{5}{36} + 7^2 \cdot \frac{6}{36}$$

$$+ 8^2 \cdot \frac{5}{36} + 9^2 \cdot \frac{4}{36} + 10^2 \cdot \frac{3}{36} + 11^2 \cdot \frac{2}{36} + 12^2 \cdot \frac{1}{36} = \frac{329}{6}.$$

If we insert this result and the expectation $\text{E}[X] = 7$, already calculated in the example 4.48, into equation (4.47), we obtain the variance

$$\text{Var}(X) = \frac{329}{6} - 7^2 = 35/6$$

and, according to definition 4.35, the standard deviation

$$\sigma(X) = \sqrt{35/6},$$

i. e. the standard deviation is around 34.5 % of the expectation.

Example 4.53 (Continuation of example 4.50)

If we use the probability density function $g_X(x)$ given by equation (4.29) to calculate the expectation $E\left[X^2\right]$, we get

$$E\left[X^2\right] = \frac{1}{L} \int_0^L x^2 dx = \frac{L^2}{3}.$$

Substituting this result and the expectation given by equation (4.41) into equation (4.47), we obtain the variance

$$Var(X) = \frac{L^2}{3} - \frac{L^2}{4} = \frac{L^2}{12}$$

and, according to definition 4.35, the standard deviation

$$\sigma(X) = \frac{L}{2\sqrt{3}}.$$

The relative standard deviation is thus equal to $1/\sqrt{3}$, i. e. about 57.7 %.

When a random quantity is considered as a measured quantity, its numerical value will generally depend on the origin of the underlying scale and the chosen unit of measurement. Any change in the value of the measured quantity due to a change in these special constants can be described by the linear transformation $f \circ X = aX + b$. In particular, if we assume $a > 0$, this transformation is also called "transformation of scale". A typical application of a transformation of scale is the transformation of temperature scales.

Example 4.54 (Transformation of temperature scales)

a) The conversion of the numerical values of the thermodynamic temperature scale (KELVIN scale) to those of the CELSIUS scale is given by

$$Y = X - 273.15.$$

In this case we have $Var(Y) = Var(X)$ and $\sigma(Y) = \sigma(X)$.

b) The conversion of the numerical values of the CELSIUS scale to those of the FAHRENHEIT scale is given by

$$Y = \frac{9}{5}X + 32.$$

In this case we have $Var(Y) > Var(X)$ and $\sigma(Y) > \sigma(X)$.

Another important application of a transformation of scale is the introduction of standardized or reduced quantities.

Example 4.55 (Standardized and reduced random quantities)
a) The standardization of a random quantity is done by the transformation

$$Y = \frac{X - E[X]}{\sigma(X)}.$$

This yields $E[Y] = 0$ and $Var(Y) = 1$. Note that standardized random quantities are of dimension number, i. e. they are pure numbers.

b) Reduced quantities are used, for example, in theoretical physics or chemistry. They are calculated by dividing the respective quantity by the corresponding reference quantity. In this way quantities of dimension number are obtained, such as in thermodynamics the reduced pressure or the reduced molar volume. Using reduced quantities facilitates the comparability of material properties.

A transformation of scale changes the expectation and generally also the variance and the standard deviation. However, in the case of a standardized quantity, this is the purpose of this transformation.

4.14 Multivariate random quantities

A random quantity is, according to its definition 4.24, at first only a scalar quantity. However, vectors or tensors whose components are random quantities can also be considered as random quantities. In principle, such a convention makes the use of terms like "random vector" or "random tensor" unnecessary.

We will call all random quantities that are not scalar quantities multivariate random quantities and write them as column vector, i. e. as $X = (X_1, \dots, X_n)^T$, regardless of the particular meaning of its components. However, the n scalar random quantities $X_1, \dots, X_n$ combined in this way can belong to quantities of the same or a different kind. Consequently, the quantity dimensions[44] of the components of such a column vector are not necessarily the same, i. e. the considered multivariate random quantity X is not actually a vector.

In the strict sense, we have to give a more general meaning to the term "multivariate random quantity", because we cannot assume that the components of a multivariate random quantity have been obtained by assigning real numbers to the outcomes of the *same* sample space Ω, but we also have to allow the possibility that they have been obtained by assigning real numbers to the outcomes of *different* sample spaces, because different random experiments have to be taken into account. These two borderline cases are schematically illustrated in Fig. 4.13, using the two components of a two-dimensional

---

44 A quantity dimension should not be confused with a dimension of a vector space in linear algebra. Quantity dimensions are related to physical properties such as length, time, mass, electric charge, etc. For more details see [8].

random quantity as an example, where in the first case, shown on the left side of Fig. 4.13 the elements of the sample space Ω must be ordered pairs (ω_1, ω_2), while in the second case, shown on the right side of Fig. 4.13, the elements of the sample spaces Ω_1 and Ω_2 are single elements ω_1 and ω_2, respectively.

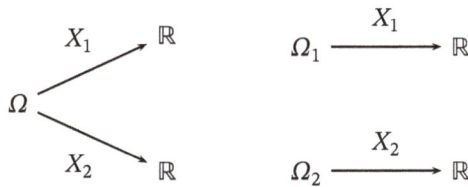

Fig. 4.13: Creation of a two-dimensional random quantity.

In the case of multivariate random quantities with more than two components, and thus with an increasing number of possible assignments, the situation can quickly become confusing. Therefore, in practical applications where it is questionable whether we can really distinguish between all possible cases, it seems more reasonable to consider only sample spaces whose elements are ordered n-tuples[45] (i. e. ordered pairs, triples, quadruples, etc.), where n denotes the number of components of the respective multivariate random quantity. We leave it open, however, whether the elements of the n-tuples are all assigned to only one or to several random experiments or have the same physical dimension.

We give two examples of multivariate random quantities that frequently occur in practical applications.

Example 4.56 (Position of an object in space)
We are interested in the position of an object in space. Since we can never measure a position exactly, we have to treat it as a random event. Thus, the sample space Ω will contain all possible positions. To each position we can assign an ordered triple of real numbers, which we consider to be the co-ordinates of the corresponding position of the object under consideration. Thus, we obtain the three-dimensional random quantity $X = (X_1, X_2, X_3)^\mathsf{T}$, which we call "position". The column vector $x = (x_1, x_2, x_3)^\mathsf{T}$ represents a realization of this random quantity, which may be obtained as the result of a measurement.

Example 4.57 (Complex valued random quantities)
Complex valued quantities are often used in electrical measurements, e. g. as complex voltages, currents, resistances or powers. When a complex quantity occurs as

---

45 In mathematics, a tuple is an ordered list of mathematical objects called the elements of the tuple. A tuple of n elements is denoted by a comma-separated list in parentheses.

a measured quantity, it can be represented by a two-dimensional random quantity $X = (X_1, X_2)^\mathsf{T}$, where the component X_1 denotes the real part and the component X_2 denotes the imaginary part. Then the complex number $x = x_1 + ix_2$ represents a realization of this random quantity. Note, however, that x is not identical to a measurement result $x = (x_1, x_2)^\mathsf{T}$, since the latter is a real vector.

From our considerations we can conclude that we can obtain a multivariate random quantity by generalizing a scalar random quantity according to the definition 4.24. This gives the following definition:

Definition 4.36 (Multivariate random quantity)

A column vector of functions $X = (X_1, \ldots, X_n)^\mathsf{T}$ which assigns a column vector of real numbers $x = (x_1, \ldots, x_n)^\mathsf{T}$ to each element ω of the sample space Ω is called a multivariate random quantity, provided that the set

$$\{\omega \in \Omega \,|\, X_1(\omega) \le x_1, \ldots, X_n(\omega) \le x_n, \, x \in \mathbb{R}^n\}$$

is an event in the set of events $\mathcal{A}$.

The additional requirement

$$\mathfrak{E}_X = \{\omega \in \Omega \,|\, X_1(\omega) \le x_1, \ldots, X_n(\omega) \le x_n, \, x \in \mathbb{R}^n\}, \qquad \mathfrak{E}_X \in \mathcal{A},$$

occurring in this definition is again the regularity condition (measurability condition) which we already know from section 4.8. Thus, a column vector of functions is a multivariate random quantity only if $\mathfrak{E}_X$ is an event of the set of events $\mathcal{A}$. In this case the event $\mathfrak{E}_X$ means that the values of the random quantities X_i—i. e. the components of the multivariate random quantity X—are not greater than given real numbers x_i. Often—especially in practical applications—this event is simply abbreviated to the denotation $(X \le x)$.

In the case of multivariate random quantities, it is not always possible or useful to distinguish between purely discrete and purely continuous quantities. If all components of a multivariate random quantity are discrete, we can certainly speak of a discrete multivariate random quantity.

Example 4.58 (Sums of the number of dots when rolling two dice)

We return to the example 4.6 in section 4.1. We simultaneously roll two ideally symmetric dice—one white and one black—and assign the number of dots on the white die to the random quantity X_1 and the number of dots on the black die to the random quantity X_2. From these two discrete random quantities we form the two-dimensional random quantity $X = (X_1, X_2)^\mathsf{T}$. In this way we obtain a discrete multivariate random quantity.

We can speak of a continuous multivariate random quantity if all of its components are continuous.

Example 4.59 (Measurement of a current-voltage characteristic)

When measuring a current-voltage characteristic the electric current is measured as a function of the voltage. We assign the random quantity X_1 to the voltage and the random quantity X_2 to the electric current. From these two continuous random quantities we form the two-dimensional random quantity $\boldsymbol{X} = (X_1, X_2)^\mathsf{T}$. This is a continuous multivariate random quantity.

However, in the case where a multivariate random quantity has both discrete and continuous components, the usual classification fails. We could then speak of a "mixed" multivariate random quantity.

Example 4.60 (Measurement of neutron spectra)

When measuring thermal neutron spectra with a proportional detector, the number of neutrons per second and their energy are recorded. We can assign the discrete random quantity X_1 to the number of counted neutrons and the continuous random quantity X_2 to their respective energies.

4.15 Multivariate distribution functions

In this section we will look at the probability distribution function of a multivariate random quantity. In order to introduce this function, we can in principle proceed similarly as in section 4.9, where we introduced the probability distribution function of a scalar random quantity.

In the previous section, when defining the multivariate random quantity $\boldsymbol{X}$, we stated that it must satisfy the regularity condition

$$\mathfrak{C}_{\boldsymbol{X}} = \{\omega \in \Omega \,|\, X_1(\omega) \leq x_1, \dots, X_n(\omega) \leq x_n, \boldsymbol{x} \in \mathbb{R}^n\}, \qquad \mathfrak{C}_{\boldsymbol{X}} \in \mathcal{A}, \qquad (4.48)$$

i. e. the event $\mathfrak{C}_{\boldsymbol{X}}$ must be a possible event of the set of events $\mathcal{A}$.

We now ask for the probability $P(\mathfrak{C}_{\boldsymbol{X}})$ of the event $\mathfrak{C}_{\boldsymbol{X}}$ given by the relation (4.48). This probability is identical to the probability

$$P\left(\bigcap_{i=1}^{n}\{X_i \leq x_i\}\right) = P(\{X_1 \leq x_1\} \cap \dots \cap \{X_n \leq x_n\}),$$

that the values of the random quantities X_i do not exceed the given values x_i. This probability is a function of $\boldsymbol{x} = (x_1, \dots, x_n)^\mathsf{T}$ and is called multivariate probability distribution function[46] (also abbreviated as multivariate distribution function or multivariate probability distribution) and denoted by[47] $G_{X_1,\ldots,X_n}(x_1,\ldots,x_n)$, often abbreviated to $G_X(x)$. Obviously, $G_{X_1,\ldots,X_n}(x_1,\ldots,x_n)$ is the joint probability distribution function of the random quantities $X_1,\ldots,X_n$. Thus, we can give the following definition:

Definition 4.37 (Multivariate probability distribution function)

Let Ω be a sample space, then the real-valued function

$$G_X(x) = G_{X_1,\ldots,X_n}(x_1,\ldots,x_n) =$$

$$P(\{\omega \in \Omega \,|\, X_1(\omega) \leq x_1,\ldots,X_n(\omega) \leq x_n\}) = P\left(\bigcap_{i=1}^{n}\{X_i \leq x_i\}\right) =$$

$$P(\{X_1 \leq x_1\} \cap \cdots \cap \{X_n \leq x_n\}), \quad (x_1,\ldots,x_n)^\mathsf{T} \in \mathbb{R}^n,$$

is called multivariate (joint) probability distribution function of the multivariate random quantity $X = (X_1,\ldots,X_n)^\mathsf{T}$.

In the case of stochastically independent random quantities $X_1,\ldots,X_n$, the construction of the multivariate probability distribution function $G_{X_1,\ldots,X_n}(x_1,\ldots,x_n)$ is particularly easy, since it follows from the proposition 4.13 about stochastic independence, the equation (4.9), and the definition 4.27 of the probability distribution function of a univariate random quantity, that

$$P\left(\bigcap_{i=1}^{n}\{X_i \leq x_i\}\right) = \prod_{i=1}^{n} P(X_i \leq x_i) = \prod_{i=1}^{n} G_{X_i}(x_i).$$

Therefore, from the definition 4.37 we obtain the following proposition:

Proposition 4.23 (Distribution function of independent quantities)

Given the joint probability distribution functions $G_{X_i}(x_i)$ $(i = 1,\ldots,n)$ of independent random quantities X_i $(i = 1,\ldots,n)$,

$$G_{X_1,\ldots,X_n}(x_1,\ldots,x_n) = \prod_{i=1}^{n} G_{X_i}(x_i) \tag{4.49}$$

is valid.

---

[46] The probability distribution function $G_X(x)$ of a scalar random variable X is therefore also called "univariate probability distribution function".

[47] In deviation from the usual denotation $F_{X_1,\ldots,X_n}(x_1,\ldots,x_n)$ in probability theory, we generally use the denotation $G_{X_1,\ldots,X_n}(x_1,\ldots,x_n)$ in this book, as it is also used in Suppl. 1 of the GUM [54, 55], to avoid any confusion with model functions.

The converse of this proposition is also true, i. e. if the joint probability distribution function of the random quantities $X_1, \ldots, X_n$ can be represented by the equation (4.49), then these random quantities are stochastically independent.

We give two examples of the application of proposition 4.23.

Example 4.61 (Multivariate distribution function when rolling two dice)

We simultaneously roll two distinguishable dice, both assumed to be ideally symmetric. We assign the number of dots on one die to the random quantity X_1 and the number of dots on the other die to the random quantity X_2. Both random quantities are—as we already know—discrete and equally distributed.

We are interested in the probability distribution function $G_{X_1,X_2}(x_1, x_2)$ of the random quantities X_1 and X_2. In order to obtain this function, we use the fact that these random quantities are stochastically independent, because the outcome of one die does not depend on the outcome of the other. Therefore, we can apply the proposition 4.23, which yields the multivariate probability distribution

$$G_{X_1,X_2}(x_1, x_2) = G_{X_1}(x_1)G_{X_2}(x_2)$$

of the number of dots, denoted by X_1 and X_2, when the two dice are rolled.

Since the probability of each of the possible outcomes of a roll of the dice is $1/6$, we obtain—according to equation (4.15)—the discrete probability distribution function

$$G_X(x) = \frac{1}{6} \sum_{k=1}^{6} \Theta(x - k).$$

The probability distribution function we are interested in is then given by

$$G_{X_1,X_2}(x_1, x_2) = \frac{1}{36} \sum_{i=1}^{6} \sum_{j=1}^{6} \Theta(x_1 - i)\Theta(x_2 - j). \tag{4.50}$$

This is a discrete multivariate probability distribution function.

Example 4.62 (Positioning precision)

During the inspection of the positioning device of an automatic placement machine, it was found that 99 % of the positions reached were within the tolerance. We would like to know the positioning precision of the positioning device.

The following information is available:

- The tolerance is given by a circular disc with diameter $T = 1\ \mu m$ and the target position as the centre.
- The directional deviation and the radial deviation from the target position are stochastically independent.
- The directional deviation is uniformly distributed and described by the probability distribution function

$$G_{X_1}(x_1) = \frac{x_1}{2\pi}, \qquad 0 \le x_1 \le 2\pi,$$

where X_1 denotes the azimuth angle.

– The radial deviation, denoted by X_2, is described by the probability distribution function

$$G_{X_2}(x_2) = 1 - \exp\left(-\frac{x_2^2}{2\sigma^2}\right), \qquad 0 \le x_2 < \infty,$$

where σ denotes the standard deviation of the radial deviation.

Applying proposition 4.23 yields the probability distribution function[48]

$$G_{X_1,X_2}(x_1, x_2) = \frac{x_1}{2\pi}\left[1 - \exp\left(-\frac{x_2^2}{2\sigma^2}\right)\right],$$

$$0 \le x_1 \le 2\pi, \quad 0 \le x_2 < \infty, \tag{4.51}$$

of the two-dimensional quantity $\mathbf{X} = (X_1, X_2)^\mathsf{T}$, which describes the probability of a deviation from the target position.

The probability that a given position is within the tolerance is given by

$$P\left(X_1 \le 2\pi, X_2 \le \frac{T}{2}\right) = G_{X_1,X_2}\left(2\pi, \frac{T}{2}\right) = 1 - \exp\left(-\frac{T^2}{8\sigma^2}\right).$$

According to the measurement results, this probability is 0.99. Therefore, based on the given tolerance $T = 1\,\mu m$, we obtain the standard deviation $\sigma = 0.16\,\mu m$. The positioning precision can be expressed numerically by the inverse of the standard deviation (see VIM, 2.15, Note 1 [3, 2]).

The following proposition summarizes the main mathematical properties of a multivariate probability distribution function:

Proposition 4.24 (Properties of a multivariate probability distribution)
A multivariate probability distribution function $G_{X_1,\dots,X_n}(x_1, \dots, x_n)$ of a random quantity $\mathbf{X} = (X_1, \dots, X_n)^\mathsf{T}$ has the following properties:

a) $G_{X_1,\dots,X_n}(x_1, \dots, x_n)$ is monotonically increasing,

b) $G_{X_1,\dots,X_n}(x_1, \dots, x_n)$ is continuous from above,

c) $\displaystyle\lim_{x_k \to -\infty} G_{X_1,\dots,X_k,\dots,X_n}(x_1, \dots, x_k, \dots, x_n) = 0, \qquad k = 1, \dots, n,$

d) $\displaystyle\lim_{x_1 \to +\infty, \dots, x_n \to +\infty} G_{X_1,\dots,X_n}(x_1, \dots, x_n) = 1,$

e) $\displaystyle\lim_{\varepsilon_1 \to 0, \dots, \varepsilon_n \to 0} \left[G_{X_1,\dots,X_n}(x_1, \dots, x_n) - G_{X_1,\dots,X_n}(x_1 - \varepsilon_1, \dots, x_n - \varepsilon_n)\right] =$

$$P(X_1 = x_1, \dots, X_n = x_n), \qquad \varepsilon_1 > 0, \dots, \varepsilon_n > 0.$$

---

48 This probability distribution function is known as the density of the RAYLEIGH distribution.

These properties are similar to those of a univariate probability distribution function. In order to make them more understandable, we give some additional explanations. Property a) means that for increasing values of its arguments $x_1, \ldots, x_n$, a multivariate probability distribution function can only increase or remain constant, i.e. if an argument x_k takes the values u and v, respectively, and u is less than v, then the value $G_{X_1, \ldots, X_k, \ldots, X_n}(x_1, \ldots, x_{k-1}, u, x_{k+1}, \ldots, x_n)$ is less than or at most equal to the value $G_{X_1, \ldots, X_k, \ldots, X_n}(x_1, \ldots, x_{k-1}, v, x_{k+1}, \ldots, x_n)$. Property b) means that the value of a multivariate probability distribution function at a jump discontinuity is always the upper value. Property c) means that no component of the multivariate random quantity $X = (X_1, \ldots, X_n)^\mathsf{T}$ can take the value $-\infty$ (note that $X_k = -\infty$, $k = 1, \ldots, n$, is the impossible event), while property d) means that every component of X can take any real value (note that $X_k < +\infty$, $k = 1, \ldots, n$, is the certain event[49]). Finally, property e) means that at a jump discontinuity a multivariate probability distribution instantly changes its value by $P(X_1 = x_1, \ldots, X_n = x_n)$. In addition, this property also means that $P(X_1 = x_1, \ldots, X_n = x_n) = 0$ if there is no jump discontinuity at the point $x = (x_1, \ldots, x_n)^\mathsf{T}$.

Property e) is especially significant for continuous multivariate probability distribution functions, because *by definition* these functions have no jump discontinuities. It follows immediately that $P(X = x) = 0$ holds for any tuple of values x of a continuous multivariate random quantity X, i.e. the probability that certain values of its components occur is always zero. On the other hand, this also means that if a multivariate random quantity takes definite values with non-zero probability, then its multivariate probability distribution function necessarily has a jump discontinuity.

4.16 Multivariate density functions

The probability density function of a multivariate random quantity X can be obtained by a generalization of the definition 4.29 of the univariate probability density function. We just have to use a multidimensional integration instead of the one-dimensional integration.

Definition 4.38 (Multivariate probability density function)
The real-valued function $g_{X_1, \ldots, X_n}(x_1, \ldots, x_n)$ is called a multivariate probability density function of the random quantity $X = (X_1, \ldots, X_n)^\mathsf{T}$ when

$$G_{X_1, \ldots, X_n}(x_1, \ldots, x_n) = \int_{-\infty}^{x_1} \cdots \int_{-\infty}^{x_n} g_{X_1, \ldots, X_n}(\xi_1, \ldots, \xi_n)\, d\xi_1 \cdots d\xi_n \qquad (4.52)$$

---

49 We do not use the expressions $X_k \leq +\infty$ ($k = 1, \ldots, n$) here, because *by definition* neither $+\infty$ nor $-\infty$ belong to the set of real numbers.

is the multivariate probability distribution function of this random quantity.

The multivariate probability density function[50] $g_{X_1,...,X_n}(x_1, ..., x_n)$—we will also use the notation $g_X(x)$—is abbreviated as multivariate density function or multivariate probability density.

By definition 4.38 a multivariate probability distribution function is traced back to its associated multivariate probability density function, if such a density function actually exists. However, this is guaranteed by a generalized differentiation theorem of H. LE-BESGUE [67], which states that a continuous monotone function has a finite derivative at each of its points, except of a countable set of points. Since a multivariate probability distribution $G_X(x)$ satisfies the conditions of this theorem, it is almost everywhere differentiable and thus its probability density function $g_X(x)$ exists. At the countable many points where $G_X(x)$ is continuous but not differentiable, we can choose arbitrary values of the probability density function without changing the probability distribution function, because the latter is only specified up to a null set (see section 4.3).

If we apply the well-known NEWTON-LEIBNIZ fundamental theorem of calculus to the equation (4.52) given in the definition 4.38 of the multivariate probability density function, we obtain

$$g_{X_1,...,X_n}(x_1, ..., x_n) = \frac{\partial^n G_{X_1,...,X_n}(x_1, ..., x_n)}{\partial x_1 \partial x_2 \cdots \partial x_n}. \tag{4.53}$$

Thus, we can calculate the multivariate probability density function from a given multivariate probability distribution function by partial differentiation, provided that the probability distribution function is differentiable. We give an example for the application of equation (4.53):

Example 4.63 (Continuation of example 4.62)
We determine the multivariate probability density function by calculating the derivatives of the multivariate probability distribution function, given by equation (4.51), with respect to x_1 and x_2, which yields

$$g_{X_1,X_2}(x_1, x_2) = \frac{x_2}{2\pi\sigma^2} \exp\left(-\frac{x_2^2}{2\sigma^2}\right), \tag{4.54}$$

$$0 \leq x_1 < 2\pi, \quad 0 \leq x_2 < \infty.$$

This is the density of a uniform distribution (rectangular distribution) within the interval $[0, 2\pi)$ with respect to X_1 and the density of a RAYLEIGH distribution within the

---

[50] In deviation from the usual denotation $f_{X_1,...,X_n}(x_1, ..., x_n)$ in probability theory we generally use in this book the denotation $g_{X_1,...,X_n}(x_1, ..., x_n)$, as it is also used in Suppl. 1 of the GUM [54, 55], in order to avoid any confusion with model functions. This is consistent with the corresponding denotation of the multivariate probability distribution function in the previous section.

half-open interval $[0, \infty)$ with respect to X_2. For every other point, this probability density function takes the value zero.

If we insert equation (4.54) into equation (4.52), we obtain the probability distribution function given by equation (4.51), as expected.

Not every multivariate real function is suitable as a multivariate probability density. The conditions of a multivariate probability density follow from the properties of a multivariate probability distribution.

Proposition 4.25 (Conditions of a multivariate probability density)

The multivariate probability density function $g_{X_1,...,X_n}(x_1, ... , x_n)$ of a multivariate random quantity $X = (X_1, ... , X_n)^\mathsf{T}$ must satisfy the following conditions:

a) $g_{X_1,...,X_n}(x_1, ... , x_n)$ is a non-negative function,

b) $g_{X_1,...,X_n}(x_1, ... , x_n)$ fulfils the normalization condition

$$\int\limits_{-\infty}^{+\infty} \cdots \int\limits_{-\infty}^{+\infty} g_{X_1,...,X_n}(x_1, ... , x_n) \mathrm{d}x_1 \cdots \mathrm{d}x_n = 1 \,.$$

The condition a) follows from the property of every probability distribution function to be a monotonically increasing function. The normalization condition b) follows from equation (4.52) and property d) in proposition 4.24. Another condition not stated in proposition 4.25, namely that a probability density function must be zero at infinity, is already implicit in the normalization condition and therefore does not need to be stated explicitly.

The definition 4.38 of the multivariate probability density function is initially only valid under the condition that the corresponding multivariate probability distribution function is continuous. However, it can be shown that this condition can be dropped by proving that a multivariate probability density function can also be specified for a discrete multivariate probability distribution function, provided that we allow a product of DIRAC's δ-functions to be used as a multivariate probability density function as well. This makes it possible to express multivariate probability density functions of multivariate random quantities with one or more discrete components by the equation (4.52). The proof, which is similar to the one given in section 4.10 but more elaborate, will not be given here.

If we represent the components of a discrete multivariate probability distribution function by the HEAVISIDE function, we are able to calculate the corresponding discrete multivariate probability density function by partial differentiation according to equation (4.53), implicitly using the relation (4.31). We demonstrate this procedure with an example:

Example 4.64 (Continuation of example 4.61)

By partial differentiation of the equation (4.50) with respect to x_1 and x_2 we obtain

$$g_{X_1,X_2}(x_1, x_2) = \frac{1}{36} \sum_{i=1}^{6} \sum_{j=1}^{6} \delta(x_1 - i)\delta(x_2 - j) =$$

$$\left(\frac{1}{6} \sum_{i=1}^{6} \delta(x_1 - i)\right)\left(\frac{1}{6} \sum_{j=1}^{6} \delta(x_2 - j)\right).$$

Thus, the probability density function is given by 36 equal δ-functions, i. e. we obtain a discrete uniform distribution, as it was to be expected.

This example shows that a multivariate probability density function of stochastically independent random variables is equal to the product of the corresponding univariate probability density functions. It is easy to prove that this statement holds in general. If we apply equation (4.53) to the equation (4.49) given in proposition 4.23, we obtain the following proposition:

Proposition 4.26 (Density function of independent random quantities)

Given the joint probability density functions $g_{X_i}(x_i)$ ($i = 1, \dots, n$) of stochastically independent random quantities X_i ($i = 1, \dots, n$),

$$g_{X_1,\dots,X_n}(x_1, \dots, x_n) = \prod_{i=1}^{n} g_{X_i}(x_i) \tag{4.55}$$

is valid.

The converse of this proposition is also true, i. e. if the joint probability density function of the random quantities $X_1, \dots, X_n$ can be represented by the equation (4.55), then these quantities are stochastically independent.

In section 4.11, we looked at the transformation of univariate random quantities. It is also possible to transform a multivariate random quantity $X = (X_1, \dots, X_n)^\mathsf{T}$ into another multivariate random quantity $Y = (Y_1, \dots, Y_m)^\mathsf{T}$ by using the transformations

$$Y_i = f_i(X_1, \dots, X_n), \qquad i = 1, \dots, m, \tag{4.56}$$

where f_i ($i = 1, \dots, m$) denote real, continuous and differentiable functions. In general, however, we cannot assume that $m = n$ is valid, i. e. that the number of components of the random quantities X and Y is equal. Unfortunately, this means that we cannot generally expect that the system of equations (4.56) to be uniquely invertible, because this is only guaranteed if $m = n$ is valid and the JACOBIan determinant [68] of the system of equations (4.56) is not equal to zero. However, if these conditions are fulfilled, we can state the following proposition:

Proposition 4.27 (Transformation of a multivariate probability density)

Let $X_1, \ldots, X_n$ be n real random quantities with a joint probability density function $g_{X_1,\ldots,X_n}(x_1, \ldots, x_n)$ and let $f_1, \ldots, f_n$ be real, continuous and differentiable functions. Then the joint probability density function of the random quantities $Y_1, \ldots, Y_n$ is given by

$$g_{Y_1,\ldots,Y_n}(y_1, \ldots, y_n) = |\det J| \, g_{X_1,\ldots,X_n}(x_1, \ldots, x_n), \qquad (4.57)$$

with

$$x_i = f_i^{-1}(y_1, \ldots, y_n), \qquad i = 1, \ldots n, \qquad (4.58)$$

and the JACOBIan determinant

$$\det J = \begin{vmatrix} \dfrac{\partial f_1^{-1}(y_1, \ldots, y_n)}{\partial y_1} & \cdots & \dfrac{\partial f_1^{-1}(y_1, \ldots, y_n)}{\partial y_n} \\ \vdots & \ddots & \vdots \\ \dfrac{\partial f_n^{-1}(y_1, \ldots, y_n)}{\partial y_1} & \cdots & \dfrac{\partial f_n^{-1}(y_1, \ldots, y_n)}{\partial y_n} \end{vmatrix},$$

where the functions f_i^{-1} ($i = 1, \ldots n$) are the inverses of the transformation functions f_i ($i = 1, \ldots n$).

In the general case $m \neq n$, which is also quite common, we can make this proposition valid by artificially introducing additional random quantities, thus forcing $m = n$. These additional random quantities must, of course, be eliminated in the final result. More details on the application of this method will be given in the next section, as some additional requirements are still missing here.

In the case of many practically important multivariate transformations, the conditions of proposition 4.27 are fulfilled. We give some examples.

Example 4.65 (Transformation to polar co-ordinates)

The following relations apply to a transformation from planar Cartesian co-ordinates (X, Y) to polar co-ordinates (R, Φ):

$$X = R \cos \Phi, \qquad Y = R \sin \Phi. \qquad (4.59)$$

The transformation is uniquely invertible and the number of components of the original random quantity and the transformed random quantity is equal, i.e. proposition 4.27 holds.

The JACOBIan determinant of the transformation is given by

$$|\det J| = \begin{vmatrix} \cos \phi & -r \sin \phi \\ \sin \phi & r \cos \phi \end{vmatrix} = r.$$

Substituting this result into equation (4.57) and using equations (4.58) and (4.59), we obtain the transformed probability density function

$$g_{R,\Phi}(r, \phi) = r \, g_{X,Y}(r \cos \phi, r \sin \phi).$$

Example 4.66 (Transformation to spherical co-ordinates)

The following relations apply to a transformation from spatial Cartesian co-ordinates (X, Y, Z) to spherical co-ordinates (R, Θ, Φ):

$$X = R \sin \Theta \cos \Phi, \qquad Y = R \sin \Theta \sin \Phi. \qquad Z = R \cos \Theta. \tag{4.60}$$

The transformation is uniquely invertible and the number of components of the original random quantity and the transformed random quantity is equal, i. e. proposition 4.27 holds.

The JACOBIAN determinant of the transformation is given by

$$|\det J| = \begin{vmatrix} \sin \theta \cos \phi & r \cos \theta \cos \phi & -r \sin \theta \sin \phi \\ \sin \theta \sin \phi & r \cos \theta \sin \phi & r \sin \theta \cos \phi \\ \cos \theta & -r \sin \theta & 0 \end{vmatrix} = r^2 \sin \theta.$$

Substituting this result into equation (4.57) and using equations (4.58) and (4.60), we obtain the transformed probability density function

$$g_{R,\Theta,\Phi}(r, \theta, \phi) = r^2 \sin \theta \, g_{X,Y,Z}(r \sin \theta \cos \phi, r \sin \theta \sin \phi, r \cos \theta).$$

Example 4.67 (Transformation to cylindrical co-ordinates)

The following relations apply to a transformation from spatial Cartesian co-ordinates (X, Y, Z) to cylindrical co-ordinates (R, Φ, Z):

$$X = R \cos \Phi, \qquad Y = R \sin \Phi, \qquad Z = Z. \tag{4.61}$$

The transformation is uniquely invertible and the number of components of the original random quantity and the transformed random quantity is equal, i. e. proposition 4.27 holds.

The JACOBIAN determinant of the transformation is given by

$$|\det J| = \begin{vmatrix} \cos \phi & -r \sin \phi & 0 \\ \sin \phi & r \cos \phi & 0 \\ 0 & 0 & 1 \end{vmatrix} = r.$$

Substituting this result into equation (4.57) and using equations (4.58) and (4.61), we obtain the transformed probability density function

$$g_{R,\Phi,Z}(r, \phi, z) = r \, g_{X,Y,Z}(r \cos \phi, r \sin \phi, z).$$

Example 4.68 (General linear transformation)

Consider a general linear (affine) transformation

$$Y = AX + b \tag{4.62}$$

of the multivariate random quantities X and Y with the same number of components, where A denotes a real quadratic matrix and b is a vector of the same dimension as X and Y. If the transformation is invertible, i. e. if

$$X = A^{-1}(Y - b)$$

is valid and the inverse matrix A^{-1} exists, proposition 4.27 holds.

The JACOBIAN determinant of the transformation is given by

$$|\det J| = |\det A^{-1}| = \frac{1}{|\det A|} \,.$$

Substituting this result into equation (4.57) and using equations (4.58) and (4.62), we obtain the transformed probability density function

$$g_Y(y) = \frac{1}{|\det A|} \, g_X\left(A^{-1}(Y - b)\right) \,.$$

This equation is a multivariate generalization of the equation (4.35).

4.17 Marginal distributions

We have seen in section 4.15 that in general it is not possible to obtain a multivariate probability distribution function from univariate probability distribution functions. This is only possible if the corresponding scalar random quantities are stochastically independent, and thus we can apply proposition 4.23. A similar statement holds for the corresponding probability density functions, as we have discussed in the previous section. In this case, an applicability of proposition 4.26 presupposes also the stochastic independence of the scalar random quantities involved, in order to be able to calculate the multivariate probability density function as a product of the corresponding univariate probability density functions. However, it is always possible to reduce a multivariate probability distribution function to a probability distribution of lower dimension. This is also true for probability density functions. The resulting probability distribution functions and probability density functions are called marginal distributions and marginal densities respectively.

The mathematical procedure used to calculate marginal distributions and marginal densities is called marginalization. Along with the BAYES-LAPLACE theorem, it is one of the most important tools of Bayesian statistics.

In the case of multivariate probability distribution functions, marginalization is easy to perform. Because of the definition 4.37 and the fact that $\{X_k < \infty\}$ is the certain event of the random quantity X_k,

$$G_{X_1,\ldots,X_{k-1},X_k,X_{k+1},\ldots,X_n}(x_1, \ldots, x_{k-1}, \infty, x_{k+1}, \ldots, x_n) =$$

$$P(\{X_1 \le x_1\} \cap \cdots \cap \{X_k \le \infty\} \cap \cdots \cap \{X_n \le x_n\}) =$$

$$P(\{X_1 \le x_1\} \cap \cdots \cap \{X_{k-1} \le x_{k-1}\} \cap \{X_{k+1} \le x_{k+1}\} \cap \cdots \cap \{X_n \le x_n\}) =$$

$$G_{X_1,\ldots,X_{k-1},X_{k+1},\ldots,X_n}(x_1, \ldots, x_{k-1}, x_{k+1}, \ldots, x_n)$$

is valid, i. e. from an n-dimensional probability distribution function we get a $(n-1)$-dimensional probability distribution function where the random quantity X_k no longer appears. Therefore, we can easily eliminate an unwanted random quantity from a multivariate probability distribution function by formally setting its value equal to ∞.

Proposition 4.28 (Marginalization of probability distribution functions)
The marginal distribution of a multivariate probability distribution function of the random quantities $X_1, \ldots, X_n$ can be obtained by marginalizing with respect to an argument x_k to eliminate the random quantity X_k:

$$G_{X_1,\ldots,X_{k-1},X_{k+1},\ldots,X_n}(x_1, \ldots, x_{k-1}, x_{k+1}, \ldots, x_n) =$$

$$G_{X_1,\ldots,X_{k-1},X_k,X_{k+1},\ldots,X_n}(x_1, \ldots, x_{k-1}, \infty, x_{k+1}, \ldots, x_n). \tag{4.63}$$

We now turn to the marginalization of multivariate probability density functions. We first give an example.

Example 4.69 (Marginal density of a two-dimensional probability density)
Let $g_{X_1,X_2}(x_1, x_2)$ be the probability density function of a two-dimensional random quantity. We are interested in the marginal density function $g_{X_1}(x_1)$.

Using the definition 4.38 of the multivariate probability density function we obtain by marginalization

$$G_{X_1,X_2}(x_1, \infty) = \int\limits_{-\infty}^{x_1} \int\limits_{-\infty}^{+\infty} g_{X_1,X_2}(x_1, x_2)\, dx_1 dx_2\,.$$

According to the definition 4.29 of the univariate probability density function we can write the equation

$$G_{X_1}(x_1) = \int\limits_{-\infty}^{x_1} g_{X_1}(x_1)\, dx_1\,.$$

Using the equation (4.63) with $n = 2$ and equating the resulting two equations yields

$$\int_{-\infty}^{x_1} g_{X_1}(x_1)\, dx_1 = \int_{-\infty}^{x_1} \left(\int_{-\infty}^{+\infty} g_{X_1,X_2}(x_1, x_2)\, dx_2 \right) dx_1$$

and thus finally

$$g_{X_1}(x_1) = \int_{-\infty}^{+\infty} g_{X_1,X_2}(x_1, x_2)\, dx_2 .$$

In this example we obtained the desired marginal density by integrating with respect to the random quantity we wish to eliminate. This procedure can be generalized and leads to the following proposition:

Proposition 4.29 (Marginalization of probability density functions)
The marginal density of a multivariate probability density function of the random quantities $X_1, \ldots, X_n$ can be obtained by a marginalization with respect to an argument x_k, in order to eliminate the random quantity X_k:

$$g_{X_1,\ldots,X_{k-1},X_{k+1},\ldots,X_n}(x_1, \ldots, x_{k-1}, x_{k+1}, \ldots, x_n) =$$

$$\int_{-\infty}^{+\infty} g_{X_1,\ldots,X_n}(x_1, \ldots, x_{k-1}, \xi_k, x_{k+1}, \ldots, x_n)\, d\xi_k . \quad (4.64)$$

This proposition can also be used to solve the problem we postponed in the previous section, namely to calculate probability density functions of transformed random quantities when the number of original and transformed random quantities is not equal.

Example 4.70 (Sum and difference of random quantities)
Let $g_{X_1,X_2}(x_1, x_2)$ be the joint probability density function of the random quantities X_1 and X_2. We are interested in the probability density function of the random quantity $Y = X_1 \pm X_2$, i. e. the sum or difference of the random quantities, respectively.

We consider the transformation equations

$$X_1 = Y \mp Z, \qquad X_2 = Z. \quad (4.65)$$

In order to be able to apply the proposition 4.27, which requires an equal number of original and transformed random quantities, we have introduced an additional random quantity denoted by Z.

The JACOBIan determinant of the transformation is given by

$$\det J = \begin{vmatrix} 1 & \mp 1 \\ 0 & 1 \end{vmatrix} = 1 \,.$$

If we insert this result into the equation (4.57) and use the equations (4.58) and (4.65), we get

$$g_{Y,Z}(y,z) = g_{X_1,X_2}(y \mp z, z)$$

and finally by marginalization according to equation (4.64)

$$g_Y(y) = \int\limits_{-\infty}^{+\infty} g_{X_1,X_2}(y \mp z, z)\, dz \,. \tag{4.66}$$

Note that if the random quantities X_1 and X_2 are stochastically independent, the result that we obtain for the difference of these quantities is their correlation integral, which is used in signal processing. For their sum, we obtain their convolution integral, which we will discuss in more detail below.

Example 4.71 (Product of random quantities)

Let $g_{X_1,X_2}(x_1, x_2)$ be the joint probability density function of the random quantities X_1 and X_2. We are interested in the probability density function of the random quantity $Y = X_1 X_2$, i. e. the product of the random quantities.

We consider the transformation equations

$$X_1 = \frac{Y}{Z}, \qquad X_2 = Z \,. \tag{4.67}$$

In order to be able to apply the proposition 4.27, which requires an equal number of original and transformed random quantities, we have introduced an additional random quantity denoted by Z.

The JACOBIan determinant of the transformation is given by

$$\det J = \begin{vmatrix} \dfrac{1}{Z} & -\dfrac{Y}{Z^2} \\ 0 & 1 \end{vmatrix} = \frac{1}{Z} \,.$$

If we insert this result into the equation (4.57) and use the equations (4.58) and (4.67), we get

$$g_{y,z}(y,z) = \frac{1}{|z|} g_{X_1,X_2}\left(\frac{y}{z}, z\right)$$

and finally by marginalization according to equation (4.64)

$$g_Y(y) = \int\limits_{-\infty}^{+\infty} \frac{1}{|z|} g_{X_1,X_2}\left(\frac{y}{z}, z\right) dz \,.$$

Example 4.72 (Quotient of random quantities)

Let $g_{X_1,X_2}(x_1, x_2)$ be the joint probability density function of the random quantities X_1 and X_2. We are interested in the probability density function of the random quantity $Y = X_1/X_2$, i. e. the quotient of the random quantities.

We consider the transformation equations

$$X_1 = YZ, \qquad X_2 = Z. \tag{4.68}$$

In order to be able to apply the proposition 4.27, which requires an equal number of original and transformed random quantities, we have introduced an additional random quantity denoted by Z.

The JACOBIan determinant of the transformation is given by

$$\det J = \begin{vmatrix} Z & Y \\ 0 & 1 \end{vmatrix} = Z.$$

If we insert this result into the equation (4.57) and use the equations (4.58) and (4.68), we obtain

$$g_{Y,Z}(y, z) = |z| \, g_{X_1,X_2}(yz, z)$$

and finally by marginalization according to equation (4.64)

$$g_Y(y) = \int_{-\infty}^{+\infty} |z| \, g_{X_1,X_2}(yz, z) \, dz \, .$$

Before concluding this section, we consider the case of sums of statistically independent quantities, which is important for statistical applications. In this case, using proposition 4.26, we can write equation (4.66) in the form

$$g_Y(y) = \int_{-\infty}^{+\infty} g_{X_1}(y - z) \, g_{X_2}(z) \, dz \, .$$

Integrals of this type are called convolutions. Convolutions are a particular kind of integral transforms. Besides its application in probability theory they are also used in many other branches of mathematics.

The convolution of probability density functions is related to the sum of random quantities, as stated in the following proposition:

Proposition 4.30 (Convolution of probability density functions)

Let $g_X(x)$ and $g_Y(y)$ be the probability density functions of two stochastically independent random quantities X and Y. Then the probability density function of the random quantity $Z = X + Y$ is given by the convolution

$$g_Z(z) = (g_X * g_Y)(z) = \int_{-\infty}^{+\infty} g_X(z - \xi) \, g_Y(\xi) \, d\xi \, .$$

We add some remarks to this proposition:

a) A convolution is a commutative operation, i. e. $f_1 * f_2 = f_2 * f_1$ is valid for any two functions f_1 and f_2.
b) The functions involved in a convolution need not necessarily be of the same type.
c) If the functions involved are of the same type, the convolution does not necessarily result in a function of the same type.

We demonstrate the application of the proposition 4.30 by two examples.

Example 4.73 (Convolution of normal distributions)
Let

$$g_{X_i}(x_i) = \frac{1}{\sigma_i\sqrt{2\pi}} \exp\left(-\frac{(x_i - \mu_i)^2}{2\sigma_i^2}\right), \qquad i = 1,2,$$

be the probability density functions of two stochastically independent, normally distributed random quantities X_1 and X_2, where μ_1 and μ_2 denote the means, and σ_2 and σ_2 the standard deviations of the corresponding normal distributions.
Applying the proposition 4.30, we first obtain the probability density function

$$g_Y(y) = \frac{1}{2\pi\sigma_1\sigma_2} \int\limits_{-\infty}^{+\infty} \exp\left(-\frac{(y - \xi - \mu_1)^2}{2\sigma_1^2} - \frac{(\xi - \mu_2)^2}{2\sigma_2^2}\right) d\xi$$

of the random quantity $Y = X_1 + X_2$. Using the identity

$$\frac{(y - \xi - \mu_1)^2}{\sigma_1^2} + \frac{(\xi - \mu_2)^2}{\sigma_2^2} =$$

$$\frac{(y - \mu_1 - \mu_2)^2}{\sigma_1^2 + \sigma_2^2} + \frac{\sigma_1^2 + \sigma_2^2}{\sigma_1^2\sigma_2^2}\left(\xi - \frac{\sigma_2^2(y - \mu_1) + \sigma_1^2\mu_2}{\sigma_1^2 + \sigma_2^2}\right)^2$$

and the substitution

$$\xi = \frac{\sigma_1\sigma_2}{\sqrt{\sigma_1^2 + \sigma_2^2}}t + \frac{\sigma_2^2(y - \mu_1) + \sigma_1^2\mu_2}{\sigma_1^2 + \sigma_2^2}$$

we obtain

$$g_Y(y) = \frac{1}{2\pi\sqrt{\sigma_1^2 + \sigma_2^2}} \exp\left(-\frac{(y - \mu_1 - \mu_2)^2}{2(\sigma_1^2 + \sigma_2^2)}\right)\int\limits_{-\infty}^{+\infty} e^{-t^2/2}dt.$$

Since the integral yields $\sqrt{2\pi}$, we finally get the probability density function

$$g_Y(y) = \frac{1}{\sigma\sqrt{2\pi}} \exp\left(-\frac{(y - \mu)^2}{2\sigma^2}\right),$$

with

$$\mu = \mu_1 + \mu_2 \qquad \text{and} \qquad \sigma = \sqrt{\sigma_1^2 + \sigma_2^2}$$

This probability density function is again the density of a normal distribution, i. e. the sum of two stochastically independent, normally distributed random quantities is also normally distributed.

We can iterate this procedure in the case of a sum of more than two independent, normally distributed random quantities. Thus, for the random quantity "arithmetic mean", i. e. for

$$Y = \overline{X} = \frac{1}{n} \sum_{i=1}^{n} X_i \, ,$$

we obtain the probability density function

$$g_Y(y) = \frac{1}{\overline{\sigma}\sqrt{2\pi}} \exp\left(-\frac{(y - \overline{\mu})^2}{2\overline{\sigma}^2}\right) ,$$

with

$$\overline{\mu} = \frac{1}{n} \sum_{i=1}^{n} \mu_i \qquad \text{and} \qquad \overline{\sigma} = \sqrt{\frac{1}{n^2} \sum_{i=1}^{n} \sigma_i^2} \, , \tag{4.69}$$

i. e. the arithmetic mean of stochastically independent, normally distributed random quantities is also normally distributed with the mean $\overline{\mu}$ and the standard deviation $\overline{\sigma}$ given by the equations (4.69).

Note that in the formula for the standard deviation $\overline{\sigma}$, the sum is *not* divided by $n(n-1)$, as it is usually done in statistics textbooks when calculating empirical standard deviations of the mean value, but rather by n^2. This is because here the parameters μ_i and σ_i ($i = 1, \dots , n$) of all probability density functions are known. If in addition all random quantities X_i are identically distributed, i. e. $\mu_i = \mu$ and $\sigma_i = \sigma$ is valid, we obtain the result

$$\overline{\mu} = \mu \qquad \text{and} \qquad \overline{\sigma} = \frac{\sigma}{\sqrt{n}} \, .$$

This is the well-known "$1/\sqrt{n}$ law" of statistics. It shows that averaging reduces the dispersion without changing the mean value.

As another example, let us consider the convolution of uniform distributions, which also occurs very often in applications of metrology.

Example 4.74 (Convolution of uniform distributions)
The probability density functions of two stochastically independent, uniformly distributed random quantities X_1 and X_2 are given by

$$g_{X_i}(x_i) = \frac{\Theta(x_i - a_i) - \Theta(x_i - b_i)}{b_i - a_i} \, , \qquad i = 1, 2 \, ,$$

where $a_1 < b_1$ and $a_2 < b_2$ is valid. Moreover, without loss of generality, we assume that $b_1 - a_1 \le b_2 - a_2$ is valid. These inequalities in turn imply the chain of inequalities $a_1 + a_2 < b_1 + a_2 \le a_1 + b_2 < b_1 + b_2$. Note that the first and last inequalities in this chain are strict, while the inequality in the middle becomes an equality if the distributions of X_1 and X_2 are equal.

Applying the proposition 4.30 yields initially the probability density function

$$g_Y(y) = \frac{1}{(b_1 - a_1)(b_2 - a_2)} \int_{-\infty}^{+\infty} \Theta(y - \xi - a_1)\Theta(\xi - a_2)d\xi$$

$$- \frac{1}{(b_1 - a_1)(b_2 - a_2)} \int_{-\infty}^{+\infty} \Theta(y - \xi - b_1)\Theta(\xi - a_2)d\xi$$

$$- \frac{1}{(b_1 - a_1)(b_2 - a_2)} \int_{-\infty}^{+\infty} \Theta(y - \xi - a_1)\Theta(\xi - b_2)d\xi$$

$$+ \frac{1}{(b_1 - a_1)(b_2 - a_2)} \int_{-\infty}^{+\infty} \Theta(y - \xi - b_1)\Theta(\xi - b_2)d\xi.$$

Since it is easy to verify that

$$\int_{-\infty}^{+\infty} \Theta(y - \xi - u)\Theta(\xi - v)d\xi = (y - u - v)_+$$

is valid, where

$$(y - c)_+ = (y - c)\Theta(y - c)$$

denotes a so-called truncated linear function,[51] we finally obtain

$$g_Y(y) = \frac{(y - a_1 - a_2)_+ - (y - b_1 - a_2)_+}{(b_1 - a_1)(b_2 - a_2)}$$

$$- \frac{(y - a_1 - b_2)_+ - (y - b_1 - b_2)_+}{(b_1 - a_1)(b_2 - a_2)}.$$

---

51 Truncated linear functions are a special case of the so-called truncated polynomials $(y - c)_+^n$, which are frequently used in the theory of spline functions.

This probability density function can also be written in the conventional form

$$g_Y(y) = \begin{cases} \dfrac{y - (a_1 + a_2)}{(b_1 - a_1)(b_2 - a_2)} & \text{if } a_1 + a_2 < y \le b_1 + a_2 \\[2mm] \dfrac{1}{b_2 - a_2} & \text{if } b_1 + a_2 \le y \le a_1 + b_2 \\[2mm] \dfrac{b_1 + b_2 - y}{(b_1 - a_1)(b_2 - a_2)} & \text{if } a_1 + b_2 \le y < b_1 + b_2 \\[2mm] 0 & \text{otherwise} \end{cases}.$$

This function[52] is shown in Fig. 4.14. It has the shape of a symmetrical trapezoid with the slope of the flanks equal to one, a base of length $(b_2 - a_2) + (b_1 - a_1)$ and a top of length $(b_2 - a_2) - (b_1 - a_1)$.

If the probability distributions of the random quantities X_1 and X_2 are equal, i. e. if $a_1 = a_2 = a$ and $b_1 = b_2 = b$, the probability density function of their sum becomes

$$g_Y(y) = \begin{cases} \dfrac{y - 2a}{(b - a)^2} & \text{if } 2a < y \le a + b \\[2mm] \dfrac{2b - y}{(b - a)^2} & \text{if } a + b \le y < 2b \\[2mm] 0 & \text{otherwise} \end{cases}.$$

This is the density of a symmetric triangular distribution.[53] It is technically used in digital data processing to intentionally add noise to digitized audio or video signals, in order to randomize quantization errors (known as dithering).

4.18 Multivariate expectations

In this section we will extend the calculation of expectations to multivariate random quantities. This problem is relatively easy to solve, based on the preparations in sections 4.12 and 4.17.

Obviously we cannot calculate the expectation of a multivariate random quantity according to the equation (4.40) in the definition 4.32, because we are now dealing with a multivariate probability density function, i. e. we have to calculate the expectation using this particular density function. However, we can generalize the equation (4.40)

---

52 This is a special case of the density of what is known as a trapezoidal distribution, which does not necessarily have flanks of equal slope.

53 Asymmetric triangular distributions also exist. The symmetric one is also known as SIMPSON's distribution, named after the English mathematician T. SIMPSON (*1710, †1761).

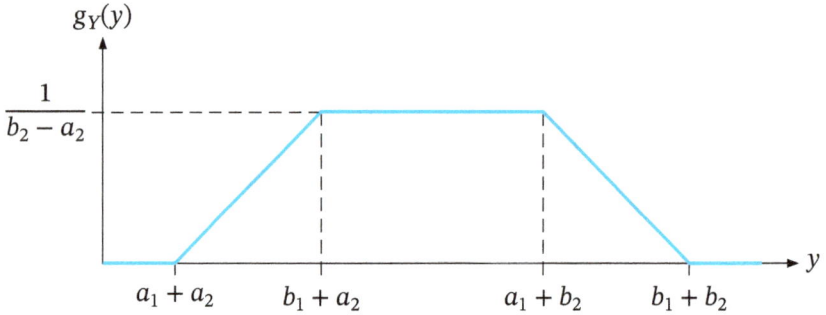

Fig. 4.14: Probability density function of the sum of two stochastically independent, uniformly distributed random quantities.

by formally replacing the random quantity X by the multivariate random quantity $\mathbf{X}$. This yields

$$E[\mathbf{X}] = \int_{-\infty}^{+\infty} \mathbf{x}\, g_X(\mathbf{x})\, d\mathbf{x}\,.$$

In a one-dimensional case, this equation corresponds to the definition 4.32 and is therefore suitable for our purpose. In order to understand its meaning clearly, we write it in the equivalent but more detailed form

$$E[X_i] = \int_{-\infty}^{+\infty} \cdots \int_{-\infty}^{+\infty} x_i\, g_{X_1,\dots,X_n}(x_1,\dots,x_n)\, dx_1 \cdots dx_n\,, \qquad i = 1,\dots,n\,, \qquad (4.70)$$

i. e. as a system of n equations. In this form it becomes evident that we are actually calculating the expectation of the n-dimensional random quantity $\mathbf{X}$ using its multivariate probability density function, and that this calculation is done component by component, i. e.

$$E[\mathbf{X}] = \begin{pmatrix} E[X_1] \\ \vdots \\ E[X_n] \end{pmatrix}\,.$$

Furthermore, if we have a closer look at the equation (4.70), we can see that it can be simplified. To do this, we first write this equation as

$$E[X_i] = \int_{-\infty}^{+\infty} x_i\, g_{X_i}(x_i)\, dx_i\,, \qquad i = 1,\dots,n\,.$$

with

$$g_{X_i}(x_i) = \int_{-\infty}^{+\infty} \cdots \int_{-\infty}^{+\infty} g_{X_1,\dots,X_n}(x_1,\dots,x_n)\, dx_1 \cdots dx_{i-1} dx_{i+1} \cdots dx_n \qquad (4.71)$$

The multiple integral (4.71) has to be taken with respect to all indices except the index i, which has to be excluded. A comparison with the equation (4.64) in proposition 4.29 shows that these integrations eliminate all components of the multivariate random quantity except the component X_i, i. e. it only gives the marginal density of this excluded component. Thus, our considerations lead to the following definition of the expectation of a multivariate random quantity:

Definition 4.39 (Expectation of a multivariate random quantity)
The expectations of the components of a multivariate random quantity with the joint probability density function $g_{X_1,...,X_n}(x_1, ... , x_n)$ are given by

$$E[X_i] = \int_{-\infty}^{+\infty} x_i\, g_{X_i}(x_i)\mathrm{d}x_i, \qquad i = 1, ... , n, \tag{4.72}$$

where $g_{X_i}(x_i)$ denotes the marginal density of the component X_i, provided that the integral is absolutely convergent.

The equation in this definition is formally equivalent to the equation (4.40) in definition 4.32. Because of this formal analogy, it is tempting to simply apply all the rules given in section 4.12 to the calculation of the expectations of the components of any multivariate random quantity, and hence to the random quantity itself. But if we try to use the definition 4.33 to calculate the expectation of a function of a multivariate random quantity, we are doomed to failure. It turns out that in this case actually the following definition has to be used:

Definition 4.40 (Expectation of a function of a multivariate quantity)
The expectation of a function $f(X_1, ... , X_n)$ of a multivariate random quantity with probability density function $g_{X_1,...,X_n}(x_1, ... , x_n)$ is given by

$$E\left[f(X_1, ... , X_n)\right] =$$
$$\int_{-\infty}^{+\infty} \cdots \int_{-\infty}^{+\infty} f(x_1, ... , x_n)\, g_{X_1,...,X_n}(x_1, ... , x_n)\, \mathrm{d}x_1 \cdots \mathrm{d}x_n, \tag{4.73}$$

provided that the integral is absolutely convergent.

Therefore, we must not carelessly extrapolate the results obtained in section 4.12 to multivariate random quantities. In the case of a function of multivariate random quantities, we notice that the equation (4.72) in definition 4.39 is only a special case of the equation (4.73) in definition 4.40, in which—due to the simplicity of the function—the marginalization was immediately possible.

Since in the case of a multivariate quantity the calculation of the expectation is still a linear operation, the following generalization of proposition 4.20 is valid:

Proposition 4.31 (Linearity of the expectation)
Let $X = (X_1, \ldots, X_n)^\mathsf{T}$ be a multivariate random quantity, a_i ($i = 1, \ldots, m$) real constants, and $f_i(X_1, \ldots, X_n)$ ($i = 1, \ldots, m$) real functions, then

$$\mathsf{E}\left[\sum_{i=1}^m a_i f_i(X_1, \ldots, X_n)\right] = \sum_{i=1}^m a_i \mathsf{E}\left[f_i(X_1, \ldots, X_n)\right] \tag{4.74}$$

is valid, provided that the expectations $\mathsf{E}\left[f_i(X)\right]$ ($i = 1, \ldots, m$) exist.

An immediate consequence of this proposition is the linear relation

$$\mathsf{E}[\mathbf{A}X + \mathbf{b}] = \mathbf{A}\,\mathsf{E}[X] + \mathbf{b}, \tag{4.75}$$

where X denotes a multivariate random quantity, $\mathbf{A}$ a real matrix, and $\mathbf{b}$ a real vector. Note that $\mathbf{A}$ is not necessarily a square matrix.

When using the equation (4.75), it is important to remember that both the matrix $\mathbf{A}$ and the vector $\mathbf{b}$ are *not* random quantities.

We get another important special case of the proposition 4.31, if we set $a_i = 1$ and $f_i(X) = X_i$ ($i = 1, \ldots, n$) in the equation (4.74). This yields the relation

$$\mathsf{E}\left[\sum_{i=1}^n X_i\right] = \sum_{i=1}^n \mathsf{E}[X_i], \tag{4.76}$$

i. e. the expectation of a sum of random quantities is equal to the sum of the expectations of these random quantities. This statement holds regardless of the underlying probability density function.

If we—instead of the expectation of a sum of random quantities—consider the expectation of a product of random quantities, we realize that in general

$$\mathsf{E}\left[\prod_{i=1}^n X_i\right] \neq \prod_{i=1}^n \mathsf{E}[X_i] \tag{4.77}$$

is valid. But if the random quantities are stochastically independent, the equality sign applies in relation (4.77), because from proposition 4.26 and definition 4.39

$$\mathsf{E}\left[\prod_{i=1}^n X_i\right] = \int_{-\infty}^{+\infty} \cdots \int_{-\infty}^{+\infty} \left(\prod_{i=1}^n x_i\right) g_{X_1, \ldots, X_n}(x_1, \ldots, x_n)\, dx_1 \ldots dx_n =$$

$$\int_{-\infty}^{+\infty} \cdots \int_{-\infty}^{+\infty} \left(\prod_{i=1}^n x_i\, g_{X_i}(x_i)\right) dx_1 \ldots dx_n = \prod_{i=1}^n \left(\int_{-\infty}^{+\infty} x_i\, g_{X_i}(x_i)\, dx_i\right) = \prod_{i=1}^n \mathsf{E}[X_i]$$

follows. Thus, we can state the proposition

Proposition 4.32 (Expectation of a product of independent quantities)
Let X_i $(i = 1, \ldots, n)$ be stochastically independent random quantities, then

$$\mathsf{E}\left[\prod_{i=1}^{n} X_i\right] = \prod_{i=1}^{n} \mathsf{E}[X_i]$$

is valid, i. e. the expectation of a product of *stochastically independent* random quantities is equal to the product of the expectations of these random quantities.

In order to demonstrate the application of the propositions stated in this section, we provide a fairly detailed example:

Example 4.75 (Oscillating random quantity)
The value of a random quantity Y (e. g. the temperature in a measuring room) oscillates around the value of a random quantity X according to the relation

$$Y = X + A \cos \Phi + B \sin \Phi, \tag{4.78}$$

where X, A, B and Φ denote random quantities. We can assume that these random quantities are stochastically independent, i. e. according to proposition 4.26 their joint probability density function is given by

$$g_{X,A,B,\Phi}(x, a, b, \phi) = g_X(x)\, g_A(a)\, g_B(b)\, g_\Phi(\phi). \tag{4.79}$$

We are interested in the expectation and the variance of Y.
According to the equation (4.73) we initially obtain from the equations (4.78) and (4.79) for the expectation of the random quantity Y the multiple integral:

$$\mathsf{E}[Y] = \int\limits_{-\infty}^{+\infty}\!\!\!\int\limits_{-\infty}^{+\infty}\!\!\!\ldots\!\!\int (x + a \cos \phi + b \sin \phi)\, g_X(x)\, g_A(a)\, g_B(b)\, g_\Phi(\phi)\, \mathrm{d}x\, \mathrm{d}a\, \mathrm{d}b\, \mathrm{d}\phi.$$

Now, in particular, we assume that the phase Φ is uniformly distributed over the interval $[0, 2\pi]$, i. e.

$$g_\Phi(\phi) = \begin{cases} \dfrac{1}{2\pi} & \text{if } 0 \le \phi \le 2\pi \\ 0 & \text{otherwise} \end{cases} \tag{4.80}$$

is valid. Using this equation, the normalization condition of probability density functions, as well as the definition 4.32 of the expectation, we can write the expectation of the random quantity Y in the form

$$\mathsf{E}[Y] = \mathsf{E}[X] + \frac{\mathsf{E}[A]}{2\pi} \int\limits_{0}^{2\pi} \cos \phi\, \mathrm{d}\phi + \frac{\mathsf{E}[B]}{2\pi} \int\limits_{0}^{2\pi} \sin \phi\, \mathrm{d}\phi.$$

Because both integrals evaluate to zero, we finally obtain $E[Y] = E[X]$, i. e. the oscillation does not contribute to the expectation of Y. Intuitively, this result is not unexpected.

In order to calculate the variance of the random quantity Y, we can also use the equation (4.73). According to the definition 4.34 of the variance and using the previous result $E[Y] = E[X]$, we obtain from equations (4.78) and (4.79)

$$Var(Y) = \int_{-\infty}^{+\infty} \int_{-\infty}^{+\infty} \int_{-\infty}^{+\infty} \int_{-\infty}^{+\infty} f(x, a, b, \phi)\, g_X(x)\, g_A(a)\, g_B(b)\, g_\Phi(\phi)\, dx\, da\, db\, d\phi,$$

with the abbreviation

$$f(x, a, b, \phi) = \left(x - E[X] + a \cos \phi + b \sin \phi\right)^2.$$

Using the equation (4.80), the normalization condition of probability density functions, the definition 4.34 of the variance, as well as the definition 4.32 of the expectation, we can write the variance of the random quantity Y in the form

$$Var(Y) = Var(X) + \frac{E\left[A^2\right]}{2\pi} \int_0^{2\pi} \cos^2\phi\, d\phi + \frac{E\left[B^2\right]}{2\pi} \int_0^{2\pi} \sin^2\phi\, d\phi$$
$$+ \frac{E[A]E[B]}{\pi} \int_0^{2\pi} \sin\phi \cos\phi\, d\phi$$

and finally, after evaluating the integrals with respect to ϕ,

$$Var(Y) = Var(X) + \frac{E\left[A^2\right] + E\left[B^2\right]}{2},$$

i. e. the variance of Y is the sum of the variance of X and an additional term which is equal to the quadratic mean of the oscillation.

Using the equation (4.47) we can transform the last equation to

$$Var(Y) = Var(X) + \frac{Var(A) + Var(B)}{2} + \frac{\left(E[A]\right)^2 + \left(E[B]\right)^2}{2}.$$

It turns out that in addition to the variance of X, the expectations and variances of the amplitudes of the oscillations are another essential contribution to the variance of Y.

This example shows that it is not always necessary to know the probability density function completely to specify the expectation and variance of a random quantity depending on several influencing quantities. Partial information is often sufficient, as in the example above, if information about the statistical independence of the influencing quantities is available. It is therefore advisable to perform the calculations in such a way that the necessary assumptions about the probability density functions involved are made only when they are really necessary. In this way one avoids introducing *a priori* assumptions—which in some cases may not even be well-founded—even if they are not relevant to the desired result.

4.19 Covariances and correlations

In this section we deal with concepts that are only relevant in the case of multivariate random quantities. As an introduction, we consider the example of a linear combination of two random quantities.

Example 4.76 (Linear combination of two random quantities)

We consider two random quantities X_1 and X_2 with a joint probability density function, where the possibility that these random quantities are stochastically dependent is not excluded. We form the linear combination

$$Y = uX_1 + vX_2, \tag{4.81}$$

where u and v denote real constants. We are interested in the expectation and variance of the random quantity Y.

Due to the linearity of expectations and with the help of the equation (4.76) we obtain from equation (4.81) the expectation of the random quantity Y as

$$\mathsf{E}[Y] = u\,\mathsf{E}[X_1] + v\,\mathsf{E}[X_2]. \tag{4.82}$$

The variance of Y follows from equations (4.81) and (4.82), the definition 4.34 of the variance, and the linearity of expectations as

$$\mathsf{Var}(Y) = \mathsf{E}\left[(Y - \mathsf{E}[Y])^2\right] =$$

$$\mathsf{E}\left[(uX_1 - \mathsf{E}[uX_1] + vX_2 - \mathsf{E}[vX_2])^2\right] =$$

$$\mathsf{E}\left[u^2(X_1 - \mathsf{E}[X_1])^2\right] + \mathsf{E}\left[v^2(X_2 - \mathsf{E}[X_2])^2\right]$$

$$+ \mathsf{E}\left[2uv(X_1 - \mathsf{E}[X_1])(X_2 - \mathsf{E}[X_2])\right] =$$

$$u^2\mathsf{Var}(X_1) + v^2\mathsf{Var}(X_2) + 2uv\,\mathsf{E}\left[(X_1 - \mathsf{E}[X_1])(X_2 - \mathsf{E}[X_2])\right]. \tag{4.83}$$

Since we do not necessarily assume the stochastic independence of the random quantities X_1 and X_2, we cannot apply the proposition 4.32 to further simplify the equation (4.83). Therefore, the variance of the random quantity Y is not just given by the variances of the random quantities X_1 and X_2, but we obtain another term, which is generally not zero.

In this example, the additional term that occurs when calculating the variance of a linear combination of two random quantities is called the covariance.

Definition 4.41 (Covariance of random quantities)

Let X_1 and X_2 be random quantities with a joint probability density function, then

$$\text{Cov}(X_1, X_2) = \text{E}\left[(X_1 - \text{E}[X_1])(X_2 - \text{E}[X_2])\right]$$

is called the covariance of these random quantities.

The covariance describes whether and how the values of two random quantities change together. If a common change is observed, we say that the respective random quantities are correlated, otherwise we say that they are uncorrelated.

Definition 4.42 (Uncorrelated random quantities)

The random quantities X_1 and X_2 are said to be uncorrelated, if their covariance is zero, i. e. if

$$\text{Cov}(X_1, X_2) = 0.$$

In the following proposition we state the properties of the covariance. They are a direct consequence of definition 4.33.

Proposition 4.33 (Properties of the covariance)

Let X, Y, and Z be random quantities and ξ a real number, then the covariance has the properties

a) symmetry

$$\text{Cov}(X, Y) = \text{Cov}(Y, X), \tag{4.84}$$

b) additivity

$$\text{Cov}(X + Y, Z) = \text{Cov}(X, Z) + \text{Cov}(Y, Z),$$
$$\text{Cov}(X, Y + Z) = \text{Cov}(X, Y) + \text{Cov}(X, Z), \tag{4.85}$$

c) homogeneity

$$\text{Cov}(\xi X, Y) = \xi \text{Cov}(X, Y),$$
$$\text{Cov}(X, \xi Y) = \xi \text{Cov}(X, Y), \tag{4.86}$$

d) shift invariance

$$\text{Cov}(X + \xi, Y) = \text{Cov}(X, Y),$$
$$\text{Cov}(X, Y + \xi) = \text{Cov}(X, Y), \tag{4.87}$$

e) reduction to expectations

$$\text{Cov}(X, Y) = \text{E}[XY] - \text{E}[X]\text{E}[Y], \tag{4.88}$$

provided that $\text{E}[XY]$, $\text{E}[X]$, and $\text{E}[Y]$ exist.

Properties a), b) and c) show that the covariance is a bilinear form which maps a pair of random quantities to the real numbers. Property d) shows that the covariance—like the variance—is insensitive to a shift, i. e. a systematic measurement error of one (or both) of the random quantities involved does not change the value of the covariance.

We also note that the variance can be considered as a special case of the covariance, because from the definitions of variance and covariance $Cov(X,X) = Var(X)$ immediately follows.

It follows from property c) that the covariance is not scale invariant, i. e. if the units of the random quantities involved are changed, the covariance changes its value. In order to obtain a scale invariant quantity, we could—analogous to the coefficient of variation (4.46)—divide the correlation by the expectations of the respective random quantities and thus obtain a kind of relative covariance. However, it is not this practice that has become established, but rather the correlation coefficient according to BRAVAIS-PEARSON, which is defined as

Definition 4.43 (Correlation coefficient)

Let X_1 and X_2 be random quantities with a joint probability density function, then

$$\rho(X_1, X_2) = \frac{Cov(X_1, X_2)}{\sqrt{Var(X_1)Var(X_2)}} \tag{4.89}$$

is called the correlation coefficient of these random quantities. The correlation coefficient is only defined, if $Var(X_1) > 0$ and $Var(X_2) > 0$ is valid.

The following proposition gives the properties of the correlation coefficient. They are a direct consequence of the definition 4.36.

Proposition 4.34 (Properties of the correlation coefficient)

Let X_1 and X_2 be random quantities and a_1, a_2, b_1, and b_2 real numbers, then the correlation coefficient has the properties

$$\rho(X_1, X_2) = \rho(X_2, X_1)$$

and

$$\rho(a_1 X_1 + b_1, a_2 X_2 + b_2) = \rho(X_1, X_2)\, \text{sgn}\, a_1\, \text{sgn}\, a_2\,.$$

The correlation coefficient only takes values in the interval $[-1, +1]$. In order to prove this assertion, we transform the equation (4.83) into

$$Var(Y) = u^2 Var(X_1) + 2uv\, Cov(X_1, X_2) + v^2 Var(X_2),$$

using the definition 4.41 of the covariance. Since $Var(Y) \geq 0$ is valid, the right side of this equation is a positive semidefinite binary quadratic form of the real variables u and

v. From the theory of quadratic forms it follows that the discriminant[54] of this quadratic form cannot be positive, i. e. we have

$$[\text{Cov}(X_1, X_2)]^2 \leq \text{Var}(X_1)\text{Var}(X_2).$$

Calculating the square root and using the definition 4.43 gives the relation

$$-1 \leq \rho(X_1, X_2) \leq +1,$$

which finally proves the assertion.

In order to recognize what it means when $\rho(X_1, X_2) = -1$ or $\rho(X_1, X_2) = +1$ is valid, we consider a linear relation $X_2 = aX_1 + b$ between the random quantities X_1 and X_2. In this case we obtain

$$\text{Var}(X_2) = a^2 \text{Var}(X_1)$$

and

$$\text{Cov}(X_1, X_2) = \text{E}\left[(X_1 - \text{E}[X_1])(X_2 - \text{E}[X_2])\right] =$$
$$\text{E}\left[(X_1 - \text{E}[X_1])(aX_1 + b - \text{E}[aX_1 + b])\right] =$$
$$a\text{E}\left[(X_1 - \text{E}[X_1])(X_1 - \text{E}[X_1])\right] = a\text{Var}(X_1).$$

According to equation (4.89) these results finally lead to

$$\rho(X_1, X_2) = \frac{a\text{Var}(X_1)}{|a|\,\text{Var}(X_1)} = \text{sgn}\,a,$$

i. e. the value of the correlation coefficient is identical to the sign of the constant a. Thus, a correlation coefficient $\rho(X_1, X_2) = -1$ or $\rho(X_1, X_2) = +1$ gives only the sign of the slope of the linear relation $X_2 = aX_1 + b$.

The correlation coefficient can only give information about the *direction* of a possible linear relation between two random quantities. It does not allow us to say how large the change in one of the random quantities is due to a change in the other random quantity.

In the case of two uncorrelated random quantities X_1 and X_2, the application of definitions 4.42 and 4.43 immediately yields $\rho(X_1, X_2) = 0$. Obviously, the same result can be obtained by setting the slope a of the linear relation $X_2 = aX_1 + b$ to zero, i. e. by excluding any linear relation between the random quantities X_1 and X_2. However, this does not exclude the possibility of a non-linear relation between these random quantities. Thus, a statement that two random quantities are uncorrelated does not necessarily mean that there is no relation between them, but rather that there is no *linear* relation. This fact is expressed by the following proposition:

---

[54] The discriminant of the quadratic form $au^2 + buv + cv^2$ is defined as $D = b^2 - 4ac$.

Proposition 4.35 (Correlation and stochastic dependence)

If two random quantities are stochastically independent, they are also uncorrelated. The converse is generally *not* true, i. e. if two random quantities are uncorrelated, they can still be stochastically dependent.

It is easy to prove the first statement of this proposition, because in the case of two independent random quantities X_1 and X_2 the relation $E[X_1 X_2] = E[X_1]E[X_2]$ immediately follows from the proposition 4.32. Therefore, we obtain $Cov(X_1 X_2) = 0$ from the equation (4.88) and finally $\rho(X_1 X_2) = 0$ from the equation (4.89). In order to prove the second statement of the proposition 4.35, it is sufficient to give an example of two uncorrelated random variables that are nevertheless stochastically dependent.

Example 4.77 (Uncorrelated, stochastically dependent random quantities)

We consider two random quantities X_1 and X_2 with the joint probability density function

$$g_{X_1,X_2}(x_1, x_2) = \frac{1}{2\pi} \exp\left(-\sqrt{x_1^2 + x_2^2}\right).$$

This function cannot be written as $g_{X_1,X_2}(x_1, x_2) = g_{X_1}(x_1) g_{X_2}(x_2)$, i. e. the random quantities X_1 and X_2 are certainly not stochastically independent.

By transforming into polar coordinates

$$X_1 = R \cos \Phi \qquad \text{and} \qquad X_2 = R \sin \Phi$$

(see example 4.65) we obtain the expectations

$$E[X_1] = \int_{-\infty}^{+\infty}\int_{-\infty}^{+\infty} x_1 g_{X_1,X_2}(x_1, x_2)dx_1 dx_2 = \frac{1}{2\pi}\int_{0}^{\infty}\int_{0}^{2\pi} r^2 \cos\varphi\, e^{-r} dr\, d\varphi = 0,$$

$$E[X_2] = \int_{-\infty}^{+\infty}\int_{-\infty}^{+\infty} x_2 g_{X_1,X_2}(x_1, x_2)dx_1 dx_2 = \frac{1}{2\pi}\int_{0}^{\infty}\int_{0}^{2\pi} r^2 \sin\varphi\, e^{-r} dr\, d\varphi = 0,$$

and

$$E[X_1 X_2] = \int_{-\infty}^{+\infty}\int_{-\infty}^{+\infty} x_1 x_2\, g_{X_1,X_2}(x_1, x_2)dx_1 dx_2 =$$

$$\frac{1}{2\pi}\int_{0}^{\infty}\int_{0}^{2\pi} r^3 \sin\varphi \cos\varphi\, e^{-r} dr\, d\varphi = 0,$$

because the integrals with respect to φ are all zero. Substituting these results into the equation (4.88) finally yields $Cov(X_1, X_2) = 0$. Consequently, the random quantities X_1 and X_2 are uncorrelated.

There is one exception to proposition 4.35, namely that uncorrelated, normally distributed random quantities are always also stochastically independent. The reason is that the joint probability density function of uncorrelated, normally distributed random quantities is separable, i. e. can be written as a product of their marginal probability density functions.

So far we have only considered two random variables. However, the conclusions drawn can be generalized to more than two random quantities. Let us first look at an example.

Example 4.78 (Variance of a linear combination of random quantities)
We consider n random quantities X_i ($i = 1, \ldots, n$) with a joint probability distribution and form the linear combination

$$Y = \sum_{i=1}^{n} a_i X_i,$$

where the coefficients a_i ($i = 1, \ldots, n$) are arbitrary real numbers. We are interested in the expectation and variance of the random quantity Y.

Due to the linearity of expectations we obtain

$$E[Y] = \sum_{i=1}^{n} a_i E[X_i].$$

Consequently, the variance is given by

$$\text{Var}(Y) = E\left[(Y - E[Y])^2\right] = E\left[\left(\sum_{i=1}^{n} a_i(X_i - E[X_i])\right)^2\right] =$$

$$E\left[\sum_{i=1}^{n}\sum_{k=1}^{n} a_i a_k(X_i - E[X_i])(X_k - E[X_k])\right] =$$

$$\sum_{i=1}^{n}\sum_{k=1}^{n} a_i a_k E\left[(X_i - E[X_i])(X_k - E[X_k])\right] =$$

$$\sum_{i=1}^{n} a_i^2 E\left[(X_i - E[X_i])^2\right] + 2\sum_{i=1}^{n}\sum_{k=1}^{i-1} a_i a_k E\left[(X_i - E[X_i])(X_k - E[X_k])\right] =$$

$$\sum_{i=1}^{n} a_i^2 \text{Var}(X_i) + 2\sum_{i=1}^{n}\sum_{k=1}^{i-1} a_i a_k \text{Cov}(X_i, X_k).$$

Therefore, we can state the following proposition:

Proposition 4.36 (Variance of a linear combination of random quantities)

Let X_i ($i = 1, \dots, n$) be random quantities with a joint probability distribution and a_i ($i = 1, \dots, n$) be real numbers, then

$$\mathrm{Var}\left(\sum_{i=1}^{n} a_i X_i\right) = \sum_{i=1}^{n} a_i^2 \mathrm{Var}(X_i) + 2 \sum_{i=1}^{n} \sum_{k=1}^{i-1} a_i a_k \mathrm{Cov}(X_i, X_k) \qquad (4.90)$$

is valid.

It follows immediately from this proposition that if *all* random quantities are uncorrelated, then

$$\mathrm{Var}\left(\sum_{i=1}^{n} a_i X_i\right) = \sum_{i=1}^{n} a_i^2 \mathrm{Var}(X_i) \qquad (4.91)$$

is valid, and if $a_i = 1$ ($i = 1, \dots, n$), equation the (4.91) becomes identical to the so-called BIENAYMÉ formula, which gives the variance of a sum of independent (and thus uncorrelated) random quantities.

We can also write the equation (4.90) in the form

$$\mathrm{Var}(\boldsymbol{a}^\mathsf{T}\boldsymbol{X}) = \boldsymbol{a}^\mathsf{T}\mathbf{C}\boldsymbol{a}, \qquad (4.92)$$

where $\boldsymbol{X}$ denotes a vector representing a multivariate random quantity and $\boldsymbol{a}$ denotes a vector of real coefficients. The matrix

$$\mathbf{C} = \begin{bmatrix} \mathrm{Var}(X_1) & \mathrm{Cov}(X_1, X_2) & \dots & \mathrm{Cov}(X_1, X_{n-1}) & \mathrm{Cov}(X_1, X_n) \\ \mathrm{Cov}(X_2, X_1) & \mathrm{Var}(X_2) & \dots & \mathrm{Cov}(X_2, X_{n-1}) & \mathrm{Cov}(X_2, X_n) \\ \vdots & \vdots & \ddots & \vdots & \vdots \\ \mathrm{Cov}(X_{n-1}, X_1) & \mathrm{Cov}(X_{n-1}, X_2) & \dots & \mathrm{Var}(X_{n-1}) & \mathrm{Cov}(X_{n-1}, X_n) \\ \mathrm{Cov}(X_n, X_1) & \mathrm{Cov}(X_n, X_2) & \dots & \mathrm{Cov}(X_n, X_{n-1}) & \mathrm{Var}(X_n) \end{bmatrix}$$

appearing in equation (4.92) is called variance-covariance matrix[55] or more often abbreviated to just covariance matrix.

Because of the property a) in the proposition 4.33, the covariance matrix is symmetric. Furthermore, since $\mathrm{Var}(\boldsymbol{a}^\mathsf{T}\boldsymbol{X}) \geq 0$, we can deduce from equation (4.92) that the covariance matrix is positive semidefinite, i. e. $\det \mathbf{C} \geq 0$.

If $\det \mathbf{C} = 0$, there is a linear relation between the components of the multivariate random quantity $\boldsymbol{X}$, i. e. there is a vector $\boldsymbol{a} \neq 0$ such that $\boldsymbol{a}^\mathsf{T}\boldsymbol{X}$ is equal to a constant and hence $\mathrm{Var}(\boldsymbol{a}^\mathsf{T}\boldsymbol{X}) = 0$. This means that at least one of the random quantities X_i ($i = 1, \dots, n$) can be expressed as a linear combination of the remaining random quantities and is therefore dispensable.

---

55 Especially in publications on measurement uncertainty the term "uncertainty matrix" is sometimes used instead of the term "variance-covariance matrix".

By dividing each row and column of the covariance matrix by the positive square root of the corresponding diagonal element, we obtain the matrix

$$\mathbf{P} = \begin{bmatrix} 1 & \rho(X_1,X_2) & \cdots & \rho(X_1,X_{n-1}) & \rho(X_1,X_n) \\ \rho(X_2,X_1) & 1 & \cdots & \rho(X_2,X_{n-1}) & \rho(X_2,X_n) \\ \vdots & \vdots & \ddots & \vdots & \vdots \\ \rho(X_{n-1},X_1) & \rho(X_{n-1},X_2) & \cdots & 1 & \rho(X_{n-1},X_n) \\ \rho(X_n,X_1) & \rho(X_n,X_2) & \cdots & \rho(X_n,X_{n-1}) & 1 \end{bmatrix}.$$

This matrix is called correlation matrix. The correlation matrix (like the covariance matrix) is also positive semidefinite, i. e. $\det \mathbf{P} \geq 0$ is valid. The case $\det \mathbf{P} = 0$ occurs if and only if $\det \mathbf{C} = 0$, i. e. when there is a linear relation between the components of a multivariate random quantity.

In numerical calculations, it is possible that—due to rounding errors—the correlation matrix (and thus also the covariance matrix) may not appear to be positive definite, although it is known that there is no linear relation between the components of a multivariate random quantity. Consequently, it is necessary to perform all calculations with sufficient numerical precision and to use appropriate algorithms to avoid this problem.

The correlation matrix also makes it possible to decide whether all components of a random quantity are pairwise uncorrelated. If this is the case, the correlation matrix is equal to the identity matrix and the covariance matrix is a diagonal matrix with the variances as diagonal elements.

4.20 Approximate estimations

Sometimes the probability distribution function of a random quantity is not, or not very well, known. Nevertheless, we want to get at least a coarse approximation of its expectation and variance in some way. In this section we will show which approximations are useful for this purpose.

We start with a proposition concerning an estimate of the upper bound of the probability that the value of a positive real function, whose expectation is known, exceeds a given value.

Proposition 4.37 (Upper bound of the probability of a function value)
Let X be a random quantity, f be a positive real function and c be a positive real constant, then

$$P\left(f(X) > c\right) \leq \frac{1}{c} E\left[f(X)\right] \tag{4.93}$$

is valid, provided that the expectation $E\left[f(X)\right]$ exists.

Proof By definition, the expectation of $f(X)$ is

$$\mathsf{E}\,[f(X)] = \int\limits_{-\infty}^{\infty} f(x)\,g_X(x)\,\mathrm{d}x\,.$$

We split the integral into two parts

$$\mathsf{E}\,[f(X)] = \int\limits_{f(x)\leq c} f(x)\,g_X(x)\,\mathrm{d}x + \int\limits_{f(x)>c} f(x)\,g_X(x)\,\mathrm{d}x\,,$$

which are both positive, because $f(x)$ is positive by assumption. Consequently, by omitting the first integral, we obtain the inequality

$$\mathsf{E}\,[f(X)] \geq \int\limits_{f(x)>c} f(x)\,g_X(x)\,\mathrm{d}x \geq c\int\limits_{f(x)>c} g_X(x)\,\mathrm{d}x = c\,\mathsf{P}\,(f(X) > c)\,.$$

Since $c > 0$, we can divide by c. This finally proves the proposition. □

From the proposition 4.37 two important inequalities of probability theory can be derived. If we set $f(X) = X$, where X denotes a positive random quantity, and $c = k\,\mathsf{E}[X]$, we immediately obtain:

Proposition 4.38 (MARKOV's inequality)
Let X be a positive random quantity with expectation $\mathsf{E}[X]$, and k a positive real constant, then

$$\mathsf{P}(X > k\,\mathsf{E}[X]) \leq \frac{1}{k}\,, \qquad k \geq 1\,,$$

is valid, provided that $\mathsf{E}[X]$ exists.

MARKOV's inequality allows us to estimate the probability that the value of a positive random quantity is greater than a multiple of its estimate, without the need to know the underlying probability density. However, this estimate is very rough, as the following example shows.

Example 4.79 (Delay of flights)
An airline offers its passengers compensation if a flight is delayed by three hours or more. An internal statistical study shows that the expected delay could be about nine minutes. However, the probability distribution describing the delay is unknown.

We represent the delay by a random quantity called X. From the known data we calculate $k = 20$. Thus, according to MARKOV's inequality, the probability that the airline will have to pay compensation is less than 5 %.

However, it is not unreasonable to assume that the delay of a flight follows an exponential distribution [69], i. e. we have

$$G_X(x) = 1 - \mathrm{e}^{-x/\mu}\,,$$

with expectation μ. Therefore, we obtain

$$P(X > k\,E[X]) = 1 - P(X \leq k\,E[X]) = 1 - G_X(k\mu) = e^{-k}.$$

Since we have $k = 20$ in this example, the latter equation yields a probability of about $2.1 \cdot 10^{-9}$. Hence, the payment of compensation for a delayed flight will in fact be extremely rare.

If we set $f(X) = |X - E[X]|^2$ and $c = k^2 \text{Var}(X)$, we obtain from the proposition 4.37 the second important inequality:

Proposition 4.39 (CHEBYSHEV's inequality)

Let X be a random quantity with expectation μ and variance σ^2, and k a real constant, then

$$P(|X - \mu| \leq k\sigma) \geq 1 - \frac{1}{k^2}, \qquad k \geq 1,$$

is valid, provided that μ and σ exist.

Proof Inserting $f(X) = |X - \mu|^2$ and $c = k^2\sigma^2$ into the inequality (4.93) and using the definition of the variance $\sigma^2 = \text{Var}(X) = E\left[|X - \mu|^2\right]$ yields

$$P\left(|X - \mu|^2 > k^2\sigma^2\right) \leq \frac{1}{k^2}.$$

Since $|X - \mu|^2 > k^2\sigma^2$ implies $|X - \mu| > k\sigma$, we conclude that

$$P(|X - \mu| > k\sigma) \leq \frac{1}{k^2}.$$

This finally leads to

$$P(|X - \mu| \leq k\sigma) = 1 - P(|X - \mu| > k\sigma) \geq 1 - \frac{1}{k^2},$$

which proves the proposition. □

CHEBYSHEV's inequality makes it easy to estimate the probability that, for a given random quantity, the deviation of its value from its estimate exceeds a multiple of its standard deviation. A further advantage of this inequality is that it can be applied without any knowledge of the underlying probability distribution of the random quantity. However, this estimation is very rough, as the following example illustrates.

Example 4.80 (Estimation of the constant k)

The probability p that the value of a random quantity X lies within the interval $[\mu - k\sigma, \mu + k\sigma]$ is

$$p = P(|X - \mu| \leq k\sigma).$$

Using CHEBYSHEV's inequality thus yields

$$k \leq \frac{1}{\sqrt{1-p}},$$

i. e. in the case of $p = 95\,\%$ we obtain $k \approx 4.47$. In contrast, if the random quantity is normally distributed, we obtain the much smaller value $k \approx 1.96$, and in the case of a uniformly distributed random quantity only $k \approx 1.65$.

4.21 Central limit theorem

In the example 4.73 we considered a sum of stochastically independent, normally distributed random quantities. It turned out that the probability distribution of the sum is again a normal distribution. In contrast, in example 4.74 we considered a sum of stochastically independent, uniformly distributed random quantities and found that in this case the probability distribution of the sum is no longer a uniform distribution. It can even be shown that with an increasing number of terms the probability distribution of a sum of stochastically independent, uniformly distributed random quantities tends to a normal distribution, as shown in Fig. 4.15 for only a few terms. In this figure, identical uniform distributions with mean zero and standard deviation one have been used. However, a very similar result would be obtained for uniform distributions with different means and standard deviations.

Further investigation of this observation shows that in most cases the probability distribution of a sum of stochastically independent random quantities tends to a normal distribution. This is known as the central limit theorem which we state here without a proof.

Proposition 4.40 (Central limit theorem)
Let X_i $(i = 1, \ldots, n)$ be stochastically independent random quantities. The random quantity

$$Z_n = \frac{\displaystyle\sum_{i=1}^{n}(X_i - \mu_i)}{\sqrt{\displaystyle\sum_{i=1}^{n}\sigma_i^2}},$$

with $E[X_i] = \mu_i$ and $Var(X_i) = \sigma_i^2 < \infty$, is asymptotically distributed according to a standardized normal distribution, i. e.

$$\lim_{n\to\infty} P(Z_n < z) = \frac{1}{\sqrt{2\pi}} \int_{-\infty}^{z} e^{-t^2/2} dt$$

is valid, provided that none of the variances σ_i is dominant.

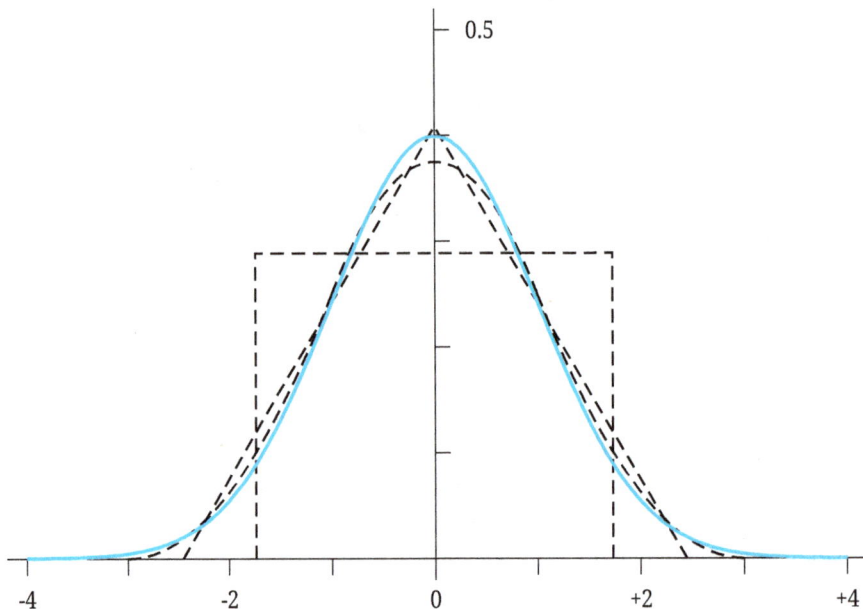

Fig. 4.15: Convolution of uniform distributions (the solid line shows the normal distribution for comparison).

This proposition does not assume any kind of probability distribution underlying the random quantities. In fact, each random quantity can have a different probability distribution with different parameters. However, there are exceptions. For example, the convolution of exponential distributions always yields an exponential distribution. This explains why, in the case of radioactive decay, the shape of the underlying distribution (an exponential distribution) does not change with the number of decaying nuclei. Also, the sum of stochastically independent, CAUCHY distributed random quantities is CAUCHY distributed [70]. Consequently, averaging measured values with an underlying CAUCHY distribution does not increase the accuracy of a particular measurement.

The central limit theorem is important in applied statistics. We can interpret it as saying that the arithmetic mean of n stochastically independent random quantities tends to a normal distribution if their number is sufficiently large. In practice, $n \geq 50$ is already sufficient to be unable to detect a significant difference from a normal distribution.

In many cases of practical importance, the central limit theorem allows a random quantity to be regarded as normally distributed, provided that it can be assumed to have been produced by an additive superposition of numerous similar influences. This explains why the normal distribution is often observed in natural phenomena. This is particularly true of random errors. However, it is always advisable to verify whether the assumption of an underlying normal distribution is actually justified.

5 Statistical methods

> It should be noted that there is no
> falsehood in interpreting any set of
> independent measurements as a random
> sample from an infinite population; …
>
> *(Ronald Aylmer Fisher, 1922)*

Statistics is a branch of applied mathematics concerned with methods of collecting, analysing, interpreting, presenting and evaluating numerical information about mass phenomena. In metrology, statistical methods are used in particular to evaluate measurement data resulting from repeated measurements. However, due to the small number of measured values in most cases, it is not possible to speak of mass phenomena in this context. Therefore, special care must be taken when using statistical methods in metrology.

In this chapter we will focus on statistical methods used in the evaluation of measurement data. In addition to conventional statistics, we will include methods of Bayesian statistics.

5.1 Populations and random sampling

In probability theory, experiments are mostly irrelevant because the emphasis is on theoretical considerations. The situation is different in more application-oriented statistics, where the data obtained from experiments and the information derived from them are the focus of interest. Here the terms "population" and "sample" play a prominent role.

In mathematical statistics we always refer to a well-defined real (or sometimes completely hypothetical) set of items, which is called population and is defined as follows (ISO 3534-1, 1.1 [38]):

Definition 5.1 (Population)
A population is the total number of items that are under consideration.

This definition says nothing about the elements of a population. Usually, the elements of a population are just objects that are carriers of one or more characteristics. A characteristic is a property inherent in or attributed to an object under consideration, which may be of a qualitative or quantitative nature (ISO 3534-2, 1.1.1 [12]). In metrology, however, we are only interested in *quantitative* characteristics.

Definition 5.2 (Quantitative characteristic)
A quantitative characteristic is a function $X : \Omega \to \mathbb{R}$ that assigns a real number $x = X(\omega)$ to each element ω of a population Ω.

https://doi.org/10.1515/9783111453712-005

A comparison of this definition with the definition 4.24 of a random quantity shows that quantitative characteristics and random quantities can be regarded as synonymous if we assume that the sample space of probability theory is a given population and understand its elements as the results of a random experiment. Such an interpretation of the term "population" is explicitly allowed in ISO 3534-1, 1.1, Note 1 [38]. Thus, it is justified to apply statistical methods to measurement results even when only a single object is involved in the measurement of a quantitative characteristic.

Since the quantitative characteristics of a population are considered as random quantities, they can be described by probability distribution functions, which are generally unknown. Whenever a very large amount of data is available, we can try to approximate the underlying distribution function by a frequency distribution. In metrology, however, this is usually not possible due to an insufficient number of measured data.

The parameters of a distribution function of a particular characteristic of a population are called population parameters (ISO 3534-2, 1.2.2 [12]).

Definition 5.3 (Population parameter)

A population parameter is a summary measure of the values of a particular characteristic of a population.

Population parameters are usually denoted by lowercase italicized Greek letters. In this chapter we will follow this convention as far as possible.

Examples of population parameters are the mean μ and the standard deviation σ. A parametrization is usually arbitrary and is done for practical reasons or because it has been found to be useful for similar problems. Different parametrizations can be used for the same distribution function, e. g. for a normal distribution, besides the mean, the standard deviation σ, the variance σ^2 or the precision $1/\sigma^2$. All three parametrizations provide equivalent descriptions of a normal distribution.

Statistical methods are particularly powerful when there are lots of data for a characteristic of interest. However, a complete data collection is practically impossible. There are many reasons for this, such as e. g. high cost or even time constraints on making a measurement. Therefore, we are generally restricted to dealing with a subset of a population, called a sample (ISO 3534-1, 1.3 [38]). During a sampling procedure (measurement), it is usually assumed that the distribution function underlying the observed values (ISO 3534-1, 1.4 [38]) of a sample is identical to that of the values of the characteristic of the population. In metrology, this can be achieved (at least approximately) by repeated measurements under well-defined repeatability or reproducibility conditions (see section 2.5 for details).

In probability theory, a sample can be mathematically described by a sample vector. This term is defined as:

Definition 5.4 (Sample vector)

A multivariate random quantity $X = (X_1, \dots, X_n)^{\mathsf{T}}$ is called sample vector, if all its components X_i ($i = 1, \dots, n$) are related to the *same* characteristic X of a population and share the *same* probability distribution function.

It follows from this definition that all components X_i ($i = 1, \dots, n$) of the sample vector necessarily have the probability density function $g_X(x)$ of the considered characteristic X of the population. Therefore, the probability that the measured values $(x_1, \dots, x_n)$ occur during a measurement is

$$P\big((X_1 \leq x_1) \cap \cdots \cap (X_n \leq x_n)\big) = \prod_{i=1}^{n} P(X_i \leq x_i),$$

i. e. the random quantities X_i ($i = 1, \dots, n$) are assumed to be stochastically independent. Thus, the sample vector is a special multivariate random quantity and its components are identically distributed and stochastically independent.[1] This assumption is widely used in statistical data evaluation because it allows most calculations to be simplified.

In probability theory, measurement data obtained by repeated measurements are considered as realization of a sample vector and are called measurement series or empirical sample. Often we just refer to "the data". The number of measured values in a measurement series is called sample size.

5.2 Statistics

In order to gain information about the parameters of the underlying population from the measured data $x = (x_1, \dots, x_n)^{\mathsf{T}}$ as a realization of a sample $X = (X_1, \dots, X_n)^{\mathsf{T}}$, many different statistics are used, such as the well-known sample mean or the sample standard deviation. In general, a statistic is defined as (ISO 3534-1, 1.8 [38]):

Definition 5.5 (Statistic)

A statistic is a fully specified function of random quantities.

In this definition the phrase "fully specified" means that a statistic depends only on the components $(X_1, \dots, X_n)$ of the sample vector. We can think of a statistic as a particular function of these components, defined independently of the probability distribution function underlying the data. For example, if that probability distribution is a normal distribution with the unknown parameters μ (mean) and σ (standard deviation), then the expressions $(X_1 + \cdots + X_n)$ or $(X_1^2 + \cdots + X_n^2)$ would be valid statistics, whereas the expressions $(X_1 + \cdots + X_n)/n - \mu$ or $(X_1^2 + \cdots + X_n^2)/\sigma^2$ do not meet the requirements of

---

[1] The abbreviation *iid* is usually used in the literature to denote this property.

a statistic because they depend on the unknown parameters μ and σ of the probability distribution function.

It is common practice to use the term statistic for both the function and the value of the function of a given sample vector.

A statistic is different from a parameter of a probability distribution, but it is usually used to calculate an approximate value of such a parameter. This is often indicated by the name of a statistic being preceded by the word "sample". In most cases, the following ordinary statistics, as defined in ISO 3534-1 [38], are used to characterize a population:

Definition 5.6 (Ordinary statistics)
Sample mean (ISO 3534-1, 1.15)

$$\overline{X} = \frac{1}{n}\sum_{i=1}^{n}X_i\,.$$

Sample variance (ISO 3534-1, 1.16)

$$S^2 = \frac{1}{n-1}\sum_{i=1}^{n}(X_i - \overline{X})^2\,.$$

Sample covariance (ISO 3534-1, 1.22)

$$S_{XY} = \frac{1}{n-1}\sum_{i=1}^{n}(X_i - \overline{X})(Y_i - \overline{Y})\,.$$

Sample correlation coefficient (ISO 3534-1, 1.23)

$$R_{XY} = \frac{S_{XY}}{\sqrt{S_X^2 S_Y^2}}\,.$$

In this defining equation, the symbols S_X^2 and S_Y^2 denote the sample variances of the random quantities X and Y, respectively.

The non-negative square root, S, of the sample variance is called sample standard deviation (ISO 3534-1, 1.17). The quotient $S/\overline{X}$ of the sample standard deviation and the sample mean (defined only for $\overline{X} > 0$) is called sample coefficient of variation (ISO 3534-1, 1.18).

In addition to these statistics, other statistics are sometimes used which are derived from the order statistics of the components of a sample vector. Order statistics are obtained by arranging the random quantities $X_1, \dots, X_n$ (the components of the sample vector) in a linear order such that

$$X_{(1)} \leq X_{(2)} \cdots X_{(n-1)} \leq X_{(n)}\,,$$

where $X_{(1)}$ is the smallest and $X_{(n)}$ the largest component. The quantities $X_{(k)}$ are called k-th order statistic (ISO 3534-1, 1.9 [38]). Order statistics are mostly used in nonparametric statistics.

The statistics derived from order statistics that are most commonly used to characterize a population are

Sample minimum
$$X_{min} = min(X_1, \ldots, X_n),$$

Sample maximum
$$X_{max} = max(X_1, \ldots, X_n),$$

Sample range (ISO 3534-1, 1.10)
$$V = X_{max} - X_{min}.$$

Sample median (ISO 3534-1, 1.13)
$$X_{med} = \begin{cases} X_{((n+1)/2)}, & \text{if } n \text{ is odd} \\ \dfrac{X_{(n/2)} + X_{(n/2+1)}}{2}, & \text{if } n \text{ is even} \end{cases}.$$

We conclude this section with two useful formulae for an iterative calculation (updating) of the sample mean
$$\overline{X}_{n+1} = \frac{n\overline{X}_n + X_{n+1}}{n+1}, \qquad n \geq 0, \tag{5.1}$$

and the sample variance
$$S^2_{n+1} = \frac{n-1}{n} S^2_n + \frac{\left(\overline{X}_n - X_{n+1}\right)^2}{n+1}, \qquad n > 0, \tag{5.2}$$

where $\overline{X}_n$ and $\overline{X}_{n+1}$ denote the sample mean for the sample size n and $(n+1)$ respectively, and S^2_n and S^2_{n+1} denote the corresponding sample variances.

5.3 Estimators

Whenever we draw a sample from a population, we assume that the measured values $(x_1, \ldots, x_n)$ are a realization of the sample vector $X = (X_1, \ldots, X_n)^\mathsf{T}$ with elements X_i ($i = 1, \ldots, n$), which are stochastically independent, identically distributed random quantities with an unknown probability density function. Therefore, we make an assumption about the underlying distribution and try to determine its parameter values in such a way as to achieve as good a fit as possible to the measured data of the sample. For this purpose, we use an estimator (ISO 3534-1, 1.12 [38]):

Definition 5.8 (Estimator)

An estimator $\hat{\Theta} = (\hat{\Theta}_1, \ldots, \hat{\Theta}_m)^\mathsf{T}$ is a statistic used in estimation of the parameter $\theta = (\theta_1, \ldots, \theta_m)^\mathsf{T}$.

The relation between the population and the sample on the one hand and the estimator $\hat{\Theta}$ and the parameter vector θ of the probability distribution function of the population on the other hand is visualized by Fig. 5.1.

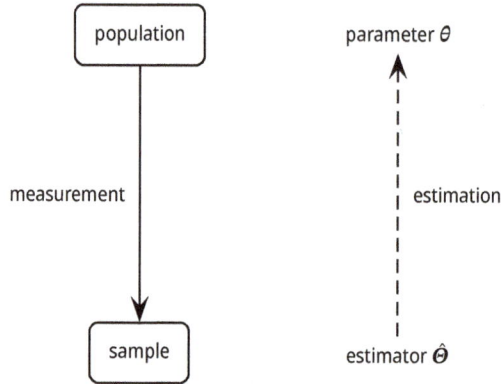

Fig. 5.1: Estimation of a parameter of a population.

Examples of estimators are the statistics introduced in the previous section, of which the sample mean $\overline{X}$ and the sample variance S^2 are the most commonly used estimators.

Whenever—for any given estimator $\hat{\Theta} = f(X_1, \ldots, X_n)$—we substitute the measured values $(x_1, \ldots, x_n)$ instead of the random quantities $(X_1, \ldots, X_n)$, we obtain the corresponding estimated value $\hat{\theta} = f(x_1, \ldots, x_n)$.

Example 5.1 (Estimating the parameters of a normal distribution)

The values $(x_1, \ldots, x_n)$ have been measured, and it is assumed that the underlying probability distribution function is a normal distribution with parameters μ (mean) and σ^2 (variance). We use the sample mean $\overline{X}$ and the sample variance S^2 as estimators. Substituting the measured values $(x_1, \ldots, x_n)$ into the formulae of these estimators yields the estimated values of the mean

$$\hat{\mu} = \overline{x} = \frac{1}{n}\sum_{i=1}^{n} x_i$$

and the variance

$$\hat{\sigma}^2 = s^2 = \frac{1}{n-1}\sum_{i=1}^{n}(x_i - \overline{x})^2$$

of the normal distribution.

It is easy to prove that in this case we obtain unbiased estimates of the parameters of the normal distribution, because—since the random quantities X_i ($i = 1, \ldots, n$) are stochastically independent and identically distributed—we get the following results for the estimators used

$$E\left[\overline{X}\right] = \frac{1}{n} \sum_{i=1}^{n} E[X_i] = \frac{1}{n} \sum_{i=1}^{n} E[X] = E[X], \tag{5.3}$$

$$\mathrm{Var}\left(\overline{X}\right) = \mathrm{Var}\left(\frac{1}{n} \sum_{i=1}^{n} X_i\right) = \frac{1}{n^2} \sum_{i=1}^{n} \mathrm{Var}(X_i) = \frac{1}{n^2} \sum_{i=1}^{n} \mathrm{Var}(X) = \frac{1}{n}\mathrm{Var}(X) \tag{5.4}$$

and

$$E\left(S^2\right) = E\left[\frac{1}{n-1}\sum_{i=1}^{n}\left(X_i - \overline{X}\right)^2\right] = \frac{1}{n-1}\sum_{i=1}^{n}E\left[X_i^2 - 2X_i\overline{X} + \overline{X}^2\right] =$$

$$\frac{1}{n-1}\left(\sum_{i=1}^{n}E\left[X_i^2\right] - \frac{2}{n}\sum_{i=1}^{n}\sum_{j=1}^{n}E[X_iX_j] + \frac{1}{n}\sum_{i=1}^{n}\sum_{j=1}^{n}E[X_iX_j]\right) =$$

$$\frac{1}{n-1}\left(\sum_{i=1}^{n}E\left[X_i^2\right] - \frac{1}{n}\sum_{i=1}^{n}\sum_{j=1}^{n}E[X_iX_j]\right) = \frac{1}{n-1}\sum_{i=1}^{n}\left(E\left[X_i^2\right] - E\left[\overline{X}^2\right]\right) =$$

$$\frac{1}{n-1}\sum_{i=1}^{n}\left(\mathrm{Var}(X_i) + (E[X_i])^2 - \mathrm{Var}\left(\overline{X}\right) - \left[E\left(\overline{X}\right)\right]^2\right) =$$

$$\frac{1}{n-1}\sum_{i=1}^{n}\left(\mathrm{Var}(X) + (E[X])^2 - \frac{1}{n}\mathrm{Var}(X) - (E[X])^2\right) = \mathrm{Var}(X).$$

Since for a normal distribution $E[X] = \mu$ and $\mathrm{Var}(X) = \sigma^2$ holds, we finally obtain $E\left(\overline{X}\right) = \mu$ and $E\left(S^2\right) = \sigma^2$, which proves the assertion.

In this example, each of the two estimators gives an unbiased value of the corresponding parameter of the underlying probability density function, regardless of the sample size. This property of an estimator is called its unbiasedness. In general, an unbiased estimator is defined as follows:

Definition 5.9 (Unbiased estimator)
Let $(X_1, \ldots, X_n)$ be a sample, then an estimator $\hat{\Theta}(X_1, \ldots, X_n)$ of the value of the parameter θ is said to be unbiased if

$$E\left[\hat{\Theta}(X_1, \ldots, X_n)\right] = \theta,$$

provided the expectation exists.

An estimator that is not unbiased is called biased estimator. The difference $E\left[\hat{\Theta}\right] - \theta$ is called bias of the estimator $\hat{\Theta}$.

If the bias of an estimator decreases with the sample size and approaches zero for an infinite sample (i. e. the estimator eventually converges to the correct value), the estimator is said to be asymptotically unbiased.

Definition 5.10 (Asymptotically unbiased estimator)
Let $(X_1, \ldots, X_n)$ be a sample, then an estimator $\hat{\Theta}(X_1, \ldots, X_n)$ of the value of the parameter θ is said to be asymptotically unbiased, if

$$\lim_{n \to \infty} \mathsf{E}\left[\hat{\Theta}(X_1, \ldots, X_n)\right] = \theta,$$

provided that the expectation exists.

Each estimator is a function of random quantities and thus a random quantity itself, *viz.* even in the case of an unbiased estimator $\hat{\Theta}$ the estimated value will show a more or less pronounced dispersion around the value θ. A measure of this dispersion is the expected mean square deviation[2]

$$D\left(\hat{\Theta}; \theta\right) = \mathsf{E}\left[\left(\hat{\Theta} - \theta\right)^2\right] = \mathsf{Var}\left(\hat{\Theta}\right) + \left(\mathsf{E}\left[\hat{\Theta}\right] - \theta\right)^2.$$

It can be seen that the expected mean square deviation is a combination of the variance of the estimator and the square of its bias. Therefore, unbiasedness by itself is still not a criterion that an estimator is better than another. This insight is reassuring because unbiased estimators exist only for a few very simple cases. Most practical problems lead only to asymptotically unbiased estimators. Moreover, for a given problem, it is quite possible that there is an estimator that gives a smaller expected mean square deviation than an unbiased estimator, i. e. unbiasedness is not an essential property.

If we want to use the expected mean square deviation as a quality criterion for an estimator, the immediate question is whether there is an estimator for which the expected mean square deviation is zero. Unfortunately, this is not the case, since it is only possible to give a non-zero lower bound on the variance of an estimator, as C. R. Rao [71] and H. Cramér [72] have proved. The result of their publications is

$$\left[\frac{d\mathsf{E}\left(\hat{\Theta}\right)}{d\theta}\right]^2 [I(\theta)]^{-1} \leq \mathsf{Var}\left(\hat{\Theta}\right). \tag{5.5}$$

The left side of this inequality is called Cramér-Rao bound. This bound depends on the Fisher information $I(\theta)$.

There are probability density functions for which the Cramér-Rao inequality (5.5) merely states that $\mathsf{Var}\left(\hat{\Theta}\right) \geq 0$, i. e. that the variance of an unbiased estimator is non-negative. This is of course a useless statement, since variances must always be non-negative. Such a result can arise because in certain cases the Fisher information $I(\theta)$

---

2 The expression $\left(\hat{\Theta} - \theta\right)^2$ is also called quadratic loss function. The expectation of a loss function is called risk function.

yields an infinite result, indicating that some conditions are violated. Actually, the FISHER information matrix cannot be calculated for an arbitrary probability density function, because special regularity conditions have to be fulfilled. We do not want to go into detail here, but we mention that, for example, such regularity conditions are not fulfilled for a uniform (rectangular) distribution.

Provided that the expected mean square deviation is used as a quality criterion for an estimator, the inequality (5.5) immediately shows that the best estimator is an unbiased estimator whose variance is equal to the CRAMÉR-RAO bound. Such an estimator is called efficient estimator. We give here the definition of the efficient estimator of a parameter vector $\theta = (\theta_1, \dots, \theta_m)^\mathsf{T}$:

Definition 5.11 (Efficient estimator)
An estimator $\hat{\Theta}$ is said to be efficient, if

$$E\left(\hat{\Theta}\right) = \theta$$

and

$$C\left(\hat{\Theta}\right) = \mathbf{I}^{-1}(\theta)$$

is valid, where $C\left(\hat{\Theta}\right)$ denotes the covariance matrix of the estimator $\hat{\Theta}$ and $\mathbf{I}(\theta)$ the FISHER information matrix .

The second equation in this definition reveals an interesting relation, namely that the covariance matrix of an efficient estimator is equal to the inverse FISHER information matrix. Correspondingly, in the case of an efficient estimator of a single parameter, its variance is equal to the inverse FISHER information. This means that as the information increases, the variance decreases and vice versa. This is an intuitively understandable result.

In the case of a sample of identically distributed random quantities, the (expected) FISHER information matrix is defined as follows

Definition 5.12 (FISHER information matrix)
Let $X = (X_1, \dots, X_n)^\mathsf{T}$ be a sample vector of identically distributed random quantities with a probability density function $g_X(x \mid \theta)$, depending on the parameter vector $\theta = (\theta_1, \dots, \theta_m)^\mathsf{T}$, then the symmetric matrix $I(\theta)$ with matrix elements

$$I_{ij} = -n\mathrm{E}\left(\frac{\partial^2 \log g_X(x \mid \theta)}{\partial \theta_i\, \partial \theta_j}\right), \qquad i, j = 1, \dots, m, \tag{5.6}$$

is called the FISHER information matrix, provided that the conditions

$$\mathrm{E}\left(\frac{\partial \log g_X(x \mid \theta)}{\partial \theta_i}\right) = 0, \qquad i = 1, \dots, m, \tag{5.7}$$

are fulfilled and all expectations exist. The expectations are calculated with respect to the probability density function $g_X(x \mid \theta)$.

We will demonstrate the calculation of the FISHER information matrix with an example whose results we will use later.

Example 5.2 (FISHER information matrix of a normal distribution)
A sample vector of stochastically independent and identically distributed random quantities is given. We assume that the underlying probability distribution is a normal distribution, described by the probability density function

$$g_X(x \mid \theta_1, \theta_2) = \frac{1}{\theta_2\sqrt{2\pi}} \exp\left(-\frac{(x - \theta_1)^2}{2\theta_2^2}\right).$$

Taking the logarithm of this probability density function yields

$$\ln g_X(x \mid \theta_1, \theta_2) = -\frac{\ln(2\pi)}{2} - \ln\theta_2 - \frac{(x - \theta_1)^2}{2\theta_2^2}$$

and by calculating the derivatives with respect to the two parameters θ_1 and θ_2

$$\frac{\partial \ln g_X(x \mid \theta_1, \theta_2)}{\partial\theta_1} = \frac{x - \theta_1}{\theta_2^2}$$

and

$$\frac{\partial \ln g_X(x \mid \theta_1, \theta_2)}{\partial\theta_2} = -\frac{1}{\theta_2} + \frac{(x - \theta_1)^2}{\theta_2^3}.$$

Taking the second derivatives of these two equations furthermore gives

$$\frac{\partial^2 \ln g_X(x \mid \theta_1, \theta_2)}{\partial\theta_1^2} = -\frac{1}{\theta_2^2},$$

$$\frac{\partial^2 \ln g_X(x \mid \theta_1, \theta_2)}{\partial\theta_1\partial\theta_2} = \frac{\partial^2 \ln g_X(x \mid \theta_1, \theta_2)}{\partial\theta_2\partial\theta_1} = -2\frac{x - \theta_1}{\theta_2^3},$$

$$\frac{\partial^2 \ln g_X(x \mid \theta_1, \theta_2)}{\partial\theta_2^2} = \frac{1}{\theta_2^2} - 3\frac{(x - \theta_1)^2}{\theta_2^4}.$$

We can verify that the conditions (5.7) are satisfied by calculating the expectations of these equations using the results $E[X - \theta_1] = 0$ and $E\left[(X - \theta_1)^2\right] = \theta_2^2$. Finally, according to the relation (5.6), we obtain the FISHER information matrix

$$I(\theta_1, \theta_2) = \frac{n}{\theta_2^2}\begin{pmatrix} 1 & 0 \\ 0 & 2 \end{pmatrix}.$$

Therefore, the FISHER information of the two parameters is given by

$$I(\theta_1) = \frac{n}{\theta_2^2} \quad \text{and} \quad I(\theta_2) = \frac{2n}{\theta_2^2}.$$

Since both parameters are unbiased, their smallest possible variances are given by the inverse of their respective FISHER information.

Using the results of this example, it can be shown that the estimator $\overline{X}$ of a sample vector consisting of stochastically independent, normally distributed components is efficient. However, this is not the case for the estimator S (sample standard deviation).

So far, the question of how to determine an estimator for a particular problem has not been considered. When dealing with this question, we need to distinguish between two cases, namely a point estimator and a region estimator (or interval estimator). The former is used to estimate the parameter value of a probability distribution function, while the latter allows a statement to be made about a region (or interval) that contains the value of a quantity with a given probability.

In the following sections we will look more closely at the construction of point estimators. We will consider the following methods:
–   method of moments,
–   maximum likelihood method,
–   method of least squares,
–   BAYESian method.

All of these methods lead to point estimators that are asymptotically efficient, i. e. for large sample sizes they are not only asymptotically unbiased, but also approximately normally distributed.

5.4   Method of moments

Expectation and variance are the most important parameters of a probability density function. They are special cases of the generalized moments of this function, which are defined as follows:

> **Definition 5.13 (Generalized moment of a probability density function)**
> Let $g_X(x \mid \theta_1, \dots, \theta_m)$ be a probability density function depending on the parameters $(\theta_1, \dots, \theta_m)$, then
> $$E\left[(X - c)^k \mid \theta\right] = \int_{-\infty}^{+\infty} (x - c)^k g_X(x \mid \theta)\, dx$$
> is called the k-th generalized moment, provided that the expectation exists. The natural number k is called the order of the moment.

If $c = 0$ in this definition, we speak simply of a moment, while if $c = E[X \mid \theta]$ we speak of a central moment. All generalized moments are functions of the parameters of the respective probability density function.

The moments of a probability density function do not necessarily exist. For example, for a CAUCHY distribution no moment exists, while for a normal distribution moments of any order can be determined.

If the moments of a probability density function exist, it can be reconstructed from them under certain conditions. This observation is the basis for the method of moments proposed by K. PEARSON at the end of the 19th century. In order to apply this method to measured data, the moments of a sample vector are equated with the corresponding moments of the underlying probability density function, which depends on the parameters to be determined, where the moments of the sample vector are defined as follows:

Definition 5.14 (Sample moment)

Let $X = (X_1, \ldots, X_n)^{\mathsf{T}}$ be a sample vector, then

$$M_k(X) = \frac{1}{n} \sum_{i=1}^{n} X_i^k$$

is called k-th sample moment.

By equating the moments of a sample vector and the corresponding moments of the probability density function depending on the parameters, we obtain a generally non-linear system of equations from which the estimators of the parameters can be determined. These estimators are called "moment estimators" and are defined as follows:

Definition 5.15 (Moment estimator)

If the moments $\mathsf{E}\left[X^k \mid \theta_1, \ldots, \theta_m\right]$ $(k = 1, \ldots, r; r \geq m)$ of a probability density function, depending on the parameters $(\theta_1, \ldots, \theta_m)$, exist and the system of equations

$$\mathsf{E}\left[X^k \mid \theta_1, \ldots, \theta_m\right] = M_k(X), \qquad k = 1, \ldots, r,$$

where $M_k(X)$ $(k = 1, \ldots, r)$ denotes the k-th moment of the sample vector X, is uniquely solvable with respect to the parameters $(\theta_1, \ldots, \theta_m)$, then the solutions $\hat{\theta}_k(X_1, \ldots, X_n)$ $(k = 1, \ldots, m)$ are called moment estimators of the parameters θ_k $(k = 1, \ldots, m)$.

Normally a system of m equations should be sufficient to determine the m moment estimators of the m parameters. However, since under certain circumstances some of the equations used to estimate the parameters may be meaningless, the condition $r \geq m$ is specified in this definition.

We demonstrate the method of moments with two examples:

Example 5.3 (Method of moments for a normal distribution)

Given a normal distribution described by the probability density function

$$g_X(x) = \frac{1}{\sqrt{2\pi\theta_2}} \exp\left(-\frac{(x - \theta_1)}{2\theta_2}\right),$$

where θ_1 denotes the mean and θ_2 denotes the variance of the distribution, we obtain the moments

$$E[X] = \theta_1 \quad \text{and} \quad E[X^2] = \theta_1^2 + \theta_2.$$

Equating these two moments with those of the sample vector $X = (X_1, \ldots, X_n)$, we obtain the system of equations

$$\theta_1 = \frac{1}{n} \sum_{i=1}^{n} X_i,$$

$$\theta_1^2 + \theta_2 = \frac{1}{n} \sum_{i=1}^{n} X_i^2.$$

Solving this system of equations yields the moment estimators

$$\hat{\Theta}_1 = \overline{X}$$

and

$$\hat{\Theta}_2 = \frac{n-1}{n} S^2$$

of the mean and variance of the normal distribution, where $\overline{X}$ denotes the sample mean and S^2 the sample variance. The estimator $\hat{\Theta}_1$ is unbiased, while the estimator $\hat{\Theta}_2$ is only asymptotically unbiased.

Example 5.4 (Method of moments for a uniform distribution)
Given a uniform distribution described by the probability density function

$$g_X(x) = \begin{cases} \dfrac{1}{\theta_2 - \theta_1} & \text{if } \theta_1 \leq x \leq \theta_2 \\ 0 & \text{otherwise} \end{cases},$$

where θ_1 and θ_2 denote the upper and lower bounds of the support of the probability density function, we obtain the moments

$$E[X] = \frac{\theta_1 + \theta_2}{2} \quad \text{and} \quad E[X^2] = \frac{\theta_1^2 + \theta_1 \theta_2 + \theta_2^2}{3}.$$

Equating these two moments with those of the sample vector $X = (X_1, \ldots, X_n)$, we obtain the system of equations

$$\frac{\theta_1 + \theta_2}{2} = \frac{1}{n} \sum_{i=1}^{n} X_i,$$

$$\frac{\theta_1^2 + \theta_1 \theta_2 + \theta_2^2}{3} = \frac{1}{n} \sum_{i=1}^{n} X_i^2.$$

Solving this system of equations under the condition $\theta_1 < \theta_2$ yields the moment estimators

$$\hat{\Theta}_1 = \overline{X} - \sqrt{\frac{3(n-1)}{n}S^2}$$

and

$$\hat{\Theta}_2 = \overline{X} + \sqrt{\frac{3(n-1)}{n}S^2}$$

of the upper and lower bounds of the uniform distribution, where $\overline{X}$ denotes the sample mean and S^2 denotes the sample variance. The two estimators $\hat{\Theta}_1$ and $\hat{\Theta}_2$ are only asymptotically unbiased, although their mean is unbiased.

5.5   Maximum likelihood method

The maximum likelihood method was developed by R. A. FISHER during the years 1912 to 1922 [73]. This method is based on considerations presented by R. A. FISHER[3] in his first publication [75]. In the following, we will have a look at some of these considerations.

If we know the probability density function $g_X(x)$ of a continuous random quantity X, the probability of obtaining a measured value x of that quantity within the infinitesimally small interval dx is

$$P(x \leq X \leq x + dx) = g_X(x)\,dx.$$

If the measurement is repeated n times under the same conditions, the probability of obtaining the measured values $x_1, \dots, x_n$ of the quantity X is

$$P(x_1 \leq X_1 \leq x_1 + dx, \dots, x_n \leq X_n \leq x_n + dx) = \left(\prod_{i=1}^{n} g_{X_i}(x_i)\right)(dx)^n. \qquad (5.8)$$

In terms of probability theory, the measured values $x_1, \dots, x_n$ represent a realization of the sample vector $\boldsymbol{X} = (X_1, \dots, X_n)^\mathsf{T}$ of a population with a given probability distribution function, where the continuous random quantities X_i ($i = 1, \dots, n$) are stochastically independent and identically distributed.

Of course, the probability of getting certain values depends on the parameters $\boldsymbol{\theta} = (\theta_1, \dots, \theta_m)^\mathsf{T}$ of the underlying probability distribution. For example, it is very unlikely to measure the values 347, 582, 413, ... , but more likely to observe the values 5, 8, 6, ... , if the population is normally distributed with the mean $\mu = 7$ and the standard deviation $\sigma = 2$. In order to express this dependence on the population parameters, we use the equation

$$P(x_1 \leq X_1 \leq x_1 + dx, \dots, x_n \leq X_n \leq x_n + dx) = \left(\prod_{i=1}^{n} g_{X_i}(x_i \mid \boldsymbol{\theta})\right)(dx)^n \qquad (5.9)$$

---

3 The same idea, however, had already been published by W. CHAUVENET [74] in 1864.

instead of the equation (5.8). This equation makes it possible to calculate the probability of observing the measured values $x_1, \dots, x_n$, given the probability density function $g_X(x \mid \theta)$ of the measurand[4] X. This probability is proportional to the function

$$l(\theta \mid \boldsymbol{x}) = \prod_{i=1}^{n} g_{X_i}(x_i \mid \theta),$$

which is called likelihood function or just likelihood. This function is defined as follows (ISO 3534-1, 1.38) [38]:

Definition 5.16 (Likelihood function)
The likelihood function is a probability density function, evaluated at the observed values, which is considered as a function of the parameters of a family of probability distribution functions.

The term "family of probability distribution functions" used in this definition denotes a set of distribution functions indexed by one or more parameters (ISO 3534-1, 2.8) [38].

It should be clear from the definition 5.16 that the likelihood function interprets the joint probability density function underlying the observed measured values (i. e. a realization of a sample vector) as a function of the unknown parameters of that density function. The difference in perspective becomes clear when we recognize that the equation (5.9) describes the probability of observing the measured values when the parameters are *known*, whereas in the case of the likelihood function it is assumed that the measured values are already given, and we are interested in the *unknown* parameters.

On the basis of the above considerations, R. A. FISHER concluded that—if only the probability density function underlying a sample is known, but not the values of its parameters—the probability of any arbitrary choice of the parameter values is determined by the likelihood function. Furthermore, he concluded that the most probable parameter values will lead to a maximum of the likelihood function.

Instead of asking for the parameter values that lead to a maximum of the likelihood function itself, we can just as well look for the parameter values that maximize the so-called log-likelihood function

$$L(\theta \mid \boldsymbol{x}) = \log l(\theta \mid \boldsymbol{x}) = \sum_{i=1}^{n} \log g_{X_i}(x_i \mid \theta). \qquad (5.10)$$

This yields the same results, because the logarithm is a monotone function, but the calculations are in most cases much more convenient.

If the measurand is represented by a discrete random quantity, we choose the likelihood function

$$l(\theta \mid \boldsymbol{x}) = \prod_{i=1}^{n} P(X_i = x_i \mid \theta),$$

---

4 The measurand is the quantity to be measured (VIM, 2.3 [3, 2])

and the log-likelihood function

$$L(\theta \,|\, \boldsymbol{x}) = \sum_{i=1}^{n} \log P(X_i = x_i \,|\, \theta), \qquad (5.11)$$

where $P(X_i = x_i \,|\, \theta)$ denotes the probability that the i-th measurement yields the value x_i. This probability depends, of course, on the parameters $\theta = (\theta_1, \dots, \theta_m)^\top$ of the underlying population.

We demonstrate the use of the maximum likelihood method to obtain estimators of the parameters of a probability distribution underlying a population by means of some typical examples. We start with an example of practical importance, where only one parameter of a probability density function of a continuous random quantity needs to be estimated.

Example 5.5 (Estimator of radioactive isotope mean lifetime)

It is well known that the decay of radioactive isotopes is a stochastic process, because the instant at which an atomic nucleus disintegrates is unknown. If we study a particular radioactive isotope, we find that the probability of a particular nucleus disintegrating can be described by an exponential distribution whose probability density function is given by

$$g_X(x \,|\, \theta) = \begin{cases} \dfrac{1}{\theta} e^{-x/\theta} & \text{if } x \geq 0 \\ 00 & \text{otherwise} \end{cases}, \qquad \theta > 0, \qquad (5.12)$$

where x denotes the elapsed time until an atomic nucleus decays, and the parameter θ denotes the mean lifetime of the nuclei. The mean lifetime is a characteristic constant of each radioactive isotope and is related to the half-life $T_{1/2}$ by the simple relation $T_{1/2} = \theta \ln 2$.

If we observe the decay of a number n of nuclei of a radioactive isotope at the instants of time[5] $x_1, \dots, x_n$, we obtain, according to equations (5.10) and (5.12), the log-likelihood function

$$L(\theta \,|\, \boldsymbol{x}) = -n \ln \theta - \frac{1}{\theta} \sum_{i=1}^{n} x_i. \qquad (5.13)$$

In calculus, it is shown that for a maximum of the log-likelihood function the necessary and sufficient conditions are

$$\left[\frac{\mathrm{d}L(\theta \,|\, \boldsymbol{x})}{\mathrm{d}\theta} \right]_{\theta = \hat{\theta}} = 0 \quad \text{and} \quad \left[\frac{\mathrm{d}^2 L(\theta \,|\, \boldsymbol{x})}{\mathrm{d}\theta^2} \right]_{\theta = \hat{\theta}} < 0 \qquad (5.14)$$

---

5 This is the elapsed time between the beginning of the observation and the observed decay.

where $\hat{\theta}$ denotes the value at which the log-likelihood function takes its maximum. Differentiating the equation (5.13) yields

$$\frac{dL(\theta\,|\,\mathbf{x})}{d\theta} = -\frac{n}{\theta} + \frac{1}{\theta^2}\sum_{i=1}^{n}x_i$$

and

$$\frac{d^2L(\theta\,|\,\mathbf{x})}{d\theta^2} = \frac{n}{\theta^2} - \frac{2}{\theta^3}\sum_{i=1}^{n}x_i\,.$$

If we insert these results in the conditions (5.14), we obtain

$$\hat{\theta} = \frac{1}{n}\sum_{i=1}^{n}x_i$$

and

$$\left[\frac{d^2L(\theta\,|\,\mathbf{x})}{d\theta^2}\right]_{\theta=\hat{\theta}} = -\frac{n}{\hat{\theta}^2} < 0\,,$$

i. e. the maximum of the log-likelihood function occurs at $\theta = \hat{\theta}$.
  Hence, the sample mean

$$\hat{\Theta} = \frac{1}{n}\sum_{i=1}^{n}X_i \tag{5.15}$$

is a maximum likelihood estimator of the mean lifetime. The expectation of this estimator is

$$E\left[\hat{\Theta}\right] = \frac{1}{n}\sum_{i=1}^{n}E[X_i] = \frac{1}{n}\sum_{i=1}^{n}E[X] = E[X] = \theta\,,$$

i. e. the estimator given by equation (5.15) is unbiased. This is, of course, due to the fact that in this example the random quantities under consideration are stochastically independent and identically distributed.

In the next example, we examine the estimation of a single parameter in the case of a sample of discrete random quantities.

Example 5.6 (Success parameter of a BERNOULLI experiment)

A BERNOULLI experiment is an experiment that has only two possible outcomes. These two outcomes are traditionally called "success" and "failure". A typical example of such an experiment is a coin flip. The random quantity of a BERNOULLI experiment is discrete with the two possible values $X = 1$ (success) and $X = 0$ (failure). If the probability of success is given by $P(X = 1) = \theta$, then the probability of failure must be $P(X = 0) = 1 - \theta$, because the sum of both probabilities must be equal to one, since there are only two possible disjoint events. The parameter θ is called the success parameter of the BERNOULLI experiment.

The two probabilities of the BERNOULLI experiment can be combined by the single expression

$$P(X = x \mid \theta) = \theta^x (1 - \theta)^{1-x}, \qquad 0 \le \theta \le 1, \tag{5.16}$$

as can be easily verified by inserting $x = 1$ and $x = 0$ respectively.

If we observe the values $x_1, \dots, x_n$ during n experiments, we can, according to equations (5.11) and (5.16), state the log-likelihood function

$$L(\theta \mid \boldsymbol{x}) = \log \theta \sum_{i=1}^{n} x_i + \log(1 - \theta) \left(n - \sum_{i=1}^{n} x_i \right). \tag{5.17}$$

In the case of a maximum of this function, the conditions (5.14) must be fulfilled, where $\hat{\theta}$ denotes the value at which the log-likelihood function takes its maximum. Differentiating the equation (5.17) gives

$$\frac{dL(\theta \mid \boldsymbol{x})}{d\theta} = \frac{1}{\theta} \sum_{i=1}^{n} x_i - \frac{1}{1-\theta} \left(n - \sum_{i=1}^{n} x_i \right)$$

and

$$\frac{d^2 L(\theta \mid \boldsymbol{x})}{d\theta^2} = -\frac{1}{\theta^2} \sum_{i=1}^{n} x_i - \frac{1}{(1-\theta)^2} \left(n - \sum_{i=1}^{n} x_i \right).$$

If we insert this result in the conditions (5.14), we obtain

$$\hat{\theta} = \frac{1}{n} \sum_{i=1}^{n} x_i$$

and

$$\left[\frac{d^2 L(\theta \mid \boldsymbol{x})}{d\theta^2} \right]_{\theta = \hat{\theta}} = -\frac{n}{\theta(1 - \theta)} < 0,$$

i. e. the maximum of the log-likelihood function occurs at $\theta = \hat{\theta}$.

Hence, the sample mean

$$\hat{\Theta} = \frac{1}{n} \sum_{i=1}^{n} X_i \tag{5.18}$$

is a maximum likelihood estimator of the success parameter of a BERNOULLI experiment, where the random quantities X_i ($i = 1, \dots, n$) can only take the values 0 or 1. This implies $\hat{\theta} = k/n$, where k denotes the number of successes in n trials. Thus, the estimated value of the success parameter of a BERNOULLI experiment is equal to the fraction of successful experiments. This result corresponds to our intuitive understanding.

Using equation (5.16), the expectation of the estimator $\hat{\Theta}$ yields

$$E[\hat{\Theta}] = \frac{1}{n} \sum_{i=1}^{n} E[X_i] = E[X] = \theta,$$

i. e. the estimator given by the equation (5.18) is unbiased. This is again due to the fact that in this example the random quantities considered are stochastically independent and identically distributed.

The first two examples dealt with probability density functions of populations characterized by only one parameter. We now turn to examples where more than one parameter is needed to obtain a complete description of the probability density function of the population. The peculiarity of these cases is the necessity to estimate all the parameters from the sampled values of a population at the same time.

Example 5.7 (Parameters of a normally distributed random quantity)

A random quantity X has been measured n times under repeatability conditions. It is assumed that this quantity is normally distributed with the probability density

$$g_X(x \mid \theta_1, \theta_2) = \frac{1}{\sqrt{2\pi\,\theta_2}} \exp\left(-\frac{(x - \theta_1)^2}{2\theta_2}\right), \tag{5.19}$$

where the parameters θ_1 (mean) and θ_2 (variance) are unknown.

We consider the sample $X = (X_1, \ldots, X_n)^\mathsf{T}$ of size n, where the random quantities X_i ($i = 1, \ldots, n$) are assumed to be stochastically independent and identically distributed with the probability density function (5.19).

If the values $x_1, \ldots, x_n$ have been observed during n repeated measurements, we obtain from the equation (5.19)—according to the equation (5.10)—the log-likelihood function

$$L(x_1, \ldots, x_n \mid \theta_1, \theta_2) = -\frac{n}{2} \ln \theta_2 - \frac{1}{2\theta_2} \sum_{i=1}^{n} (x_i - \theta_1)^2, \tag{5.20}$$

where the constant term $(-n/2 \cdot \ln 2\pi)$ has been omitted because it does not change the position of the maximum of the log-likelihood function. This corresponds to the usual practice of omitting all additive terms of the log-likelihood function that do not depend on at least one of the parameters.

It is well known from calculus that a maximum of the log-likelihood function must satisfy the necessary conditions

$$\left[\frac{\partial L(\theta \mid x)}{\partial \theta_i}\right]_{\theta = \hat{\theta}} = 0, \qquad i = 1, \ldots, m, \tag{5.21}$$

where $\hat{\theta} = (\hat{\theta}_1, \ldots, \hat{\theta}_n)$ denotes the parameter values for which this maximum is assumed. Furthermore, the HESSE matrix $\mathbf{H}$ with elements

$$H_{ij} = \left[\frac{\partial^2 L(\boldsymbol{\theta} \mid \boldsymbol{x})}{\partial \theta_i\, \partial \theta_j} \right]_{\boldsymbol{\theta} = \hat{\boldsymbol{\theta}}}$$

has to be negative definite.

Differentiating the equation (5.20) yields

$$\frac{\partial L(x_1, \dots, x_n \mid \theta_1, \theta_2)}{\partial \theta_1} = \frac{1}{\theta_2} \sum_{i=1}^{n} (x_i - \theta_1) \tag{5.22}$$

and

$$\frac{\partial L(x_1, \dots, x_n \mid \theta_1, \theta_2)}{\partial \theta_2} = -\frac{n}{2\theta_2} + \frac{1}{2\theta_2^2} \sum_{i=1}^{n} (x_i - \theta_1)^2, \tag{5.23}$$

and further differentiation of these equations leads to

$$\frac{\partial^2 L(x_1, \dots, x_n \mid \theta_1, \theta_2)}{\partial \theta_1^2} = -\frac{n}{\theta_2}, \tag{5.24}$$

$$\frac{\partial^2 L(x_1, \dots, x_n \mid \theta_1, \theta_2)}{\partial \theta_2^2} = \frac{n}{2\theta_2^2} - \frac{1}{\theta_2^3} \sum_{i=1}^{n} (x_i - \theta_1)^2, \tag{5.25}$$

$$\frac{\partial^2 L(x_1, \dots, x_n \mid \theta_1, \theta_2)}{\partial \theta_1\, \partial \theta_2} = \frac{\partial^2 L(x_1, \dots, x_n \mid \theta_1, \theta_2)}{\partial \theta_2\, \partial \theta_1} = -\frac{1}{\theta_2^2} \sum_{i=1}^{n} (x_i - \theta_1). \tag{5.26}$$

Using the equations (5.22) and (5.23), the condition (5.21) yields

$$\hat{\theta}_1 = \frac{1}{n} \sum_{i=1}^{n} x_i \tag{5.27}$$

and

$$\hat{\theta}_2 = \frac{1}{n} \sum_{i=1}^{n} \left(x_i - \hat{\theta}_1 \right)^2. \tag{5.28}$$

Substituting these results into equations (5.24), (5.25) and (5.26) we obtain the HESSE matrix

$$\mathbf{H} = \begin{pmatrix} -\dfrac{n}{\hat{\theta}_2} & 0 \\[2mm] 0 & -\dfrac{2n}{\hat{\theta}_2^2} \end{pmatrix}.$$

This matrix is obviously negative definite because it is a diagonal matrix with only negative diagonal elements. Thus, the log-likelihood function has its maximum at $\theta_1 = \hat{\theta}_1$ and $\theta_2 = \hat{\theta}_2$. Consequently, it follows from equations (5.27) and (5.28) that the statistics

$$\hat{\Theta}_1 = \frac{1}{n} \sum_{i=1}^{n} X_i$$

and

$$\hat{\Theta}_2 = \frac{1}{n} \sum_{i=1}^{n} (X_i - \hat{\Theta}_1)^2 ,$$

i.e. the sample mean and the mean squared deviation are maximum likelihood estimators of the parameters θ_1 and θ_2 of the probability density function (5.19). This result is well known and widely used.

Using the probability density function (5.19) yields the expectations

$$E[\hat{\Theta}_1] = \frac{1}{n} \sum_{i=1}^{n} E[X_i] = E[X] = \theta_1 ,$$

and

$$E[\hat{\Theta}_2] = E\left[\frac{1}{n} \sum_{i=1}^{n} (X_i - \hat{\Theta}_1)^2 \right] = \frac{1}{n} \sum_{i=1}^{n} E\left[X_i^2 - 2X_i\hat{\Theta}_1 + \hat{\Theta}_1^2 \right] =$$

$$\frac{1}{n} \left(\sum_{i=1}^{n} E[X_i^2] - \frac{2}{n} \sum_{i=1}^{n} \sum_{j=1}^{n} E[X_iX_j] + \frac{1}{n} \sum_{i=1}^{n} \sum_{j=1}^{n} E[X_iX_j] \right) =$$

$$\frac{1}{n} \left(\sum_{i=1}^{n} E[X_i^2] - \frac{1}{n} \sum_{i=1}^{n} \sum_{j=1}^{n} E[X_iX_j] \right) = \frac{1}{n} \sum_{i=1}^{n} \left[E\left(X_i^2\right) - E\left(\hat{\Theta}_1^2\right) \right] =$$

$$\frac{1}{n} \sum_{i=1}^{n} \left(\text{Var}(X_i) + [E[X_i]]^2 - \text{Var}\left(\hat{\Theta}_1\right) - [E[\hat{\Theta}_1]]^2 \right) =$$

$$\frac{1}{n} \sum_{i=1}^{n} \left(\text{Var}(X_i) + [E[X]]^2 - \text{Var}\left(\frac{1}{n} \sum_{j=1}^{n} X_j \right) - [E[X]]^2 \right) =$$

$$\frac{1}{n} \sum_{i=1}^{n} \left(\text{Var}(X_i) - \frac{1}{n^2} \sum_{j=1}^{n} \text{Var}(X_j) \right) =$$

$$\frac{1}{n} \sum_{i=1}^{n} \left(\text{Var}(X) - \frac{1}{n^2} \sum_{j=1}^{n} \text{Var}(X) \right) = \frac{n-1}{n} \text{Var}(X) = \frac{n-1}{n} \theta_2$$

of the estimators $\hat{\Theta}_1$ and $\hat{\Theta}_2$, i.e. the estimator $\hat{\Theta}_1$ is unbiased for a sample of stochastically independent and identically distributed random quantities, while this is not true for the estimator $\hat{\Theta}_2$, which is only asymptotically unbiased.

If the probability density function of a continuous random quantity under consideration is not differentiable, we cannot proceed as in the previous examples. In such a case, the maximum of the likelihood function must be determined directly. We demonstrate this with an example.

Example 5.8 (Parameters of a uniformly distributed quantity)

A random quantity X has been measured n times under repeatability conditions. It is assumed that this quantity is uniformly distributed with the probability density

$$g_X(x \mid \theta_1, \theta_2) = \begin{cases} \dfrac{1}{\theta_2 - \theta_1} & \text{if } \theta_1 \leq x \leq \theta_2 \\ 0 & \text{otherwise} \end{cases}, \tag{5.29}$$

where the parameters θ_1 and θ_2 are unknown.

We consider the sample $X = (X_1, \dots, X_n)^\mathsf{T}$ of size n, where the random quantities X_i ($i = 1, \dots, n$) are assumed to be stochastically independent and identically distributed with the probability density function (5.29).

If the values $x_1, \dots, x_n$ have been observed during n repeated measurements, we obtain from equation (5.29)—according to equation (5.10)—the log-likelihood function

$$l(x_1, \dots, x_n \mid \theta_1, \theta_2) = \begin{cases} \dfrac{1}{(\theta_2 - \theta_1)^n} & \text{if } \theta_1 \leq x_i \leq \theta_2 \ (i = 1, \dots, n) \\ 0 & \text{otherwise} \end{cases}.$$

The next step is to interpret the condition $\theta_1 \leq x_i \leq \theta_2$ ($i = 1, \dots, n$) in this function. This condition requires that *all* measured values must be within the closed interval $[\theta_1, \theta_2]$. This is certainly fulfilled if

$$\theta_1 \leq \min(x_1, \dots, x_n) \quad \text{and} \quad \max(x_1, \dots, x_n) \leq \theta_2, \tag{5.30}$$

i. e. if the smallest measured value is not less than θ_1 and the largest measured value is not greater than θ_2.

With this condition, we can now formulate the conditions for a maximum of the likelihood function. This maximum occurs if and only if the difference $(\theta_2 - \theta_1)$ takes a minimum, i. e. if the equality signs are valid in the conditions (5.30).

Therefore, we can conclude that the statistics

$$\hat{\Theta}_1 = X_{\min} \quad \text{and} \quad \hat{\Theta}_2 = X_{\max}$$

are maximum likelihood estimators of the parameters θ_1 and θ_2 of the probability density function (5.29). Thus, estimating the parameters of a uniform distribution is a typical application of an order statistic.

In order to calculate the expectations of the estimators $\hat{\Theta}_1$ and $\hat{\Theta}_2$, we need their probability density functions. Since the random quantities X_i ($i = 1, \dots, n$) are stochastically independent, we have

$$1 - G_{\hat{\Theta}_1}(z) = P\left(\hat{\Theta}_1 > z\right) =$$

$$P(X_1 > z, \dots, X_n > z) = \prod_{i=1}^{n} \left(1 - G_{X_i}(z)\right) = \left(1 - G_X(z)\right)^n$$

and

$$G_{\hat{\Theta}_2}(z) = P(\hat{\Theta}_2 \le z) = P(X_1 \le z, \dots, X_n \le z) = \prod_{i=1}^{n} G_{X_i}(z) = (G_X(z))^n.$$

Differentiating these equations yields

$$g_{\hat{\Theta}_1}(z) = n(1 - G_X(z))^{n-1} g_{X|\Theta_1,\Theta_2}(z) \tag{5.31}$$

and

$$g_{\hat{\Theta}_2}(z) = n(G_X(z))^{n-1} g_{X|\Theta_1,\Theta_2}(z). \tag{5.32}$$

From the equation (5.29) we obtain by integration

$$G_X(z) = \begin{cases} 0 & \text{if } z \le \theta_1 \\ \dfrac{z - \theta_1}{\theta_2 - \theta_1} & \text{if } \theta_1 \le z \le \theta_2 \\ 1 & \text{if } \theta_2 \le z \end{cases}.$$

Inserting this result into the equations (5.31) and (5.32), we obtain the probability density functions

$$g_{\hat{\Theta}_1}(z) = \begin{cases} \dfrac{n(\theta_2 - z)^{n-1}}{(\theta_2 - \theta_1)^n} & \text{if } \theta_1 \le z \le \theta_2 \\ 0 & \text{otherwise} \end{cases}$$

and

$$g_{\hat{\Theta}_2}(z) = \begin{cases} \dfrac{n(z - \theta_1)^{n-1}}{(\theta_2 - \theta_1)^n} & \text{if } \theta_1 \le z \le \theta_2 \\ 0 & \text{otherwise} \end{cases}$$

of the estimators $\hat{\Theta}_1$ and $\hat{\Theta}_2$. Using these density functions, we obtain the expectations of the respective estimators as

$$E[\hat{\Theta}_1] = \int_{-\infty}^{\infty} z g_{\hat{\Theta}_1}(z) dz = \int_{\theta_1}^{\theta_2} \frac{nz(\theta_2 - z)^{n-1}}{(\theta_2 - \theta_1)^n} dz = \frac{n\theta_1 + \theta_2}{n + 1}$$

and

$$E[\hat{\Theta}_2] = \int_{-\infty}^{\infty} z g_{\hat{\Theta}_2}(z) dz = \int_{\theta_1}^{\theta_2} \frac{nz(z - \theta_1)^{n-1}}{(\theta_2 - \theta_1)^n} dz = \frac{n\theta_2 + \theta_1}{n + 1},$$

i. e. the estimators $\hat{\Theta}_1$ and $\hat{\Theta}_2$ are only asymptotically unbiased.
In addition, we can say that

$$E\left[\frac{\hat{\Theta}_1 + \hat{\Theta}_2}{2}\right] = \frac{E[\hat{\Theta}_1] + E[\hat{\Theta}_2]}{2} = \frac{\theta_1 + \theta_2}{2} = E[X]$$

is valid, i. e. it turns out that the mean of the sample minimum and the sample maximum is an unbiased estimator of the expectation $E[X]$ of the population, even though the estimators $\hat{\Theta}_1$ and $\hat{\Theta}_2$ are themselves biased.

Furthermore, it turns out that

$$E\left[\frac{V}{2\sqrt{3}}\right] = \frac{E[\hat{\Theta}_2] - E[\hat{\Theta}_1]}{2\sqrt{3}} = \frac{n-1}{n+1}\frac{\theta_2 - \theta_1}{2\sqrt{3}} = \frac{n-1}{n+1}\sqrt{Var(X)}$$

is an asymptotically unbiased estimator of the standard deviation $\sqrt{Var(X)}$ of the population.

5.6   Least squares method

The least squares method was developed independently by the mathematicians and astronomers A. M. LEGENDRE [76] and C. F. GAUSS [66] more than 200 years ago, and is thus the oldest statistical estimation method at all. To understand this method, we will follow the considerations of C. F. GAUSS [14], which are still valid today.

Suppose a quantity denoted by X is measured, having a value denoted by θ, and we obtain a measured value denoted by x. Then we know from experience that in general the value x does not coincide with the value θ, i. e. there will be a deviation $d = x - \theta$. This deviation can be positive or negative. If we take measurements not only once but repeatedly, this fact applies to each of the measurements.

We assume that the measurements are made under repeatability conditions, so we can assume that all the measured values x_i ($i = 1, \ldots, n$) have the same precision, and thus we obtain the deviations

$$d_i = x_i - \theta, \qquad i = 1, \ldots, n.$$

Adding all these deviations and dividing by the number of measurements gives the mean deviation.

$$\bar{d} = \frac{1}{n}\sum_{i=1}^{n} x_i - \theta = \bar{x} - \theta.$$

This mean deviation will generally be different from zero, but we can always make it zero by correcting all the measured values by an appropriate common value. Consequently, the mean deviation is not suitable as a measure of the dispersion of the measured values. Therefore, C. F. GAUSS proposed to use the mean square deviation as a suitable measure of dispersion, which is proportional to the sum of the squared deviations. We therefore consider the expression

$$q^2 = \sum_{i=1}^{n}(x_i - \theta)^2.$$

By a simple calculation, this expression can be transformed to

$$q^2 = (n - 1)s^2 + n(\bar{x} - \theta)^2,$$

with the abbreviations

$$\overline{x} = \frac{1}{n} \sum_{i=1}^{n} x_i$$

and

$$s^2 = \frac{1}{n-1} \sum_{i=1}^{n} (x_i - \overline{x})^2 .$$

It turns out that the sum of the squares of the deviations, q^2, consists of two positive terms. Thus, this sum will become smaller if the second term becomes zero. This can be achieved by setting the parameter θ to the value

$$\hat{\theta} = \frac{1}{n} \sum_{i=1}^{n} x_i , \qquad (5.33)$$

such that we obtain the minimum

$$q_{\min}^2 = (n-1)s^2 .$$

We can therefore conclude that—according to the least squares method—the value $\hat{\theta}$ can be considered as the best estimate of the value θ of the quantity X, provided that there is no systematic measurement error in the measured values that affects this estimate.

If we drop the requirement that all measurements must be of equal precision, we can take this into account by multiplying each squared deviation by a weighting factor, which is considered to be a measure of the precision of that particular measurement, before adding them all. Usually the weighting factors are the inverse variances, which must therefore be known for each measured value. In this way we obtain the expression[6]

$$\chi^2 = \sum_{i=1}^{n} \frac{(x_i - \theta)^2}{\sigma_i^2}$$

as the sum of the weighted squared deviations, where σ_i^2 denotes the variance of the i-th measured value x_i. This expression can be rewritten as

$$\chi^2 = (n-1)\frac{s^2}{\sigma^2} + n\frac{(\overline{x} - \theta)^2}{\sigma^2} ,$$

with the abbreviations

$$\overline{x} = \frac{1}{n} \sum_{i=1}^{n} \left(\frac{\sigma}{\sigma_i}\right)^2 x_i ,$$

$$s^2 = \frac{1}{n-1} \sum_{i=1}^{n} \left(\frac{\sigma}{\sigma_i}\right)^2 (x_i - \overline{x})^2 ,$$

---

[6] The term χ^2 is conventionally used in the case of a sum of squared deviations weighted by inverse variances. The quantity χ^2 has the dimension number.

and

$$\sigma^2 = \left(\frac{1}{n} \sum_{i=1}^{n} \frac{1}{\sigma_i^2} \right)^{-1} .$$

The expression χ^2 consists of two positive terms. Therefore, we can minimize it by choosing the value

$$\hat{\theta} = \frac{1}{n} \sum_{i=1}^{n} \left(\frac{\sigma}{\sigma_i} \right)^2 x_i$$

for the parameter θ, such that we obtain the minimum

$$\chi_{\min}^2 = (n-1) \frac{s^2}{\sigma^2} .$$

If all the measured values happen to be of the same precision, i. e. if we have $\sigma_i = \sigma$, the previous result will be reproduced.

We have seen that the least squares method can be applied even if the assumption of equal precision of the measured data does not hold. Since it is not the purpose of this book to deal with the theory of the least squares method, we will not give a justification for this statement here, but refer to the specialist literature[7] dealing with this method.

The least squares method generally starts with a model equation, which in turn leads to a sum of squared deviations of the measured values from the underlying model, divided by the respective variances (providing they are known), if deemed necessary. This sum is then minimized by fitting the unknown values of the model parameters. The values thus obtained are considered to be the best estimates.

If, in an equation for a given parameter θ obtained by applying the least squares method, all the measured values x_i are replaced by the corresponding random quantities X_i and the best estimate $\hat{\theta}$ of the parameter is replaced by the quantity $\hat{\Theta}$, we obtain the so-called Last Squares Estimator of the parameter in question. For example, equation (5.33) gives the least squares estimator

$$\hat{\Theta} = \frac{1}{n} \sum_{i=1}^{n} X_i$$

of the measurand X, provided that all components of the respective sample vector $X = (X_1, \dots, X_n)$ are identically distributed, i. e. all measured values that are realizations of the sample vector have the same precision. Since this particular estimator is the arithmetic mean, it is unbiased.

In order to demonstrate the application of the least squares method, we consider an example known as linear regression[8] in statistics. As linear regression is one of the

---

7 The books of F. R. HELMERT [77] or W. W. JOHNSON [78], which are still available as reprints from some publishers, are classics in the field.
8 Note that linear regression is *not* identical with the determination of a best-fit straight line, because in regression the independent quantity is not considered to be random. In the case of a best-fit straight line,

best known and most studied methods in statistics, we will mainly present probabilistic aspects and the stochastic modelling in this example.

Example 5.9 (Linear regression)

The values y_i ($i = 1, \dots, n$) have been measured and we know or assume from the experimental conditions that the linear model

$$Y = \theta_1 + \theta_2 X$$

exists between the measurand Y and a control variable X that varies during the measurement. The unknown values of the two parameters θ_1 and θ_2 are to be determined using the least squares method.

We consider the measured values y_i ($i = 1, \dots, n$) as a realization of the sample vector $Y = (Y_1, \dots, Y_n)^\mathsf{T}$ and use the stochastic model

$$Y_i = \theta_1 + \theta_2 X_i + D_i, \qquad i = 1, \dots, n,$$

of the components of this vector, where the unknown deviations from the assumed linear relation are given by the random quantities D_i ($i = 1, \dots, n$). We do not consider the quantities X_i as random, as is common in linear regression. The parameters θ_1 and θ_2 are constant in this model.

We start by looking at the estimator of the mean deviation

$$\bar{D} = \bar{Y} - \theta_1 - \theta_2 \bar{X},$$

with the abbreviations

$$\bar{X} = \frac{1}{n} \sum_{i=1}^{n} X_i$$

and

$$\bar{Y} = \frac{1}{n} \sum_{i=1}^{n} Y_i.$$

If we form the expectation of this estimator and solve for $\mathsf{E}\left(\bar{Y}\right)$, we obtain—under the conditions given above—the linear equation

$$\mathsf{E}[\bar{Y}] = \mathsf{E}[\bar{D}] + \theta_1 + \theta_2 \bar{X}.$$

It turns out that the intercept of this line is not only determined by the parameter θ_1, but also by the mean expected deviation $\mathsf{E}[\bar{D}]$. Thus, we can interpret this expected deviation as a systematic error of the sample, i. e. $\mathsf{E}[\bar{D}] = 0$ means that no systematic error is included in the measured values. Whether this is actually the case or not, we

---

however, both the independent and dependent quantities have to be considered as random. Suitable fitting algorithms are given for example in [79, 80].

cannot deduce from the measured values, i. e. we have to assume that the estimate $\hat{\theta}_1$, which is yet to be determined, could generally include a systematic error. We need more information to decide whether this is the case or not. Therefore, in linear regression, the intercept is burdened with the possibility of a latent systematic error. On the other hand, if we already know that the intercept has to take a certain value (e. g. zero), this gives us the chance to identify a possible systematic error.

Now we consider the estimator of the sum of squares of the deviations

$$Q^2 = \sum_{i=1}^{n} (Y_i - \theta_1 - \theta_2 X_i)^2 .$$

By a simple calculation we can change this expression to

$$Q^2 = (n-1)S_{YY}\left(1 - R_{XY}^2\right)$$
$$+ (n-1)S_{XX}\left(\frac{S_{XY}}{S_{XX}} - \theta_2\right)^2 + n(\bar{Y} - \theta_2\bar{X} - \theta_1)^2 ,$$

with the abbreviations

$$S_{XX} = \frac{1}{n-1}\sum_{i=1}^{n}(X_i - \bar{X})^2 ,$$

$$S_{YY} = \frac{1}{n-1}\sum_{i=1}^{n}(Y_i - \bar{Y})^2 ,$$

$$S_{XY} = \frac{1}{n-1}\sum_{i=1}^{n}(X_i - \bar{X})(Y_i - \bar{Y}) ,$$

and

$$R_{xy} = \frac{S_{XY}}{\sqrt{S_{XX}S_{YY}}} .$$

It turns out that Q^2 consists of three terms, all of which are positive. Therefore, a minimum of this expression can be obtained by letting the last two terms become zero, i. e. by choosing the estimators of the parameters as follows

$$\hat{\theta}_2 = \frac{S_{XY}}{S_{XX}}$$

and

$$\hat{\theta}_1 = \bar{Y} - \hat{\theta}_2\bar{X} .$$

This finally gives the expression

$$Q_{\min}^2 = (n-1)S_{YY}\left(1 - R_{XY}^2\right)$$

of the least squares error estimator. The estimated value $\hat{r}_{xy}^2$ of R_{XY}^2 provides information about the degree of agreement of the data with the assumed linear relation.

The value of this so-called correlation coefficient, which is provided by most statist-ics software used for linear regression, is one measure of the quality of the model. In the case of $\hat{r}_{xy}^2 = 0$ there is no linear relation, while $\hat{r}_{xy}^2 = 1$ indicates that all measured data points lie on a straight line.

For further calculations we assume that the quantities D_i are stochastically inde-pendent and identically distributed. Based on this assumption we get $E[D_i] = E[D]$, $\text{Var}(D_i) = \text{Var}(D)$ and $\text{Cov}(D_i, D_j) = 0$. Using these conditions and the equation (5.3), we can determine the values of the estimators $\hat{\Theta}_1$ and $\hat{\Theta}_2$, which after a short calcu-lation yields

$$E\left(\hat{\Theta}_2\right) = \theta_2$$

and

$$E\left(\hat{\Theta}_1\right) = \theta_1 + E(D) .$$

From this it follows—under the above assumptions—that the estimator $\hat{\Theta}_2$ is un-biased, while for the estimator $\hat{\Theta}_1$ this is true only if $E(D) = 0$ is valid, i. e. if no systematic error is present.

Finally, we calculate the variance and covariance of the estimators. Using the equation (5.4) we obtain, after a short calculation, the results

$$\text{Var}\left(\hat{\Theta}_2\right) = \frac{\text{Var}(D)}{(n-1)S_{XX}} ,$$

$$\text{Var}\left(\hat{\Theta}_1\right) = \frac{\text{Var}(D)}{(n-1)S_{XX}} \left(\frac{n-1}{n}S_{XX} + \overline{X}^2\right)$$

and

$$\text{Cov}\left(\hat{\Theta}_1, \hat{\Theta}_2\right) = -\frac{\text{Var}(D)}{(n-1)S_{XX}} \overline{X}.$$

All three expressions depend on the unknown variance $\text{Var}(D)$, which is a measure of the random deviation of the measured points from the regression line. In order to determine this variance, we calculate the expectation of $Q_{\min}^2$, which yields

$$E\left(Q_{\min}^2\right) = (n-2)\text{Var}(D).$$

This result shows that the expectation of the minimum of the mean square deviation is proportional to the variance of the measured data.

The proportionality constant $(n-2)$, which appears in the last equation of this example, is equal to the number of degrees of freedom. The general definition of this term is given by:

Definition 5.17 (Number of degrees of freedom)
The number of degrees of freedom is equal to the number of measured values re-duced by the number of parameters to be determined.

It can be shown that the procedure shown in the example above can be generalized in such a way that the least squares method can also be applied to non-linear problems. It has already been pointed out that measured data of unequal precision can be taken into account. Constraints or known correlations can also be included. Even the assumption, usually made in regression estimation, that the control variables are not regarded as random can be dropped. All that matters is that a functional relation describing the deviation of the measured data from a given model can be stated fairly well. However, probabilistic aspects should also be considered in order to assess the quality of the results. The example has shown how this can be done in principle.

A final comment at the end of this section: The view of some authors that the least squares method is merely a special case of the maximum likelihood method with an underlying normal distribution of the data is not correct. We have seen that nowhere in this section the probability distribution function underlying the data has been used. Indeed, different methods may give similar or even the same results under certain circumstances, but this does not justify the conclusion that one method is actually a special case of the other. The least squares method can be justified by probabilistic reasoning without the need for a likelihood function. It can even be shown that, in the case of linear problems, the least squares method yields estimators with the smallest possible variance, without the need for any assumption about the underlying probability distribution function. The proof of this property of the least squares method, now known as the GAUSS-MARKOV theorem, was already given by C. F. GAUSS [14] and recovered by A. A. MARKOFF [81].

5.7 BAYESian methods

BAYESian methods can be traced back to the publications of T. BAYES [49] and P. S. LAPLACE [50] and are thus much older than so-called "classical statistics", which was strongly influenced by R. A. FISHER. Both statistical methods have many similarities, but there is an essential difference between the BAYESian method and the statistical estimation methods introduced in the previous sections. In order to understand this difference, we return to the considerations in the previous section that led to the maximum likelihood method.

The maximum likelihood method is based on the assumption that the measured values $x_1, \ldots, x_n$ are realizations of the sample vector $X = (X_1, \ldots, X_n)^\mathsf{T}$ of a population. In addition, the continuous random quantities X_i ($i = 1, \ldots, n$) are assumed to be statistically independent and identically distributed. It is further assumed that the family of probability density functions $g_X(x \mid \theta)$ is known, and that the probability of obtaining certain measured values depends on the parameters $\theta = (\theta_1, \ldots, \theta_m)^\mathsf{T}$ of the underlying population. It is claimed that under these conditions the probability of measuring the values $x_1, \ldots, x_n$ is given by the expression

$$P(x_1 \leq X_1 \leq x_1 + dx, \ldots, x_n \leq X_n \leq x_n + dx) = l(\theta \,|\, x)(dx)^n$$

with the likelihood function

$$l(\theta \,|\, x) = \prod_{i=1}^{n} g_{X_i}(x_i \,|\, \theta),$$

which is interpreted as a function of the unknown parameter θ, given the observed data x. The argument for the applicability of the maximum likelihood method is then that those parameters are most likely to give a maximum of the likelihood function. *De facto*, however, this is not necessarily the case, as the likelihood function alone does not allow any probability statements to be made. The use of the maximum likelihood method in clinical diagnostics, for example, would be tantamount to interpreting the results of medical examinations without considering a particular prevalence, and thus potentially leading to erroneous conclusions with the corresponding consequences (for more details see example 4.27).

When evaluating measured data, we should take into account—similar to how a physician generally does not ignore the prevalence of a serious illness—the probability of an occurrence of certain parameter values, in order to avoid wrong conclusions. In obvious cases, of course, an unreasonable measurement result can easily be recognized, such as if in elementary particle physics, where usually only small mass values occur, a negative mass value would be obtained. However, it is much better to consider the prior knowledge about possible parameter values already during the evaluation of the measured data. This is exactly what is done when using the Bayesian method.

It is an agreed implicit understanding of the maximum likelihood method that all parameters θ are generally *constants*, which are however unknown. On the contrary, when using the Bayesian method, the parameters are always considered as *random quantities*. Thus, it is possible to assign a probability density function $g_\theta(\theta)$ to the parameters θ—the so-called prior[9] —which describes the probability of the parameter values. The prior probability density depends only on the parameters, but not on the measured data, because it describes our *a priori* knowledge (i. e. *before* we are aware of any measured data) about the parameters.

When we speak in the context of the Bayesian method of *a priori* knowledge, we are not referring to a chronology, in the sense of knowledge that already exists before the start of the measurement. Any knowledge obtained during or after the measurement, or even at a much later time, can also be considered *a priori* knowledge, provided that it is not directly related to the measurement being performed. Typical examples of *a priori* knowledge are scientific fundamentals such as physical theories, results published

---

9 This is an abbreviation of the term prior probability density function.

in peer-reviewed journals, manuals or databases, results stated in calibration certificates, but also hypotheses to be verified or rejected, and so-called expert knowledge of
metrologists.

If we multiply the prior probability density by the likelihood[10] and then normalize
the resulting expression, we obtain a probability density function, which is briefly called
the posterior.[11] Thus, we have the relation (BAYES-LAPLACE theorem)

$$g_\Theta(\theta \mid x) = C\, l(\theta \mid x)\, g_\Theta(\theta),$$

where C denotes the normalization constant. The posterior probability density describes
our *a posteriori* knowledge (i. e. after we have obtained the measured data) about the
parameter values. The likelihood includes the information contained in the data as a
supplement to the *a priori* knowledge.

In this context, it should not go unmentioned that the BAYESian approach is considered inadmissible in classical statistics. However, the objections raised are of a more
philosophical nature and can therefore be considered largely irrelevant to a practical
application of the BAYESian method.

Using the posterior probability density obtained by applying the BAYES-LAPLACE
theorem, we are able to define the term BAYESian estimator[12]

Definition 5.18 (BAYESian estimator)
The conditional expectation

$$\hat{\Theta} = E[\Theta \mid X] = \int_{-\infty}^{+\infty} \theta\, g_\Theta(\theta \mid x)\, d\theta$$

formed with the posterior probability density $g_\Theta(\theta \mid x)$ is a BAYESian estimator of
the random quantity Θ (parameter) given the sample X, provided the integral is
absolutely convergent.

This definition shows very well the difference with the estimation methods of conventional statistics. Since the BAYESian estimator is a conditional expectation with respect to
the posterior probability density, which represents *all* the available information about
the parameters, it also takes into account the knowledge that already exists before a
measurement is made.

---

10 In BAYESian statistics the likelihood function is usually just called the likelihood.
11 This is an abbreviation of the term posterior probability density function.
12 It should be noted that this definition of a BAYESian estimator represents only *one* possibility, since
other BAYESian estimators exist, such as e. g. the posterior mode or the posterior median. However, the
definition given here is widely used.

To evaluate the performance of a BAYESian estimator as a function of the sample size (especially if we are interested in a comparison with other estimators), we need its expectation, which is defined as:

Definition 5.19 (Expectation of the BAYESian estimator)

The expression

$$E[\hat{\Theta}] = \int\limits_{-\infty}^{+\infty} \hat{\Theta} g_X(x) \, dx$$

formed with the probability density function $g_X(x)$ of the sample X is the expectation of a BAYESian estimator of the random quantity Θ (parameter) given the sample X, provided the integral is absolutely convergent.

Choosing a "correct" likelihood is one of the problems that BAYESian and conventional (orthodox) statistics have in common. This is part of a stochastic modelling of the measurement, which is necessary to allow evaluation of the measured data. However, BAYESian statistics also requires a prior probability density as an additional component of the stochastic model. This requirement is usually considered more problematic, because while assumptions about the likelihood can often be justified empirically (at least in principle), this is only true to a certain extent for a prior probability density. We will not enter into this discussion here, but will limit ourselves to a brief discussion of some of the methods commonly used to specify prior probability densities in stochastic modelling.

Often we can state the prior probability density based on our experience alone, if we know sufficiently well the quantity to be measured (the measurand) and the measurement system used. In particular, if we are dealing with a measurement that has already been made many times, we can reuse the prior probability density that has been used before.

If we have no or insufficient knowledge of the measurand before the measurement, we can use the so-called uninformative prior proposed by H. JEFFREYS [36, 82], which is usually regarded as an attempt to model complete ignorance. For each parameter, this prior probability density is proportional to the positive square root of the respective diagonal element of the expected FISHER information matrix (also called FISHER information for short), ignoring factors that do not depend on the parameter under consideration. This approach also includes an implicit *a priori* assumption of the stochastic independence of the parameters.

The prior probability density according to H. JEFFREYS has the important property of being invariant with respect to a reparametrization and can be regarded as a generalization of the so-called indifference principle which was already proposed by P. S. LAPLACE. However, some authors find it problematic that an uninformative prior probability density often does not satisfy the normalization requirement, which is mandatory for any probability density function, and thus call such a prior probability density an improper prior. In most cases, however, this deficiency is not essential for a practical application of such a prior, since the resulting posterior probability density is generally normalizable. Moreover, in all cases of practical importance, it can be shown that an improper prior is only a borderline case of a properly normalizable prior probability density.

In the following, we will demonstrate the application of a prior according to H. JEF-FREYS by an example of some practical importance, where the estimators of the parameters mean and standard deviation of a normal distribution are jointly constructed:

Example 5.10 (Repeated measurements)

A random quantity X has been measured n times under repeatability conditions. No prior knowledge exists about this random quantity, but we assume that the measured values are normally distributed, i.e. they are a realization of a sample with the probability density function

$$g_X(x \mid \theta_1, \theta_2) = \frac{1}{\theta_2 \sqrt{2\pi}} \exp\left(-\frac{(x - \theta_1)^2}{2\theta_2^2}\right), \tag{5.34}$$

where θ_1 denotes the mean and θ_2 denotes the standard deviation of the normal distribution function.

In example 5.2, we have already calculated the FISHER information matrix of the normal distribution function. From this calculation it follows that the diagonal elements are constant with respect to the parameter θ_1 and proportional to θ_2^{-2} with respect to the parameter θ_2. Therefore, according to H. JEFFREYS, we obtain the prior probability density

$$g_{\Theta_1, \Theta_2}(\theta_1, \theta_2) \propto \frac{1}{\theta_2}.$$

The probability density function (5.34) leads to the likelihood

$$l(\theta_1, \theta_2 \mid \boldsymbol{x}) = \theta_2^{-n} \exp\left(-\frac{1}{2\theta_2^2} \sum_{i=1}^{n}(x_i - \theta_1)^2\right),$$

where, as usual, a multiplicative constant that does not depend on the parameters has been suppressed.

According to the BAYES-LAPLACE theorem, the product of the prior probability density and the likelihood yields the posterior probability density.

$$g_{\Theta_1, \Theta_2}(\theta_1, \theta_2 \mid \boldsymbol{x}) = C\theta_2^{-(n+1)} \exp\left(-\frac{1}{2\theta_2^2} \sum_{i=1}^{n}(x_i - \theta_1)^2\right),$$

which can be transformed into the form

$$g_{\Theta_1, \Theta_2}(\theta_1, \theta_2 \mid \boldsymbol{x}) = C\theta_2^{-(n+1)} \exp\left(-\frac{(n-1)s^2 + n(\theta_1 - \bar{x})^2}{2\theta_2^2}\right),$$

with the abbreviations

$$\bar{x} = \frac{1}{n} \sum_{i=1}^{n} x_i$$

and

$$s^2 = \frac{1}{n-1} \sum_{i=1}^{n} (x_i - \overline{x})^2 .$$

Finally, integrating the posterior probability density function with respect to the parameters θ_1 and θ_2 yields the normalization constant

$$C = \frac{\sqrt{\frac{2n}{\pi}} \left(\frac{(n-1)s^2}{2} \right)^{(n-1)/2}}{\Gamma \left(\frac{n-1}{2} \right)} .$$

The posterior probability density function is thus a product of the density function of a normal distribution with respect to θ_1 and the density function of an inverse gamma distribution with respect to θ_2^2.

Marginalizing the posterior probability density with respect to the two parameters θ_1 and θ_2 yields the marginal distribution functions

$$g_{\Theta_1}(\theta_1 \,|\, \boldsymbol{x}) = \frac{\Gamma \left(\frac{n}{2} \right)}{\Gamma \left(\frac{n-1}{2} \right)} \sqrt{\frac{n}{\pi(n-1)s^2}} \left(1 + \frac{n(\theta_1 - \overline{x})^2}{(n-1)s^2} \right)^{-n/2}$$

and

$$g_{\Theta_2}(\theta_2 \,|\, \boldsymbol{x}) = 2 \frac{\left(\frac{(n-1)s^2}{2} \right)^{(n-1)/2}}{\Gamma \left(\frac{n-1}{2} \right)} \theta_2^{-n} \exp \left(-\frac{(n-1)s^2}{2\theta_2^2} \right) .$$

These marginal distribution functions show that the random quantities

$$T = \frac{\Theta_1 - \overline{X}}{S} \sqrt{n} \quad \text{and} \quad Z = \frac{(n-1)S^2}{\Theta_2^2}$$

have a t distribution (also called Student distribution) and a χ^2 distribution, respectively, with $(n-1)$ degrees of freedom.

Using the posterior probability density we obtain, according to the definition 5.18, the estimators

$$\hat{\Theta}_1 = \overline{X} \quad \text{and} \quad \hat{\Theta}_2^2 = \frac{n-1}{n-3} S^2$$

of the parameters θ_1 and θ_2. Calculating the expectations of these estimators according to the definition 5.19 yields

$$E\left(\hat{\Theta}_1 \right) = \theta_1 \quad \text{and} \quad E\left(\hat{\Theta}_2^2 \right) = \frac{n-1}{n-3} \theta_2^2 .$$

Thus, it turns out that the estimator $\hat{\Theta}_1$ is unbiased, while the estimator $\hat{\Theta}_2$ is only asymptotically unbiased.

The factor $(n-1)/(n-3)$ that appears in this example is called the BAYESian correction. This factor increases the variance and approaches one as the sample size increases.

A method of determining the prior probability density based on an information-theoretical approach was proposed by E. T. JAYNES [83, 84, 85]. This method is known as the principle of maximum entropy because it recommends choosing the prior probability density in such a way that—based on the available knowledge—the information entropy, introduced by C. E. SHANNON [86] in 1948, assumes a maximum. This information entropy can be seen as a measure of the expected information contained in a given probability distribution.

The easiest way to understand the relationship between information and probability is to imagine an unexpected event. Then it becomes clear that an event with a very low probability has a higher information content than a very probable event that happens every day. Journalists learn in their training that an event such as *"Dog bites man!"* would make an uninteresting newspaper story, while a very rare event such as *"Man bites dog!"* is certainly worthy of a headline.

The principle of maximum entropy is beyond the scope of this book. We refer the reader to the publications of E. T. JAYNES. However, we shall mention at least two examples of the application of this method which are of practical importance. For instance, if we know or believe that the value of a quantity lies between two bounds and that it is very unlikely that this quantity takes values outside the interval defined by these bounds, we can choose a uniform distribution with these bounds as the prior probability density. On the other hand, if we have a quantity value and its associated uncertainty from one source or another (e. g. from a calibration certificate), and no other knowledge is available, we can choose as a prior probability density a normal distribution with the given quantity value as the mean and its associated standard uncertainty as the standard deviation.

We demonstrate by an example the application of the principle of maximum entropy in a case where prior knowledge is available due to a calibration certificate, and consequently we can use the probability density of a normal distribution as a prior probability density.

Example 5.11 (Calibration of a measurement system)
We calibrate a length measuring device using a material measure (e. g. a gauge block). We consider the length of the material measure to be a random quantity, which we denote by X. This quantity is measured n times under repeatability conditions. We assume that the measured values are normally distributed, i. e. they are a realization of a sample with the probability density function

$$g_X(x \mid \theta, \sigma) = \frac{1}{\sigma\sqrt{2\pi}} \exp\left(-\frac{(x-\theta)^2}{2\sigma^2}\right),$$

where θ denotes the mean and σ denotes the standard deviation of a normal distribution function. This probability density function leads to the likelihood

$$l(\theta, \sigma \,|\, \boldsymbol{x}) = \sigma^{-n} \exp\left(-\frac{1}{2\sigma^2} \sum_{i=1}^{n} (x_i - \theta)^2\right),$$

where, as usual, a multiplicative constant that does not depend on the parameters has been suppressed. A calibration certificate is available for the material measure, stating its assigned value w and the associated standard uncertainty u. These two values represent our prior knowledge regarding the length of the measured object. Using the principle of maximum entropy we can assign a normal distribution to this prior knowledge, i. e. the prior probability density function is given by

$$g_\Theta(\theta) = \frac{1}{u\sqrt{2\pi}} \exp\left(-\frac{(\theta - w)^2}{2u^2}\right).$$

It is permissible here to replace the standard deviation of the normal distribution by the standard uncertainty u, since these two terms denote the same quantity (see the GUM or the next chapter for details).

When calibrating a measurement system, the variance is often known. Thus, in the following we can assume that the standard deviation σ of the data is given. This decision is mainly motivated by the fact that without such an assumption we would not be able to obtain an analytical solution to the problem, but would have to rely on numerical calculations.

According to the BAYES-LAPLACE theorem, the product of the prior probability density and the likelihood yields, under the conditions given above, the posterior probability density

$$g_\Theta(\theta \,|\, \boldsymbol{x}) = C \exp\left(-\frac{1}{2\sigma^2} \sum_{i=1}^{n} (x_i - \theta)^2 - \frac{(\theta - w)^2}{2u^2}\right),$$

where C denotes the normalization constant. This equation can be transformed, with some computational effort, into

$$g_\Theta(\theta \,|\, \boldsymbol{x}) = \sqrt{\frac{nu^2 + \sigma^2}{2\pi u^2 \sigma^2}} \exp\left[-\frac{nu^2 + \sigma^2}{2u^2 \sigma^2}\left(\theta - \frac{n\bar{x}u^2 + w\sigma^2}{nu^2 + \sigma^2}\right)^2\right], \tag{5.35}$$

with the abbreviation

$$\bar{x} = \frac{1}{n} \sum_{i=1}^{n} x_i.$$

Since the posterior probability density turns out to be the density function of a normal distribution, we obtain the estimator

$$\hat{\Theta} = \frac{n\bar{X}u^2 + w\sigma^2}{nu^2 + \sigma^2}$$

of the parameter θ. This estimator is asymptotically unbiased. The value of the parameter θ is a weighted mean of the sample mean and the value stated in the calibration certificate, both weighted by their respective variances.

When calibrating a measurement system, we are normally only interested in the deviation $d = \theta - w$ from the value of the material standard.[13] The estimator of this deviation is given by the expression

$$\hat{D} = \frac{nu^2}{nu^2 + \sigma^2} \left(\overline{X} - w \right) .$$

In the case of a large sample size n we obtain the relation $\hat{D} = \overline{X} - w$, i. e. this estimator is also asymptotically unbiased.

Taking into account that u is usually much smaller than σ and n is not particularly large, $\hat{D} < \overline{X} - w$ follows, i. e. as a general rule systematic measurement errors are overestimated when calibrating a measurement system.

We now come to another important method, introduced by H. RAIFFA and R. SCHLAIFER [87] in conjunction with BAYESian decision theory, which can be used to construct a prior probability density function. Such a prior is usually called a conjugated prior, because the method always yields a posterior that belongs strictly to the same family of probability density functions as the prior itself. Prior and posterior probability are therefore called conjugated probability densities with respect to the likelihood.

In addition to the parameters to be estimated—which are always considered as random quantities in BAYESian statistics—a conjugated probability density depends on further parameters, commonly called hyperparameters. However, these parameters are *not* random quantities, but rather serve as a more detailed specification of the prior probability density. In general, there are at least two hyperparameters for each of the parameters to be estimated. In this way, the prior can be very flexibly adapted to the available information.

In the following, we will demonstrate the use of conjugated prior probability density functions using exactly the same examples that were used to demonstrate the use of the maximum likelihood method, in order to allow for a direct comparison of these methods.

Example 5.12 (Continuation of example 5.5)
We again take the probability density function

$$g_X(x \mid \theta) = \begin{cases} \dfrac{1}{\theta} e^{-x/\theta} & \text{if } x \geq 0 \\ 0 & \text{otherwise} \end{cases} , \qquad \theta > 0 ,$$

---

13 This is the usual way to determine a systematic deviation.

for granted. This yields the likelihood

$$
l(\theta \mid \boldsymbol{x}) = \begin{cases} \dfrac{1}{\theta^n} \exp\left(-\dfrac{1}{\theta} \sum_{i=1}^{n} x_i\right) & \text{if } \theta > 0 \\ 0 & \text{otherwise} \end{cases}.
$$

As conjugated prior probability density of θ, we choose

$$
g_\Theta(\theta; \alpha, \beta) = \begin{cases} \dfrac{\beta^\alpha}{\Gamma(\alpha)} \theta^{-(1+\alpha)} e^{-\beta/\theta} & \text{if } \theta > 0 \\ 0 & \text{otherwise} \end{cases}, \qquad \alpha > 0, \quad \beta > 0.
$$

This is an inverse gamma distribution. According to the Bayes-Laplace theorem the product of the prior probability density and the likelihood yields the posterior probability density

$$
g_\Theta(\theta \mid \boldsymbol{x}) = \begin{cases} \dfrac{\tilde{\beta}^{\alpha+n}}{\Gamma(\alpha + n)} \theta^{-(1+\alpha+n)} e^{-\tilde{\beta}/\theta} & \text{if } \theta > 0 \\ 0 & \text{otherwise} \end{cases}, \qquad \alpha > 0, \quad \beta > 0,
$$

with

$$
\tilde{\beta} = \beta + \sum_{i=1}^{n} x_i.
$$

Thus, the posterior probability density is again an inverse gamma distribution.

Using this posterior probability density, we obtain according to the definition 5.18 the Bayesian estimator

$$
\hat{\Theta} = \frac{\beta + n\overline{X}}{\alpha + n - 1}
$$

of the parameter θ. The expectation of this estimator is given by

$$
\mathsf{E}[\hat{\Theta}] = \frac{\beta + n\theta}{\alpha + n - 1},
$$

i. e. it is asymptotically unbiased.

Example 5.13 (Continuation of example 5.6)

We again start with the relation

$$
\mathsf{P}(X = x \mid \theta) = \theta^x (1 - \theta)^{1-x}, \qquad 0 \le \theta \le 1,
$$

which describes the probability of a Bernoulli experiment. This leads to the likelihood

$$
l(\theta \mid \boldsymbol{x}) = \theta^{n\overline{x}} (1 - \theta)^{n - n\overline{x}}, \qquad 0 \le \theta \le 1,
$$

with

$$
\overline{x} = \frac{1}{n} \sum_{i=1}^{n} x_i.
$$

As the conjugated prior probability density of θ, we choose

$$g_\Theta(\theta; \alpha, \beta) = \frac{\Gamma(\alpha + \beta)}{\Gamma(\alpha)\Gamma(\beta)} \theta^{\alpha-1}(1 - \theta)^{\beta-1}, \qquad 0 \leq \theta \leq 1.$$

This is a beta distribution. According to the BAYES-LAPLACE theorem the product of the prior probability density and the likelihood yields the posterior probability density

$$g_\Theta(\theta \mid x) = \frac{\Gamma(n + \alpha + \beta)}{\Gamma(n\bar{x} + \alpha)\Gamma(n - n\bar{x} + \beta)} \theta^{n\bar{x}+\alpha-1}(1 - \theta)^{n-n\bar{x}+\beta-1},$$

$$0 \leq \theta \leq 1.$$

Thus, the posterior probability density is again a beta distribution.

Using this posterior probability density we obtain according to the definition 5.18 the BAYESian estimator

$$\hat{\Theta} = \frac{\alpha + n\bar{X}}{n + \alpha + \beta}$$

of the parameter θ. The expectation of this estimator is given by

$$\mathsf{E}[\hat{\Theta}] = \frac{\alpha + n\theta}{n + \alpha + \beta},$$

or equivalently by

$$\mathsf{E}[\hat{\Theta}] = \frac{\alpha + k}{n + \alpha + \beta},$$

if we denote by k the number of successes during a BERNOULLI experiment with n trials. This shows that the estimator is asymptotically unbiased.

To get an unbiased estimator, we would need $\alpha = \beta = 0$. In this case we would get the prior probability density suggested by J. B. S. HALDANE [88], which leads to the posterior probability density

$$g_\Theta(\theta \mid x) = \frac{(n - 1)!}{(k - 1)!(n - k - 1)!} \theta^{k-1}(1 - \theta)^{n-k-1}, \qquad 0 \leq \theta \leq 1.$$

This posterior cannot be normalized in the case of $k = 0$ and $k = n$, i.e. these definitely possible cases have to be excluded. This restriction does not favour a practical application of the prior probability density of J. B. S. HALDANE.

Example 5.14 (Continuation of example 5.7)
We again use the probability density function

$$g_X(x \mid \theta_1, \theta_2) = \frac{1}{\sqrt{2\pi\theta_2}} \exp\left(-\frac{(x - \theta_1)^2}{2\theta_2}\right)$$

of a normal distribution, where θ_1 denotes the mean and θ_2 denotes the variance. This leads to the likelihood

$$l(\theta_1, \theta_2 \mid \boldsymbol{x}) = \frac{1}{(2\pi\theta_2)^{n/2}} \exp\left(-\frac{(n-1)s^2 + n(\theta_1 - \overline{x})^2}{2\theta_2}\right),$$

with

$$\overline{x} = \frac{1}{n} \sum_{i=1}^{n} x_i$$

and

$$s^2 = \frac{1}{n-1} \sum_{i=1}^{n} (x_i - \overline{x})^2.$$

As the conjugated prior probability density, we choose the density of a so-called normal gamma distribution given by

$$g_{\theta_1, \theta_2}(\theta_1, \theta_2; \lambda, \nu, \alpha, \beta) =$$

$$\sqrt{\frac{\nu}{2\pi}} \frac{\beta^\alpha}{\Gamma(\alpha)} \theta_2^{-(2\alpha+3)/2} \exp\left(-\frac{2\beta + \nu(\theta_1 - \lambda)^2}{2\theta_2}\right), \qquad \alpha > 0, \quad \beta > 0.$$

This probability density function is the product of a normal distribution with respect to the parameter θ_1 and a inverse gamma distribution with respect to the parameter θ_2. The choice of this particular prior probability density is understandable because the likelihood is already proportional to the probability density function of a normal gamma distribution. A comparison with the likelihood shows that the purpose of the parameters λ and ν is to describe the prior knowledge of the mean, and that of the parameters α and β is to describe the prior knowledge of the variance.

Using this posterior probability density we obtain according to the definition 5.18 the Bayesian estimator

$$g_{\theta_1, \theta_2}(\theta_1, \theta_2 \mid \boldsymbol{x}) =$$

$$\sqrt{\frac{\nu + n}{2\pi}} \frac{\tilde{\beta}^{(n+2\alpha)/2}}{\Gamma\left(\frac{n}{2} + \alpha\right)} \theta_2^{-(n+2\alpha+3)/2} \exp\left(-\frac{2\tilde{\beta} + (\nu + n)\left(\theta_1 - \tilde{\lambda}\right)^2}{2\theta_2}\right),$$

with the abbreviations

$$\tilde{\beta} = \beta + \frac{n\nu}{\nu + n} \frac{(\lambda - \overline{x})^2}{2} + \frac{(n-1)s^2}{2}$$

and

$$\tilde{\lambda} = \frac{\nu\lambda + n\overline{x}}{\nu + n}.$$

Thus, the posterior probability density is again a normal gamma distribution.

Marginalizing the posterior probability density yields the marginal distribution functions

$$g_{\Theta_1}(\theta_1 \mid \boldsymbol{x}) = \sqrt{\frac{\nu + n}{2\pi\tilde{\beta}}} \frac{\Gamma\left(\frac{n+1}{2} + \alpha\right)}{\Gamma\left(\frac{n}{2} + \alpha\right)} \left(1 + \frac{(\nu + n)(\theta_1 - \tilde{\lambda})^2}{2\tilde{\beta}}\right)^{-(n+2\alpha+1)/2}$$

and

$$g_{\Theta_2}(\theta_2 \mid \boldsymbol{x}) = \frac{\tilde{\beta}^{(n+2\alpha)/2}}{\Gamma\left(\frac{n}{2} + \alpha\right)} \theta_2^{-(n+2\alpha+2)/2} e^{-\tilde{\beta}/\theta_2}$$

of the parameters θ_1 and θ_2. Using these results we obtain, according to the definition 5.18, the BAYESian estimators

$$\hat{\Theta}_1 = \frac{\nu\lambda + n\overline{X}}{\nu + n}$$

and

$$\hat{\Theta}_2 = \frac{1}{2\alpha + n - 2}\left(2\beta + \frac{n\nu}{\nu + n}(\lambda - \overline{X})^2 + (n - 1)S^2\right) \tag{5.36}$$

of the parameters θ_1 and θ_2. Both estimators are asymptotically unbiased.

This example shows clearly that the application of the BAYESian method can be considered as a kind of learning process, because the results—a combination of prior knowledge and the information obtained from the measurement—can be interpreted very well in this case. The estimator $\hat{\Theta}_1$ in this example shows that the resulting mean is a weighted average of the mean λ known before the measurement and the mean $\overline{x}$ obtained from the measured data, where the weights are given by the respective sample sizes ν and n. This is obviously a generalization of the equation (5.1) for an actualization of the mean value (update) which would be obtained in the case of $\nu = 1$. As the sample size n increases, the prior knowledge becomes less important compared to the information provided by the measurement. In the case of a small sample size, however, our prior knowledge becomes essential and will prevent us from drawing potentially incorrect conclusions from a result based on only a few measurements. On the other hand, if our prior knowledge is confirmed by the measurement result, there will be no change in the mean. If this is not the case, it is certainly reasonable to give greater weight to the measured values only if they are confirmed by many repetitions under the same or similar conditions.

The estimator $\hat{\Theta}_2$ in example 5.14 can be considered as a generalization of the equation (5.2) for an actualization of the variance that would be obtained in the case of $\alpha = 1$, $\beta = 0$ and $\nu = 1$. It is obvious that as the sample size n increases, the variance already known before the measurement, which in this example is proportional to the parameter β, becomes less important compared to the information about the variance obtained by the measurement. Since the estimator $\hat{\Theta}_2$ is asymptotically unbiased, it is guaranteed to converge to the variance of the distribution in the limit of an infinite sample size, but this convergence is not necessarily uniform. This is due to the middle term in the equation (5.36), which is proportional to the squared deviation of the mean obtained by the measurement from the mean already known. If this deviation is large, the variance may be increased and not, as we would normally expect, reduced by the information obtained from the measurement. In such a case, the discrepancy between the prior knowledge and the measurement result increases the uncertainty. This is usually a motivation to carry out further investigations to explain this unexpected deviation. Learning does not always mean reducing our uncertainty, but in certain cases it can also give rise to doubt.

To conclude this section, we will again consider sampling from a population with an underlying uniform (rectangular) distribution. This time, however, the aim is to show how to deal with uncertain boundaries of this distribution by using a conjugated prior probability density.

Example 5.15 (Continuation of example 5.8)
We start with the probability density function

$$g_X(x \mid \theta_1, \theta_2) = \begin{cases} \dfrac{1}{\theta_2 - \theta_1} & \text{if } \theta_1 \leq x \leq \theta_2 \\ 0 & \text{otherwise} \end{cases}$$

of a rectangular distribution. This leads to the likelihood

$$l(\theta_1, \theta_2 \mid \boldsymbol{x}) = \begin{cases} \dfrac{1}{(\theta_2 - \theta_1)^n} & \text{if } \theta_1 \leq x_i \leq \theta_2, \quad i = 1, \dots, n \\ 0 & \text{otherwise} \end{cases}.$$

As the conjugated prior probability density we choose a modified Pareto distribution

$$g_{\Theta_1, \Theta_2}(\theta_1, \theta_2; \alpha, \beta_1, \beta_2) = \begin{cases} \alpha(\alpha + 1)\dfrac{(\beta_2 - \beta_1)^\alpha}{(\theta_2 - \theta_1)^{\alpha+2}} & \text{if } \theta_1 < \beta_1 \text{ and } \theta_2 > \beta_2 \\ 0 & \text{otherwise} \end{cases},$$

where $\alpha > 0$ and $\beta_1 < \beta_2$ is assumed. This prior probability density describes our prior knowledge about the two boundaries θ_1 and θ_2, where β_1 and β_2 denote the boundaries known in advance. The parameter α characterizes the vagueness of our prior knowledge. In the limit $\alpha \to \infty$, the prior probability density converges to Dirac's δ-function $\delta(\theta_1 - \beta_1)\delta(\theta_2 - \beta_2)$, i. e. this limit describes the case where the boundaries of the rectangular distribution are known exactly.

According to the BAYES-LAPLACE theorem the product of the prior probability density and the likelihood yields the posterior probability density

$$g_{\Theta_1,\Theta_2}(\theta_1,\theta_2\,|\,\boldsymbol{x}) =$$

$$\begin{cases} (n+\alpha)(n+\alpha+1)\dfrac{(\gamma_2-\gamma_1)^{n+\alpha}}{(\theta_2-\theta_1)^{n+\alpha+2}} & \text{if } \theta_1 < \gamma_1 \text{ and } \theta_2 > \gamma_2 \\ 0 & \text{otherwise} \end{cases}$$

with the abbreviations

$$\gamma_1 = \min\,(\beta_1, x_{\min}) \qquad \text{and} \qquad \gamma_2 = \max\,(\beta_2, x_{\max})\,.$$

The two conditions $\theta_1 < \gamma_1$ and $\theta_2 > \gamma_2$ take into account that the constraints $\theta_1 \leq x_i \leq \theta_2$ ($i = 1, \dots, n$) exist for the measurement values, i. e. all measured values have to be within the boundaries of the rectangular distribution. If the boundaries β_1 and β_2 are too tight due to the prior knowledge, they have to be changed appropriately, such that all measured values lie within the boundaries, otherwise they remain unchanged. Consequently, the values γ_1 and γ_2 are the new bounds of the parameters θ_1 and θ_2 resulting from the information obtained from the measurement. The function value $g_{\Theta_1,\Theta_2}(\gamma_1, \gamma_2\,|\,\boldsymbol{x})$ denotes the posterior mode.

The posterior probability density is again a modified PARETO distribution. Marginalization yields the marginal distribution functions

$$g_{\Theta_1}(\theta_1\,|\,\boldsymbol{x}) = (n+\alpha)\dfrac{(\gamma_2-\gamma_1)^{n+\alpha}}{(\gamma_2-\theta_1)^{n+a+1}}\,, \qquad \theta_1 < \gamma_1\,,$$

and

$$g_{\Theta_2}(\theta_2\,|\,\boldsymbol{x}) = (n+\alpha)\dfrac{(\gamma_2-\gamma_1)^{n+\alpha}}{(\theta_2-\gamma_1)^{n+a+1}}\,, \qquad \theta_2 > \gamma_2\,,$$

of the parameters θ_1 and θ_2. Using these results we obtain, according to the definition 5.18, the BAYESian estimators

$$\hat{\Theta}_1 = \dfrac{(n+\alpha)\min\,(\beta_1, X_{\min}) - \max\,(\beta_2, X_{\max})}{n+\alpha-1}$$

and

$$\hat{\Theta}_2 = \dfrac{(n+\alpha)\max\,(\beta_2, X_{\max}) - \min\,(\beta_1, X_{\min})}{n+\alpha-1}$$

of the parameters θ_1 and θ_2. Thus, the estimator of the centre of the interval (expectation of the rectangular distribution) is given by

$$\dfrac{\hat{\Theta}_1 + \hat{\Theta}_2}{2} = \dfrac{\min\,(\beta_1, X_{\min}) + \max\,(\beta_2, X_{\max})}{2}\,.$$

This is obviously a reasonable result, taking into account prior knowledge only if it does not conflict with the measured data.

5.8 Interval and region estimation

In the previous sections we have only dealt with so-called point estimators. These estimators assign an estimated value to the parameters of the probability distribution of a population, using all the information obtained from the measurement. The result of the measurement is therefore exactly one value for each of the parameters to be estimated. However, these estimated values do not usually correspond to the values of the parameters of the population. In order to be able to make a statement about the reliability (accuracy) of the values obtained in this way, interval estimators are usually used.

Whereas a point estimator is used to find the best estimate of a measurand that is in agreement with the measured data, an interval estimator is used to construct, on the basis of the available measured data, an interval that contains the most likely[14] value of the measurand under consideration. Therefore, an interval estimator is an important extension to a point estimator. One could even argue that interval estimators alone are sufficient, and point estimators are unnecessary, but this is certainly debatable.

The interval estimator traditionally used in conventional statistics is the confidence interval introduced by J. NEYMAN [89]. This interval is defined as (see ISO 3534-1, 1.28 [38]):

Definition 5.20 (Confidence interval)
The confidence interval is the interval estimator $(\hat{T}_0, \hat{T}_1)$ of the parameter θ with the statistics $\hat{T}_0$ and $\hat{T}_1$ as interval limits, such that

$$P\left(\hat{T}_0 < \theta < \hat{T}_1\right) \geq 1 - \alpha \tag{5.37}$$

holds, where $(1 - \alpha)$ denotes the confidence level or confidence coefficient. The value α is called level of significance.

It is necessary to add some comments to this definition:
- The interval estimator is given by the two estimators of the interval boundaries $\hat{T}_0$ and $\hat{T}_1$.
- The interval boundaries $\hat{T}_0$ and $\hat{T}_1$ are *random quantities* because they are by definition statistics and thus functions of random quantities.
- The parameter θ in this definition is *not* a random quantity, but rather a constant value that is unknown.
- The interval boundaries $\hat{T}_0$ and $\hat{T}_1$ do not depend on the parameter θ, but they are functions of the confidence level $(1 - \alpha)$.

---

14 We use the term "likely" here because, as we will see later, an interval estimator does not allow a clear statement of a probability in any case.

For reasons of objectivity, the confidence level $(1 - \alpha)$ has to be chosen in advance (i. e. before any measurement data are obtained) and is therefore a known value. This value reflects the percentage of cases where, in a long series of repeated measurements (i. e. in the case of a large sample) made under identical conditions, the confidence interval contains the value of the parameter. It does *not* mean that this interval contains the value of the parameter with *probability* $(1 - \alpha)$, because it either contains the value or it does not.

In order to better understand the last statement, let us have a closer look at the underlying relationships. For any measurement, the estimated values $\hat{t}_0$ and $\hat{t}_1$ of the sampling functions $\hat{T}_0$ and $\hat{T}_1$ are known values because the interval $(\hat{t}_0, \hat{t}_1)$ represents one of the possible realizations of the confidence interval $(\hat{T}_0, \hat{T}_1)$ by the measurement. Since $\hat{T}_0$ and $\hat{T}_1$ are random quantities, the estimated values $\hat{t}_0$ and $\hat{t}_1$ will change from one measurement to the next. After each measurement, however, they are known values. But since the parameter θ is constant but unknown, we do not know whether the statement $\hat{t}_0 < \theta < \hat{t}_1$ is true or false. We only know that only *one* of these two possibilities can be true. So it is meaningless to talk about the probability $P(\hat{t}_0 < \theta < \hat{t}_1)$ that the statement $\hat{t}_0 < \theta < \hat{t}_1$ is true, because either it is or it is not true. On the other hand, we can give a probability $P(\hat{T}_0 < \theta < \hat{T}_1)$ that the statement $\hat{T}_0 < \theta < \hat{T}_1$ is true, because $\hat{T}_0$ and $\hat{T}_1$ are random quantities. If numerous measurements are made, the statement $\hat{T}_0 < \theta < \hat{T}_1$ will be true "on average" or "in the long run" about $(1 - \alpha)$ of the time.

Therefore, the probability in the relation (5.37) only tells us how often a confidence interval contains the value of the parameter θ if we have taken many samples from a population and calculated the confidence interval for each one. This probability cannot be applied to a single sample. A statement like "the value of the measurand is within the confidence interval with a probability of 95 %" is wrong. In contrast, however, the statement "95 % of a large number of confidence intervals contain the value of the measurand" corresponds to the definition of a confidence interval.

Having clarified the meaning of a confidence interval, we now turn to the calculation of this interval. Let us first consider a simple example.

Example 5.16 (Confidence interval for the mean of a normal distribution)

We start with a population based on a normal distribution with mean μ and standard deviation σ. If we take a sample of size n from a population with this distribution, the best estimator of the mean is given by the sample mean $\overline{X}$.

Since a normal distribution is symmetrical about its mean μ, it makes sense to construct the confidence interval symmetrically about the estimator $\overline{X}$ of the parameter μ, i. e. we insert into equation (5.37)

$$\hat{T}_0 = \overline{X} - U \quad \text{and} \quad \hat{T}_1 = \overline{X} + U,$$

where the quantity U must be such that, for a given confidence level $(1 - \alpha)$, the condition

$$P\left(\overline{X} - U < \mu < \overline{X} + U\right) \geq 1 - \alpha \tag{5.38}$$

is satisfied. However, since the mean μ cannot be a random quantity in conventional statistics, this condition is replaced by the equivalent condition

$$P\left(\mu - U < \overline{X} < \mu + U\right) \geq 1 - \alpha, \tag{5.39}$$

where for each sample vector $\boldsymbol{X} = (X_1, \ldots, X_n)^\mathsf{T}$ the quantity U is uniquely determined. Since the random quantities X_i ($i = 1, \ldots, n$) are all normally distributed with the mean μ and the standard deviation σ, the sample mean $\overline{X}$ is normally distributed with the mean μ and the standard deviation $\sigma/\sqrt{n}$ (see example 4.73). If we now apply the linear transformation

$$\overline{X} = \mu + \frac{\sigma}{\sqrt{n}} Z, \tag{5.40}$$

the resulting random quantity Z becomes standard normally distributed[15] (see example 4.43). If we additionally use $U = k\sigma/\sqrt{n}$, we obtain the new condition

$$P\left(-k < Z < +k\right) \geq 1 - \alpha,$$

where the underlying probability distribution function is a standard normal distribution. Because of the relation between probability and probability distribution function, this condition can also be written as

$$\frac{1}{\sqrt{2\pi}} \int_{-k}^{+k} e^{-z^2/2} dz \geq 1 - \alpha,$$

where we took advantage of the fact that the random quantity Z is normally distributed by default. If we also take into account that the density of a normal distribution is symmetric, we can transform this equation to

$$\alpha \geq \mathrm{erfc}\left(\frac{k}{\sqrt{2}}\right),$$

where

$$\mathrm{erfc}(z) = \frac{2}{\sqrt{\pi}} \int_{z}^{\infty} e^{-\xi^2} d\xi$$

defines the complementary error function [90]. Since $\mathrm{erfc}(z)$ is monotonous, its inverse function exists, so that a value k can be uniquely assigned to each value α

---

15 A random quantity is called standard normally distributed if it is distributed according to a normal distribution with mean zero and standard deviation one.

(for example, for $\alpha = 0.05$ we get the value $k \geq 1.96$). Using the smallest resulting value of k, we finally obtain the estimators

$$\hat{T}_0 = \overline{X} - k(\alpha)\frac{\sigma}{\sqrt{n}} \qquad \text{and} \qquad \hat{T}_1 = \overline{X} + k(\alpha)\frac{\sigma}{\sqrt{n}}$$

of the bounds of the confidence intervals for the mean of a normal distribution, provided that the standard deviation σ is known.

In this case, the length of the confidence interval $\hat{T}_1 - \hat{T}_0$ is independent[16] of the measured data. Apart from the significance level α, it depends only on the sample size n and the standard deviation σ of the normal distribution, and decreases with decreasing standard deviation and increasing sample size. For $\sigma = 0$ and $n \to \infty$ respectively, the confidence interval contracts to the middle of the interval given by $\overline{X}$, just as we would expect.

A closer examination of the calculations performed to obtain the confidence interval shows that an essential step in the progress was the transition from the inequality (5.38) to the inequality (5.39), because only in this way it became possible to use the probability distribution function of the sample mean $\overline{X}$ for the calculation. The double inequalities that occur as arguments of the probabilities in conditions (5.38) and (5.39) are illustrated in Fig. 5.2.

The upper part of the figure shows the unknown confidence interval centred on the mean μ, while the lower part shows the confidence intervals centred on the respective sample means $\overline{X}$ available from the measured data. It is obvious that the mean μ lies within the interval $(\overline{X} - U, \overline{X} + U)$, as required by condition (5.38), whenever the respective sample mean $\overline{X}$ lies within the interval $(\mu - U, \mu + U)$, as required by condition (5.39). However, the transformation of the double inequality in condition (5.38) into the double inequality in condition (5.39) does not generally justify considering both conditions as equivalent. In fact, this is the case in the example because the probability density function depends only on $|\overline{X} - \mu|$, but this is not necessarily true in general.

Another important step in the calculation of the confidence interval in the example 5.16 was the introduction of the quantity

$$Z = \frac{\overline{X} - \mu}{\sigma}\sqrt{n}$$

by the linear transformation (5.40). The quantity Z is a so-called pivotal quantity. A pivotal quantity is a random quantity that is a function of the observed sample vector of a measurand and the unknown and unobservable parameters, but whose probability distribution function does not depend on the parameters and is therefore fully known.

---

16 In general, both the position of the centre and the length of the confidence interval depend on the measured data.

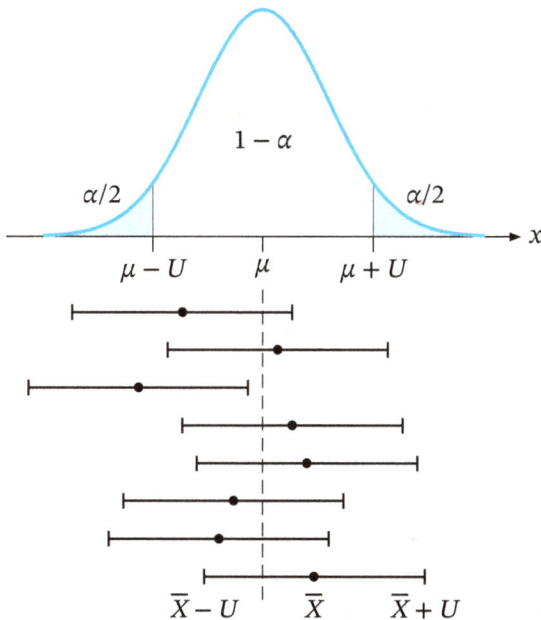

Fig. 5.2: Confidence intervals of eight different samples of the same sample size of a normally distributed random quantity X with mean μ and known standard deviation.

The method used in the example 5.16 to calculate the confidence interval is known as pivot method. This method can be used whenever a pivotal quantity can be found for the parameter and the double inequality (5.37) in the definition 5.20 of the confidence intervals can be transformed so that the pivotal quantity appears as the middle term of the double inequality.

We demonstrate the use of the pivot method for another case of practical importance by continuing the example 5.16.

Example 5.17 (Continuation of example 5.16)
When calculating the confidence interval in the example 5.16, we assumed that the standard deviation σ of the normal distribution is known. Since this is generally not the case, we will now drop this assumption.

We start with the condition (5.39). In order to replace the sample mean $\overline{X}$ by the pivotal quantity T, we apply a linear transformation this time, given by

$$\overline{X} = \mu + \frac{S}{\sqrt{n}} T \quad \text{and} \quad T = \frac{\overline{X} - \mu}{S}\sqrt{n},$$

where S denotes the standard deviation of the sample. If we now additionally use $U = kS/\sqrt{n}$, the relation (5.39) becomes

$$P(-k < T < +k) \geq 1 - \alpha. \tag{5.41}$$

Since we have assumed that the random quantity X is normally distributed, the pivotal quantity T is distributed according to STUDENT's t-distribution[17] with $(n-1)$ degrees of freedom, as proved by W. S. GOSSET. [91]. The t-distribution with ν degrees of freedom is given by the probability density function

$$g_T(t;\nu) = \frac{\Gamma\left(\frac{\nu+1}{2}\right)}{\sqrt{\nu\pi}\,\Gamma\left(\frac{\nu}{2}\right)}\left(1+\frac{t^2}{\nu}\right)^{-(\nu+1)/2}.$$

Because of the relation between probability and probability distribution function, condition (5.41) can also be written as

$$\int_{-k}^{+k} g_T(t;n-1)dt \geq 1-\alpha,$$

where $g_T(t;n-1)$ denotes the probability density function of a t-distribution with $(n-1)$ degrees of freedom.

Since every probability distribution function is monotone, there exists its inverse function, such that to each value α a value k can be uniquely assigned (for example, for $\alpha = 0.05$ and $n = 10$ we obtain the value $k \geq 2.262$). Using the smallest resulting value of k, we finally obtain the estimators

$$\hat{T}_0 = \overline{X} - k(\alpha, n-1)\sqrt{\frac{S^2}{n}} \quad \text{and} \quad \hat{T}_1 = \overline{X} + k(\alpha, n-1)\sqrt{\frac{S^2}{n}}$$

of the boundaries of the confidence intervals for the mean of a normal distribution when the standard deviation σ is unknown.

In this case, apart from the significance level α and the sample size n, the length of the confidence interval also depends on the sample variance S^2 and decreases with increasing sample size.

The dependence of the length of the confidence interval on the sample variance implies that a large dispersion of the measurements will lead to a large confidence interval. It also means that the length of the confidence interval is not constant, as in the previous example where the standard deviation σ was assumed to be known, but is generally different for each sample.

This concludes our discussion of the confidence interval. Hopefully it has become clear to the reader where the difficulties with this interval estimator lie. The main problem can be summarized by the following statement: A confidence interval can be thought

---

[17] The so-called STUDENT's t-distribution was first derived by W. S. GOSSET, who had to publish under the pseudonym "Student" to conceal the fact that his employer, the Guinness brewery, was already using statistical methods for quality control at the time he was writing.

of as an interval produced by a random process which in $(1 - \alpha)$ of all cases contains the value of the parameter of interest and in α of all cases does not. But we do not know (and will never be able to know) which of the two cases applies to a confidence interval associated with our measurement. This is an unsatisfactory situation and makes the confidence interval unsuitable for metrological applications.

We now turn to the term "coverage interval". The coverage interval is based on BAYESian statistics. In metrology, it has replaced the confidence interval. The *Guide to the Expression of Uncertainty in Measurement (GUM)* does not yet contain the term "coverage interval", only the term "statistical coverage interval" is mentioned (GUM, C.2.30 [5, 4]). However, the latter term has the meaning of statistical tolerance interval (ISO 3534-1, 1.26 [38]), which is different from both the coverage interval and the confidence interval. The GUM rejects the use of the terms "confidence interval" and "confidence level" as inappropriate, but does not clearly state which terms should be used instead (GUM, 6.2.2 [5, 4]). This only happens in the Suppl. 1 of the GUM [54, 55], where the terms "coverage interval" and "coverage probability" are introduced. The coverage interval is defined as (GUM Suppl. 1, 3.12) [54, 55]:

interval containing the value of a quantity with a stated probability, based on the information available

and the coverage probability as (GUM Suppl. 1, 3.13) [54, 55]:

probability that the value of a quantity is contained within a specified coverage interval

The *International Vocabulary of Metrology—Basic and General Concepts and Associated Terms (VIM)* also contains a definition of the coverage interval (VIM:2008, 2.36 [3, 2, 1]), which is essentially the same as that given in Suppl. 1 of the GUM.

From the two definitions given in the Suppl. 1 of the GUM and in the VIM, the following formal definition of the coverage interval can be derived:

Definition 5.21 (Coverage interval)
The coverage interval is an interval (t_0, t_1) of the parameter Θ, such that

$$P(t_0 < \Theta < t_1) = \int_{t_0}^{t_1} g_\Theta(\theta \,|\, \boldsymbol{x}) \, d\theta \geq 1 - \alpha \qquad (5.42)$$

holds, where $g_\Theta(\theta \,|\, \boldsymbol{x})$ denotes the posterior probability density of the parameter Θ given the data $\boldsymbol{x} = (x_1, \ldots, x_n)^\mathsf{T}$. The value $(1 - \alpha)$ is called coverage probability.

This definition shows that for the coverage interval, in contrast to the confidence interval, the interval boundaries t_0 and t_1 are not random quantities, but rather constant values, which of course have to be determined from the respective measurement result. On the other hand—according to BAYESian statistics—the parameter θ is now considered as a random quantity, because we do not have sufficient information about its value. Additional information can only be obtained from the measurement. Once the measured values are known, the corresponding probability density function (the posterior probability density) of the parameter can be determined, and thus its expectation, which is its best estimate, can be calculated. Subsequently, the coverage interval can be determined using the *same* probability density function.

The coverage interval can be characterized as an interval containing all values that with a given probability can reasonably be attributed to a measurand. This probability statement is valid for any *single* measurement and does not require a frequency interpretation as is required for the confidence interval. The coverage interval thus corresponds to an interval that most metrologists intuitively associate with the term "confidence interval", namely that it allows a statement to be made about the reliability of the value of a measurand. However, it is important to be aware that there are essential differences between the confidence interval of orthodox statistics and the coverage interval, which are by no means purely philosophical, as is often claimed. Unfortunately, the definitions of the coverage interval as given in the Suppl. 1 of the GUM or in the VIM are not very helpful, because the phrase "... *based on the information available*" gives no reference to the posterior probability density and could well be misinterpreted in the sense of orthodox statistics as information based solely on the measurement.

We now give an example of how a coverage interval can be calculated. In order to allow a direct comparison with the results obtained for the confidence interval in the examples above, we calculate the coverage interval again for the mean of a normal distribution.

Example 5.18 (Coverage interval for the mean of a normal distribution)

In the example 5.10 we considered the case where a random quantity X is measured several times under repeatability conditions, assuming no prior knowledge of the measurand. The application of BAYESian statistics yielded the probability density function

$$g_M(\mu \mid x) = \frac{\Gamma\left(\frac{n}{2}\right)}{\Gamma\left(\frac{n-1}{2}\right)} \sqrt{\frac{n}{\pi(n-1)s^2}} \left(1 + \frac{n(\mu - \overline{x})^2}{(n-1)s^2}\right)^{-n/2} \tag{5.43}$$

for the mean μ of a normal distribution with empirical mean

$$\overline{x} = \frac{1}{n} \sum_{i=1}^{n} x_i$$

and the empirical standard deviation

$$s^2 = \frac{1}{n-1} \sum_{i=1}^{n} (x_i - \overline{x})^2$$

of the measured values $(x_1, \ldots, x_n)$. Using this probability density function, we obtain under the constraint[18] $n > 2$

$$E[\mu] = \bar{x}$$

as the best estimate of the mean μ of the normal distribution.

In order to calculate the coverage interval of the mean μ of the normal distribution, we have to determine—according to the definition 5.21—the interval boundaries t_0 and t_1 from the condition

$$P(t_0 < \mu < t_1) = \int_{t_0}^{t_1} g_M(\mu \mid \boldsymbol{x})\,d\mu = 1 - \alpha, \qquad (5.44)$$

where the coverage probability $(1 - \alpha)$ is given. For comparability with the calculation of the confidence interval, we choose interval boundaries that are symmetrical to the best estimate $\bar{x}$. Therefore, we set

$$t_0 = \bar{x} - k\sqrt{\frac{s^2}{n}} \quad \text{and} \quad t_1 = \bar{x} + k\sqrt{\frac{s^2}{n}}.$$

If we use these interval boundaries, insert the probability density function (5.43) into the condition (5.44) and apply the substitutions

$$t = \frac{\mu - \bar{x}}{s}\sqrt{n} \quad \text{and} \quad \mu = \bar{x} + \frac{s}{\sqrt{n}}t$$

we obtain the condition

$$\frac{\Gamma\!\left(\frac{n}{2}\right)}{\Gamma\!\left(\frac{n-1}{2}\right)} \frac{1}{\sqrt{(n-1)\pi}} \int_{-k}^{+k} \left(1 + \frac{t^2}{n-1}\right)^{-n/2} dt = 1 - \alpha$$

for the value k. It shows that for a given coverage probability $(1 - \alpha)$, the value k is determined by a t-distribution with $(n - 1)$ degrees of freedom. This gives us the interval estimator

$$\hat{T}_0 = \bar{X} - k(\alpha, n - 1)\sqrt{\frac{S^2}{n}} \quad \text{and} \quad \hat{T}_1 = \bar{X} + k(\alpha, n - 1)\sqrt{\frac{S^2}{n}}$$

for the boundaries of the coverage interval. This result is in total agreement with the result obtained for the confidence interval in example 5.17.

---

18 In the case of $n \leq 2$ the expectation does not exist.

The result of this example, that there is no *numerical* difference between the boundaries of the confidence interval and those of the coverage interval when both intervals are calculated for the mean of a normal distribution, appears *prima facie* unexpected. It can be shown (see E. T. JAYNES [92]) that under certain circumstances the confidence interval and the coverage interval can coincide for *strictly mathematical* reasons. However, this should not lead to the conclusion that confidence interval and coverage interval are the same anyway, because there are a sufficient number of cases where the two intervals do not coincide, or where the confidence interval is obviously unreasonable while the coverage interval can be considered meaningful [92].

Confidence interval and coverage interval will generally not coincide for the simple reason that the calculation of the coverage interval is based on the posterior probability density, whereas the calculation of the confidence interval uses only the probability density function of the sample (i. e. only the information contained in the measured data, represented by a frequency distribution). This often results—under otherwise identical conditions—in a coverage interval being shorter than the confidence interval, because more information is available due to the prior probability density, which generally leads to a smaller uncertainty. We illustrate this with an example.

Example 5.19 (Coverage interval for a measurement system calibration)
In the example 5.11 we considered the calibration of a measurement system using a material measure whose value w and associated standard uncertainty u could be taken from its calibration certificate. From the equation (5.35) we obtain the probability density function

$$g_D(d \mid x) = \sqrt{\frac{nu^2 + \sigma^2}{2\pi u^2 \sigma^2}} \exp\left[-\frac{nu^2 + \sigma^2}{2u^2\sigma^2}\left(d - \frac{nu^2(\bar{x} - w)}{nu^2 + \sigma^2}\right)^2\right]$$

for the deviation $d = \theta - w$ of the value of the material measure, where

$$\bar{x} = \frac{1}{n}\sum_{i=1}^{n} x_i$$

denotes the empirical mean of the measured values and σ the standard deviation associated with the measurements performed during the calibration, which is assumed to be known. Using this probability density function yields

$$\mathsf{E}[D] = \frac{nu^2}{nu^2 + \sigma^2}\left(\bar{x} - w\right)$$

and

$$\mathrm{Var}(D) = \frac{u^2\sigma^2}{nu^2 + \sigma^2}$$

as the best estimate of the deviation of the value of the material measure and the associated variance.

In addition, we are now also interested in the coverage interval of the deviation d. For this purpose—according to the definition 5.21—we must determine the interval boundaries t_0 and t_1 from the condition

$$P\left(t_0 < d < t_1\right) = \int_{t_0}^{t_1} g_D(d \,|\, \boldsymbol{x}) \, d\mu = 1 - \alpha,$$

where the coverage probability $(1 - \alpha)$ is given. Again, if we require a coverage interval that is symmetric with respect to the expectation, we can choose

$$t_0 = \frac{nu^2(\bar{x} - w)}{nu^2 + \sigma^2} - k \frac{u\sigma}{\sqrt{nu^2 + \sigma^2}}$$

and

$$t_1 = \frac{nu^2(\bar{x} - w)}{nu^2 + \sigma^2} + k \frac{u\sigma}{\sqrt{nu^2 + \sigma^2}}.$$

If we use these interval boundaries, insert the probability density function (5.43) into the condition (5.44) and use the substitutions

$$t = \frac{\sqrt{nu^2 + \sigma^2}}{u\sigma}\left(d - \frac{nu^2(\bar{x} - w)}{nu^2 + \sigma^2}\right)$$

and

$$d = \frac{nu^2(\bar{x} - w)}{nu^2 + \sigma^2} + \frac{u\sigma}{\sqrt{nu^2 + \sigma^2}}\, t,$$

we obtain the condition

$$\frac{1}{\sqrt{2\pi}} \int_{t_0}^{t_1} e^{-t^2/2} dt = 1 - \alpha$$

for the value k. This condition is equivalent to the relation

$$\alpha = \operatorname{erfc}\left(\frac{k}{\sqrt{2}}\right),$$

which allows us to calculate the value k for a given value α.

Thus, we have a coverage interval of length

$$t_1 - t_0 = k(\alpha)\frac{2u\sigma}{\sqrt{nu^2 + \sigma^2}},$$

which is independent of the measured values. This interval is by the factor

$$\sqrt{\frac{nu^2}{nu^2 + \sigma^2}}$$

shorter than the confidence interval calculated for the same problem. In the limit $n \to \infty$ this factor approaches one, i. e. for a large sample size $n \gg (\sigma/u)^2$ prior knowledge is practically irrelevant, and the information obtained from the measurement becomes dominant.

We now turn to another aspect to consider when calculating coverage intervals. In the previous two examples, we made the decision to construct the coverage interval symmetrically with respect to the expectation of the parameter. However, we could have made any other decision about the interval boundaries. So we can conclude that the definition 5.21 does not uniquely determine the coverage interval. Obviously, more than one coverage interval is possible for a given coverage probability. Suppl. 1 of the GUM gives two possible specifications of the coverage interval, the *probabilistic symmetric coverage interval* and the *shortest coverage interval*.

The probabilistic symmetric coverage interval is the coverage interval for a quantity such that the probability that the quantity is less than the smallest value in the interval is equal to the probability that the quantity is greater than the largest value in the interval (GUM Supplement 1, 3.15 [54, 55]). This corresponds to the formal definition:

Definition 5.22 (Probabilistic symmetric coverage interval)
The probabilistic symmetric coverage interval is an interval (t_0, t_1) of the parameter Θ such that

$$P(\Theta < t_0) = \int_{-\infty}^{t_0} g_\Theta(\theta \,|\, x) \, d\theta \leq \frac{\alpha}{2}$$

and

$$P(\Theta > t_1) = \int_{t_1}^{\infty} g_\Theta(\theta \,|\, x) \, d\theta \leq \frac{\alpha}{2}$$

where $g_\Theta(\theta \,|\, x)$ denotes the posterior probability density of the parameter Θ given the data $x = (x_1, \ldots, x_n)^\mathsf{T}$ and $(1 - \alpha)$ the coverage probability.

Of course, this definition corresponds to the definition 5.21 of the coverage interval. However, unlike the latter definition, it gives separate conditions for the interval boundaries t_0 and t_1, such that the coverage interval becomes unambiguous. For a symmetric and unimodal[19] posterior probability density (like e. g. a normal distribution), the probab-

---

19 A probability density function is called unimodal if it has only one maximum, bimodal if it has two maxima, and multimodal if it has many maxima.

ilistic symmetric coverage interval corresponds to the expanded uncertainty[20] of the GUM.

The shortest coverage interval is defined as the coverage interval for a quantity with the shortest length among all coverage intervals for that quantity with the same coverage probability (GUM Supplement 1, 3.16 [54, 55]). In general, we are interested in the shortest possible coverage interval because, for a given coverage probability, a shorter coverage interval obviously corresponds to a higher reliability of the estimated value of a measurand.

The question now is how to determine the shortest possible coverage interval. We will deal with this problem in a more general way, i. e. not for a single parameter, but for a parameter vector. In this case we can no longer speak of an interval, but of a region. The smallest possible region that contains the values of a parameter vector with a given probability is called its credible region [93]. The shortest possible coverage interval is thus a special case for a single parameter.

We start with the requirement that the credible region must correspond to a given probability. The mathematical specification of this requirement is analogous to the definition 5.21

$$P(\Theta \in \mathfrak{B}) = \int_{\mathfrak{B}} g_{\Theta}(\theta \,|\, x)\,d\theta \geq 1 - \alpha, \tag{5.45}$$

where $\mathfrak{B}$ denotes the credible region for a given probability $(1 - \alpha)$ and $g_{\Theta}(\theta \,|\, x)$ the posterior probability density of the parameter vector $\Theta = (\Theta_1, \dots, \theta_m)^{\mathsf{T}}$ given the data $x = (x_1, \dots, x_n)^{\mathsf{T}}$. The m-dimensional integral is to be understood in the sense of LE-BESGUE[21] and applies to the region $\mathfrak{B}$. The relation (5.45) can be seen as a multivariate generalization of the relation (5.42). The region $\mathfrak{B}$ defined by this relation is the credible region for a given probability $(1 - \alpha)$.

In general, there are many regions $\mathfrak{B}$ which satisfy the relation (5.45). However, we are only interested in those regions which contain the most probable values of the parameter vector Θ. This requirement can be formulated mathematically by the expression

$$\mathfrak{B} = \left\{ \theta \,|\, g_{\Theta}(\theta \,|\, x) \geq K \right\}, \tag{5.46}$$

i. e. $\mathfrak{B}$ contains all values that do not fall below a certain threshold K of the probability density. The function $g_{\Theta}(\theta \,|\, x) = K$ defines a (not necessarily connected) hyper-surface within the m-dimensional parameter space, which separates the values belonging to $\mathfrak{B}$ from those values which do not belong to $\mathfrak{B}$, but to the complement $\mathfrak{B}^c$. A region

---

20 This is a very unfortunate and clumsy name, because this value is actually *not* an uncertainty, but rather half the length of a symmetric coverage interval centred on the best estimate. This essential difference often leads to misunderstandings.

21 The class of functions which are integrable in the sense of LEBESGUE is wider than that of ordinary integrable functions (according to RIEMANN) and allows, for example, the use of the δ-function of DIRAC inside the integrand, which is not allowed in the case of ordinary integrals.

determined by the equation (5.46) is called an HPD region (highest probability density region) [94].

We will now prove that an HPD region is also the smallest possible region. For this we assume that besides the HPD region $\mathcal{B}$ there is another region $\overline{\mathcal{B}}$ which also satisfies the condition (5.45). We do not exclude the possibility that both regions overlap, so we perform a partition of the region $(\mathcal{B} \cup \overline{\mathcal{B}})$ into three disjoint parts $(\mathcal{B} \setminus \overline{\mathcal{B}})$, $(\mathcal{B} \cap \overline{\mathcal{B}})$ and $(\overline{\mathcal{B}} \setminus \mathcal{B})$, as shown in Fig. 5.3 by a VENN diagram.

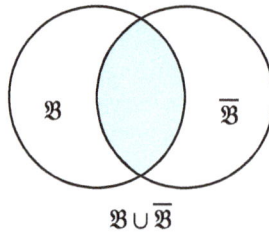

$$\mathcal{B} \cup \overline{\mathcal{B}}$$

Fig. 5.3: Illustration of the relations between sets: The left circle represents the set $\mathcal{B}$, the right circle the set $\overline{\mathcal{B}}$. The set intersection $(\mathcal{B} \cap \overline{\mathcal{B}})$ is represented by a grey area, the set difference $(\mathcal{B} \setminus \overline{\mathcal{B}})$ by the left white area and the set difference $(\overline{\mathcal{B}} \setminus \mathcal{B})$ by the right white area.

Obviously the equations $\mathcal{B} = (\mathcal{B} \setminus \overline{\mathcal{B}}) \cup (\mathcal{B} \cap \overline{\mathcal{B}})$ and $\overline{\mathcal{B}} = (\overline{\mathcal{B}} \setminus \mathcal{B}) \cup (\mathcal{B} \cap \overline{\mathcal{B}})$ are valid. Since the two regions are disjoint, the addition theorem of probability theory yields

$$P\left(\boldsymbol{\Theta} \in \mathcal{B}\right) = P\left(\boldsymbol{\Theta} \in (\mathcal{B} \setminus \overline{\mathcal{B}})\right) + P\left(\boldsymbol{\Theta} \in (\mathcal{B} \cap \overline{\mathcal{B}})\right)$$

and

$$P\left(\boldsymbol{\Theta} \in \overline{\mathcal{B}}\right) = P\left(\boldsymbol{\Theta} \in (\overline{\mathcal{B}} \setminus \mathcal{B})\right) + P\left(\boldsymbol{\Theta} \in (\mathcal{B} \cap \overline{\mathcal{B}})\right) .$$

Consequently, due to the assumption $P\left(\boldsymbol{\Theta} \in \mathcal{B}\right) = P\left(\boldsymbol{\Theta} \in \overline{\mathcal{B}}\right)$ we obtain

$$P\left(\boldsymbol{\Theta} \in (\mathcal{B} \setminus \overline{\mathcal{B}})\right) = P\left(\boldsymbol{\Theta} \in (\overline{\mathcal{B}} \setminus \mathcal{B})\right) ,$$

or rather

$$\int_{\mathcal{B} \setminus \overline{\mathcal{B}}} g_{\boldsymbol{\Theta}}(\boldsymbol{\theta} \,|\, \boldsymbol{x}) \, d\boldsymbol{\theta} = \int_{\overline{\mathcal{B}} \setminus \mathcal{B}} g_{\boldsymbol{\Theta}}(\boldsymbol{\theta} \,|\, \boldsymbol{x}) \, d\boldsymbol{\theta} . \tag{5.47}$$

But we have $\mathcal{B} \setminus \overline{\mathcal{B}} \subseteq \mathcal{B}$ and $\overline{\mathcal{B}} \setminus \mathcal{B} \subseteq \mathcal{B}^c$, i.e. due to equation (5.46) $g_{\boldsymbol{\Theta}}(\boldsymbol{\theta} \,|\, \boldsymbol{x}) \geq K$ is valid for all $\boldsymbol{\theta} \in \mathcal{B} \setminus \overline{\mathcal{B}}$ and $g_{\boldsymbol{\Theta}}(\boldsymbol{\theta} \,|\, \boldsymbol{x}) < K$ is valid for all $\boldsymbol{\theta} \in \overline{\mathcal{B}} \setminus \mathcal{B}$. Thus, the left integrand in equation (5.47) is greater than the right integrand. The values of these two integrals, however, are equal and $g_{\boldsymbol{\Theta}}(\boldsymbol{\theta} \,|\, \boldsymbol{x})$— like any probability density function— is non-negative. Therefore, $(\mathcal{B} \setminus \overline{\mathcal{B}})$ must be less than $(\overline{\mathcal{B}} \setminus \mathcal{B})$ and therefore, due to the equations $\mathcal{B} = (\mathcal{B} \setminus \overline{\mathcal{B}}) \cup (\mathcal{B} \cap \overline{\mathcal{B}})$ and $\overline{\mathcal{B}} = (\overline{\mathcal{B}} \setminus \mathcal{B}) \cup (\mathcal{B} \cap \overline{\mathcal{B}})$, it follows that $\mathcal{B}$ is less than $\overline{\mathcal{B}}$. But the region $\overline{\mathcal{B}}$ was chosen arbitrarily, i. e. the region $\mathcal{B}$ is the smallest possible.

The question now is whether the region $\mathfrak{B}$ is already completely determined by the relations (5.45) and (5.46). Unfortunately the answer is no, because ironically the simplest probability density function, namely the density of the uniform distribution, does not allow a unique determination of $\mathfrak{B}$. This statement can be understood as follows. When

$$\mu(\mathfrak{G}) = \int_{\mathfrak{G}} d\theta$$

denotes the content (length, area, volume, etc.) of an m-dimensional region $\mathfrak{G}$, the m-dimensional uniform distribution is given by the probability density function

$$g_\Theta(\theta \,|\, x) = \begin{cases} \dfrac{1}{\mu(\Omega)} & \text{for } \theta \in \Omega \\ 0 & \text{otherwise} \end{cases}, \tag{5.48}$$

where $\Omega \subseteq \mathbb{R}^m$ denotes the region of admissible values of the parameter vector Θ. This expression can be verified immediately by inserting it into the normalization condition

$$\int_{\mathbb{R}^m} g_\Theta(\theta \,|\, x)\, d\theta = 1,$$

which holds for any m-dimensional probability density function. If we now insert equation (5.48) into equation (5.45), we obtain

$$\mu(\mathfrak{B}) = (1 - \alpha)\,\mu(\Omega)$$

and therefore $\mu(\mathfrak{B}) \leq \mu(\Omega)$. Thus, only the content $\mu(\mathfrak{B})$ of the region $\mathfrak{B}$ is determined, but not its location. Therefore, we need another condition to specify the location of $\mathfrak{B}$. In the case of a uniform distribution this can only be—due to the relation (5.45)—the condition $\mathfrak{B} = \Omega$. This equation is easiest to understand by looking at the one-dimensional case. The shortest coverage interval cannot be uniquely stated for the rectangular (uniform) distribution of a parameter, because all parameter values within the boundaries of the rectangular density function have the same probability of occurring. Any restriction to a smaller interval would give undue preference to a particular parameter value and thus conflict with a uniform distribution of values.

In addition to the location, the shape of the region $\mathfrak{B}$ cannot be uniquely determined for a uniform distribution. The relation (5.46) does not help here, because in the case of a uniform distribution the probability density only takes on a constant value, by which the value of K is already fixed. In order to visualize the situation, the reader could imagine a sack of flour ($\mathfrak{B}$) being thrown into a box (Ω) into which it fits completely. The sack could take any shape compatible with the boundary conditions (it could even burst, since we do not require $\mathfrak{B}$ to be connected) without changing its volume.

In the case of a uniform distribution, we do not need to care whether the shape of $\mathfrak{B}$ is uniquely specified, because due to the requirement $\mathfrak{B} = \Omega$ this distribution is already determined by Ω. But in the case of any other probability density function with regions of constant probability density, it may happen that for certain values of α the region $\mathcal{B}$ cannot be uniquely determined. In such a situation we have two choices, either to abandon the uniqueness requirement or to try to enforce it somehow. A pragmatic solution is to abandon the uniqueness requirement, because on the one hand any additional requirement necessary to ensure uniqueness is not without a certain arbitrariness, and on the other hand we can assume that in practice the relations (5.45) and (5.46) will usually lead to a unique solution. However, if this is not the case, a selection can be made from the possible regions $\mathcal{B}$ on the basis of appropriate criteria.

With these preliminary considerations in mind, we can now give a definition of a credible region:

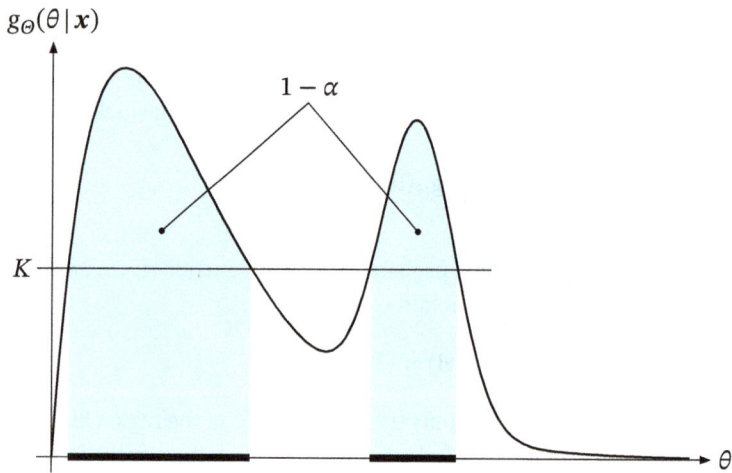

Fig. 5.4: Credible region (indicated by thick black lines on the θ axis) of a bimodal posterior probability density function.

Definition 5.23 (Credible region)

Let $g_\Theta(\theta \mid x)$ be the posterior of the parameter vector $\Theta = (\Theta_1, \ldots, \Theta_m)^\mathsf{T}$ given the data $x = (x_1, \ldots, x_n)^\mathsf{T}$ and $(1 - \alpha)$ a given probability, then the smallest, not necessarily connected, m-dimensional region

$$\mathcal{B} = \left\{ \theta \mid g_\Theta(\theta \mid x) \geq K \right\},$$

such that K satisfies the condition

$$\int_{\mathcal{B}} g_\Theta(\theta \mid x) \, d\theta \geq 1 - \alpha,$$

is called the credible region of the parameter vector Θ with coverage probability $(1 - \alpha)$.

In the one-dimensional case, i. e. if we consider only a single parameter, the credible region will usually become an interval, provided that the posterior probability density is unimodal. However, in the case of a multimodal posterior probability density, the credible region may—under certain circumstances—consist of several intervals. Such a case is illustrated in Fig. 5.3 for an example of a bimodal posterior probability density.

6 Measurement uncertainty concepts

> ... after the observer has done his part, it is up to the
> geometer to assess the uncertainty of the observations
> and the quantities derived from them by calculation
> according to strict mathematical principles, and most
> importantly, where the quantities associated with the
> observations can be derived from them by various
> combinations, to prescribe the method that leaves as
> little uncertainty as possible to be feared.
>
> *(Carl Friedrich Gauss, 1821)*

During the last two centuries the methods developed by C. F. Gauss, P. S. Laplace and their contemporaries have been successfully used to calculate the uncertainty of measurement. Most of these methods, formerly known as the theory of errors, are still valid after the introduction of some new methods published in the *Guide to the expression of uncertainty in measurement (GUM)*. In order to facilitate a better understanding of the similarities and differences between the traditional method and the methods of the GUM, both approaches are critically compared in this chapter.

6.1 The traditional method

The traditional method, developed about two centuries ago by C. F. Gauss[1] [66, 14] and P. S. Laplace [21], was initially used mainly to evaluate astronomical data. The application of these methods was very successful from the outset[2] and since then has provided a mathematically rigorous basis for the evaluation of measurement data in science and engineering.

We shall begin with considerations of the direct measurement of a measurand, following essentially the ideas of C. F. Gauss [14], which led to the traditional method of evaluating measurement data.

Suppose we measure the length of an object (e. g. the length of an iron rod) and obtain a measured value which we denote by x. We assume that the length of the rod has a fixed value that does not change over time, which we denote by w. We know from

---

[1] The work of C. F. Gauss was probably inspired by the book of J. H. Lambert, which contained a chapter on the theory of the reliability of observations and experiments [95].

[2] The calculations of C. F. Gauss enabled the dwarf planet Ceres—which had been lost after its discovery by G. Piazzi in January 1801—to be found again by F. X. v. Zach in December 1801.
P. S. Laplace was able to determine the mass of Saturn with an uncertainty—as we now know—of less than 1 %, using probabilistic methods he developed.

https://doi.org/10.1515/9783111453712-006

experience that repeated measurements usually do not give the same measured values. Therefore, we have to assume that the measured values do not generally agree with the unknown length value (even if this were the case by chance, we would remain ignorant of this fact). Therefore, for values obtained from repeated measurements under the same or similar conditions, we use the approach

$$x_i = w + b + e_i, \qquad i = 1, \dots, n, \tag{6.1}$$

where x_i denotes the i-th measurement, e_i the associated random error, b the systematic error and n the number of repeated measurements.

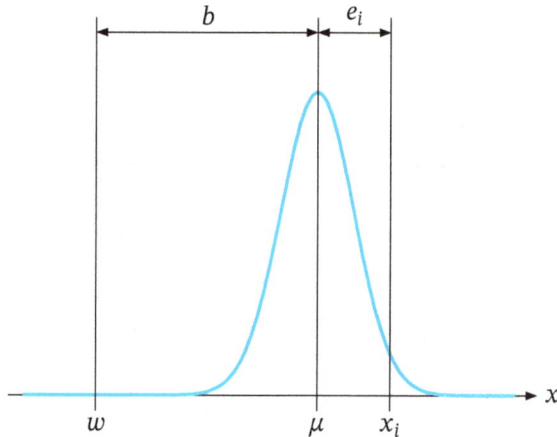

Fig. 6.1: Systematic error and random error.

The separation of measurement errors into two components of different nature was already common in the time of C. F. GAUSS [66]. The systematic error b is the constant (or better deterministic) component[3] of the error, which describes a constant shift of the measured values with respect to the true value w of the measurand. In contrast, the random error is the variable component of the error that describes the random variability of the measurement values. It can be assumed that in repeated measurements positive and negative errors e_i occur with almost the same frequency. Since experience shows that larger absolute error values occur less frequently than smaller values, a symmetrical dispersion of the measured values around the value

$$\mu = w + b$$

---

3 Not only a constant error, but any error that obeys a known physical law, such as a drift or hysteresis, can also be considered a systematic error.

results. This dispersion can be described by a probability density function of the random error e, as shown in Fig. 6.1, which was first proposed by C. F. Gauss [66] in 1809 and is therefore known today as the Gaussian (normal) distribution.

We give a first interim statement about our considerations so far. The traditional method assumes that

- each quantity has exactly one value, called the true value,
- this value is invariable with time, and
- measurement errors inevitably exist.

The deviations from the true value that occur during a measurement consist of a constant component (systematic error) and a variable component (random error). The systematic error is assumed to be known. The random error can be described sufficiently well by a Gaussian distribution.

The aim of each measurement is to obtain an estimate that is as close as possible to the true value. Let us have a closer look at how this estimate is determined by the traditional method.

If we sum over the equation (6.1) with respect to i and then divide by the number of measured values, we get

$$\overline{x} = \frac{1}{n} \sum_{i=1}^{n} x_i = w + b + \frac{1}{n} \sum_{i=1}^{n} e_i , \qquad i = 1, \dots , n .$$

Since we have assumed that the absolute values of the positive and negative random errors e_i occur on average with equal frequency, we can hope that, for a sufficiently large number of measurements, the sum on the right-hand side of this equation (i. e. the arithmetic mean of the random errors) will quickly approach zero. When this happens, we have

$$\overline{x} \approx w + b ,$$

i. e. the arithmetic mean $\overline{x}$ of the measured values is a suitable estimate of the true value w of the measurand, provided we can assume that there is no systematic error, i. e. $b = 0$ is valid. If the systematic error is different from zero, but is known, we can correct the measurement result accordingly.

Our considerations have therefore led to the conclusion that the arithmetic mean of the measured values, possibly corrected by the value b of a known systematic error, is a suitable estimate of the true value of the measurand. Hence

$$\hat{x} = \overline{x} - b \qquad (6.2)$$

holds, where $\hat{x}$ denotes the estimate of the measurand.

In general, the assumption that the mean of the random errors is zero will not hold, i. e. the estimated value $\hat{x}$ will not be the true value w of the measurand. The question is how close these two values are, i. e. how uncertain the estimate is.

Since we assume—as C. F. GAUSS did in 1821—that there is no systematic error, or that it has already been corrected, the uncertainty of the estimate $\hat{x}$ can only depend on the random errors e_i. For this uncertainty we obtain, by subtracting $\overline{x}$ on both sides in equation (6.1), using equation (6.2) and solving for e_i, the relation

$$e_i = (x_i - \overline{x}) + (\hat{x} - w), \qquad i = 1, \dots, n.$$

The measurement errors e_i are thus composed of two parts, namely the deviations of the measured values from their arithmetic mean $\overline{x}$ and the deviation of the estimated value $\hat{x}$ of the measurand from its true value w, which we assume to be negligibly small as the number of measurements increases.

Since the uncertainty of the measured values is essentially reflected by their dispersion, any measure of uncertainty should be based on it. However, only absolute values are relevant, since positive and negative errors, if not insignificant, contribute similarly. Furthermore, for any meaningful measure of uncertainty we must require that all observed values are averaged to avoid giving too much weight to extremely large or small variations in measurements.

If we take the absolute value on both sides of the last equation and then average over all measured values, we obtain

$$\frac{1}{n}\sum_{i=1}^{n}|e_i| = \frac{1}{n}\sum_{i=1}^{n}|(x_i - \overline{x}) + (\hat{x} - w)| \leq \frac{1}{n}\sum_{i=1}^{n}|x_i - \overline{x}| + |\hat{x} - w|,$$

where the triangle inequality has been used on the right-hand side. The left-hand side of this expression gives a possible measure of the uncertainty which corresponds to that originally proposed by P. S. LAPLACE. However, this is not the only way to specify an uncertainty measure. Indeed, if we do not take the absolute values, but rather square the values, we obtain the relation

$$\frac{1}{n}\sum_{i=1}^{n}e_i^2 = \frac{1}{n}\sum_{i=1}^{n}(x_i - \overline{x})^2 + (\hat{x} - w)^2.$$

The left-hand side of this equation, which is the mean squared error, corresponds to the uncertainty measure proposed by C. F. GAUSS.[4]

This gives us two different measures of measurement uncertainty, both of which are equally valid. Other measures can also be found where the latter statement is true.

---

4 Perinde ut integrale $\int x\varphi x.dx$, seu valor medius ipsius x, erroris constantis vel absentiam vel praesentiam et magnitudinem docet , integrale

$$\int xx\varphi x.dx$$

ab $x = -\infty$ usque ad $x = +\infty$ extensum (seu valor medius quadrati xx] aptissimum videtur ad incertitudinem observationum in genere definiendam et dimetiendam, ita ut e duobus observationum systematibus, quae quoad errorum facilitatem inter se differunt, eae praecisione praestare censeantur, in quibus integrale $\int xx\varphi x.dx$ valorem minorem obtinet. [14]

Thus, it follows that any measure of uncertainty is ultimately based on the prevailing conventions and is determined solely by *agreement*. This fact was already pointed out by C. F. GAUSS (see the corresponding footnote on page 32).

The present convention corresponds to the definition given by C. F. GAUSS, which is based on the mean square error. According to this convention, the value of the sample variance

$$s^2 = \frac{1}{n-1} \sum_{i=1}^{n} (x_i - \overline{x})^2$$

is a measure of the dispersion of the measured values. The factor $1/(n-1)$ in front of the sum, used instead of the factor $1/n$, ensures that for an infinite number of measurements the sample variance s^2 converges to the variance σ^2 of the GAUSSIAN distribution, as we have shown in the example 5.1.

The uncertainty $u(x_i)$ of a *single value* x_i is defined as the positive square root of the sample variance, i. e.

$$u(x_i) = s,$$

where s is the empirical standard deviation. The uncertainty of the arithmetic mean of a measurement is correspondingly given by

$$u(\overline{x}) = \frac{s}{\sqrt{n}},$$

because the value of the empirical variance of the sample mean is smaller by a factor $1/n$, as we have already shown in the example 5.1.

Since, according to the equation (6.2), $\hat{x}$ and $\overline{x}$ differ only by the constant systematic error b, their variances and thus also their uncertainties are equal. Therefore, we obtain the result $u(\hat{x}) = u(\overline{x})$.

We summarize our previous considerations:

Definition 6.1 (Result of a direct measurement)
A random quantity (measurand) X is measured n times under unchanging or similar conditions. The best estimate of the true value of the measurand is given by

$$\hat{x} = \overline{x} - b,$$

[Just as the integral $\int x\varphi(x)\mathrm{d}x$, or the mean value of x, tells us the absence or presence and magnitude of the constant error, the integral

$$\int x^2\varphi(x)\mathrm{d}x$$

extended from $x = -\infty$ to $x = +\infty$ (or the mean value of the square x^2) seems most appropriate to define and measure the uncertainty of observations in general, so that from two systems of observations, which differ from each other in terms of the probability of error, those are considered to be of higher precision, in which the integral $\int x^2\varphi(x)\mathrm{d}x$ obtains a smaller value.]

where b denotes the known systematic error, and

$$\bar{x} = \frac{1}{n} \sum_{i=1}^{n} x_i$$

is the arithmetic mean of the measured values x_i $(i = 1, \ldots, n)$.

The measurement uncertainty associated with the best estimate (called the standard uncertainty) is given by

$$u(\hat{x}) = \sqrt{\frac{1}{n(n-1)} \sum_{i=1}^{n} (x_i - \bar{x})^2}.$$

We now turn to the question of how to evaluate the measured data when we are *not* able to measure a measurand directly. For this purpose we consider the general case that a measurand Y is an arbitrary function of several different quantities $X_1, \ldots, X_n$ which can be measured directly, i. e. we assume the model equation

$$Y = f(X_1, \ldots, X_n) \tag{6.3}$$

for the quantity Y we are interested in.

Since it is assumed that the quantities $X_1, \ldots, X_n$ can be measured directly, we can determine their best estimates $\hat{x}_1, \ldots, \hat{x}_n$ as well as the measurement uncertainties $u(\hat{x}_1), \ldots, u(\hat{x}_n)$ associated with these values from the respective measurement data according to the equations given in the definition 6.1.

Therefore, it is not unreasonable to consider that

$$\hat{y} = f(\hat{x}_1, \ldots, \hat{x}_n)$$

as the best estimate of the true value of the measurand Y. In order to make a statement about the uncertainty of this estimate, we linearize the equation (6.3) at the operating point $(X_1, \ldots, X_n)$, as shown in section 3.5. Thus, by omitting the remainder, we obtain

$$Y = \hat{y} + \sum_{i=1}^{n} c_i (X_i - \hat{x}_i),$$

with the sensitivity coefficient

$$c_i = \left[\frac{\partial f(X_1, \ldots, X_n)}{\partial X_i} \right]_{X_1=\hat{x}_1, \ldots, X_n=\hat{x}_n}, \qquad i = 1, \ldots, n.$$

Next, we calculate the variance of Y according to the proposition 4.36. This yields

$$\text{Var}(Y) = \sum_{i=1}^{n} c_i^2 \text{Var}(X_i) + 2 \sum_{i=1}^{n} \sum_{j=i+1}^{n} c_i c_j \text{Cov}(X_i, X_j).$$

In the last step, we use the relation between the uncertainties and the variances, which finally results in

$$u^2(\hat{y}) = \sum_{i=1}^{n} c_i^2 u^2(\hat{x}_i) + 2 \sum_{i=1}^{n} \sum_{j=i+1}^{n} c_i c_j u(x_i, x_j),$$

where we have also replaced $\mathrm{Cov}(X_i, X_j)$ by $u(x_i, x_j)$ to facilitate the comparison with the formula stated in the GUM. We have found the well-known formula[5] for the propagation of measurement uncertainties.

We summarize our result:

Definition 6.2 (Propagation of measurement uncertainties)

Let Y be a random quantity that is not directly measurable, but depends on n directly measurable random quantities $X_1, \ldots, X_n$ according to an explicit function $Y = f(X_1, \ldots, X_n)$. Then the best estimate of Y is given by

$$\hat{y} = f(\hat{x}_1, \ldots, \hat{x}_n)$$

and the associated combined standard uncertainty by

$$u(\hat{y}) = \sqrt{\sum_{i=1}^{n} c_i^2 u^2(\hat{x}_i) + 2 \sum_{i=1}^{n} \sum_{j=i+1}^{n} c_i c_j u(x_i, x_j)},$$

with the empirical covariance

$$u(x, y) = \frac{1}{n(n-1)} \sum_{k=1}^{n} (x_k - \bar{x})(y_k - \bar{y})$$

and the sensitivity coefficients

$$c_i = \left[\frac{\partial f(X_1, \ldots, X_n)}{\partial X_i} \right]_{X_1 = \hat{x}_1, \ldots, X_n = \hat{x}_n}, \qquad i = 1, \ldots, n,$$

where $\hat{x}_i$ denotes the best estimate of the random quantity X_i and $u(\hat{x}_i)$ the associated standard uncertainty.

In order to make a statement about the reliability of the determined estimate of a measurand, the traditional method usually uses the confidence interval of conventional statistics, which has already been discussed in detail in section 5.8. Therefore, there is no need to return to it here.

Let us return to the systematic error. In our previous considerations we have always assumed that this error is either not present or it is known, so that a correction of a

---

5 This formula is also called the "law of error propagation" in many textbooks.

measurement result is possible. Unfortunately, in practice these assumptions are usually not met.

A systematic error may be unknown and may be included in the measurement result even though we cannot recognize it. Repeated measurements under the same or similar conditions are—as we have seen—not suitable for detecting systematic errors. Only additional experiments and a careful calibration of the measurement system can reveal systematic errors.

An experimental determination of a systematic error will generally only give an interval $(b - d, b + d)$ of possible values. The same is true if the systematic error is determined by an evaluation of known factors influencing the measurement, assuming a Gaussian distribution. For these two cases C. EISENHART [96] has suggested using the value of the centre of the interval, b, as before to correct the measurement result, but to consider half the width of the interval, d, as an additional contribution to the measurement uncertainty. Because of the fundamental difference between systematic and random errors, C. EISENHART proposed that the measurement result for a measurand X should be expressed in the form

$$(x - b) \pm [k_{95}\, u(x) + d],$$

where k_{95} must be determined in such a way that a 95 % confidence level is achieved. Consequently, until the publication of the *Guide to the expression of uncertainty in measurement (GUM)*, this procedure was common practice in the National Metrology Institutes of the United States (NBS, now NIST) and—similarly—the United Kingdom (NPL). Today, however, both institutes follow the internationally recognized recommendations of the GUM.

At the end of this section it should be mentioned that the least squares method, already discussed in section 5.8, was of course an integral part of the traditional method from the beginning.

6.2   The methods of the GUM

The introduction of the *Guide to the expression of uncertainty in measurement (GUM)* [5, 4] established an international harmonization of the evaluation of measurement data and the calculation of associated measurement uncertainties. This had become necessary due to the globalization of trade and the worldwide use of the International System of Units (SI), so that measurement results obtained in different countries could be easily compared.

The methods described in the GUM always presuppose a mathematical model of measurement and assume that the value of a quantity to be measured can be uniquely characterized by a single value. However, according to the GUM—unlike the traditional method—the aim of an evaluation is not to determine an estimate of the value of the measurand that is as close as possible to its "true" value, but rather to assign an interval of reasonable values to the measurand. This interval is represented by one of its values, the "measurement value", and its associated "measurement uncertainty". Both values together form the "complete measurement result".

The measurand considered is the output quantity of a mathematical model of the measurement. This model expresses the functional dependence of the output quantity on one or more input quantities. It describes both the measurement procedure and the evaluation method. In general, the model is given by one or more equations, but it can also take the form of an algorithm, which could be implemented, for example, by a computer program. The model must be sufficiently complex to achieve the required accuracy of the measurement result. Details of modelling are given in 3. For an evaluation, all quantities appearing in the model equations, except the model constants, are considered as random quantities. These include not only the measurand and the input quantities, but also all quantities[6] that directly or indirectly affect the measurement result. The input quantities can be divided into the following three classes, depending on how their values and associated measurement uncertainties are determined:

a) Quantities whose values, together with their associated measurement uncertainties, have been determined from values obtained during a measurement. These values may result from a single measurement or from repeated measurements under the same or similar conditions. They can include corrections of reported values as well as corrections for prevailing environmental conditions such as e. g. ambient temperature, barometric pressure or humidity.

b) Quantities whose values, together with their associated measurement uncertainties, have been determined *not* from values obtained during a measurement, but in some other way. These values can be taken, for example, from the calibration certificate of a material measure, or from the certificate of a reference material, or from handbooks or databases.

c) Quantities whose values, together with their associated measurement uncertainties, have been estimated on the basis of available knowledge and experimental experience (expert knowledge).

In the traditional method, the quantities listed under b) and c) are not used to calculate the measurement results.

Until the introduction of the GUM, not only were systematic and random errors treated differently, but they were also reported separately. The GUM does not distinguish between systematic and random errors, but introduces the two different evaluation methods

type A evaluation by statistical methods, and

type B evaluation by other (non-statistical) methods.

---

6 In the GUM, a distinction is made between input quantities and influence quantities. The latter are quantities that influence the quantity actually measured (VIM, 2.52, Note 2 [3, 2]).

The type A evaluation is used when the value of a quantity is determined by repeated measurements under the same or similar conditions. The value and its associated type A standard uncertainty are calculated according to the statistical methods discussed in chapter 5. The resulting formulae are identical to those given in the definition 6.1 in the previous section. Thus, the best estimate of the quantity under consideration is the arithmetic mean and its associated standard uncertainty is the empirical standard deviation.

The Type A evaluation also includes the possibility to determine the estimate of a quantity and its associated standard uncertainty by the least squares method. An important example is the determination of the parameters of a calibration curve for a measuring instrument.

Thus, there is no difference between the measurement results obtained by the GUM methods and by the traditional method when the information on the measurand is obtained only by repeated measurements or by the least squares method, provided that the systematic error is negligibly small.

The type B evaluation is used whenever the value of a quantity cannot be determined by a type A evaluation. The best estimate of the quantity and its associated type B standard uncertainty is determined by methods based on the available information, not necessarily derived from the current measurement, but originating from other sources.

The best estimates and their associated standard uncertainties obtained from a Type B evaluation are in no way less reliable than those obtained from a Type A evaluation. On the contrary, they may be more reliable if only very few measurement values are available for a Type A evaluation. This is because the arithmetic mean and the empirical standard deviation are good estimates only if the sample size (i. e. the number of measured values) is sufficiently large.

The methods used by a type B evaluation are of a different nature and include, at least to some extent, some methods of BAYESian statistics, which we discussed in chapter 5. This does not mean, however, that the methods of the GUM are based solely on BAYESian statistics. This is unfortunately not the case, although in principle it would be possible to base a theory of measurement uncertainty solely on BAYESian statistics, as K. WEISE and W. WÖGER [97, 98, 99] have demonstrated.

If Type B evaluations are used in conjunction with Type A evaluations, it must be ensured in the measurement model that uncertainty contributors already included in the empirical standard deviation are not included again in a Type B evaluation.

Once the estimates and their associated standard uncertainties have been determined for all input quantities, the next step is to calculate the estimate of the output quantity. It is assumed that the input values have already been corrected for known significant systematic effects. If this is not the case, the necessary corrections have to be included in the model of the measurement as additional input quantities.

In order to get the best estimate $\hat{y}$ of the measurand, the best estimates $(\hat{x}_1, \dots, \hat{x}_n)$ of the input quantities $(X_1, \dots, X_n)$ are inserted into the model equation(s). Thus, if the model equation

$$Y = f(X_1, \ldots, X_n)$$

is given, the best estimate of the output quantity Y is given by

$$\hat{y} = f(\hat{x}_1, \ldots, \hat{x}_n).$$

If the best estimates of the input quantities have been determined using a type A evaluation, there is another way to calculate the best estimate of the output quantity. In this case, however, it is necessary that the number of measured values is equal for all input quantities $(X_1, \ldots, X_n)$. If this condition is fulfilled, the best estimate of the output quantity Y can be calculated according to the expression

$$\hat{y} = \frac{1}{N} \sum_{k=1}^{N} f(x_{1,k}, \ldots, x_{n,k}), \tag{6.4}$$

where $x_{i,k}$ denotes the k-th measured value of the input quantity X_i and N denotes the number of measured values.

In the case of a linear model, the two ways of calculating the best estimate of the output quantity are equivalent, i.e. they lead to the same result. In the case of a non-linear model, however, different values will be obtained. In the GUM it is recommended (GUM, 4.1.4, [5, 4]) to use the possibility given by equation (6.4).

In order to obtain the standard uncertainty of the output quantity Y, the so-called *combined standard uncertainty* has to be calculated, taking into account all standard uncertainties determined by both Type A and Type B evaluations. For this purpose, the uncertainty propagation law derived by the linearization of the model equation is used. Since the corresponding formulae for the propagation of the uncertainties are identical to those shown in the previous section, the equations given in the definition 6.2 are the same as those stated in the GUM (GUM, 5, [5, 4]).

When calculating the combined standard uncertainty, the standard uncertainties determined by both Type A and Type B evaluations are treated as completely equivalent. However, this approach is not without controversy because the standard uncertainties determined by a Type A evaluation are based on frequency distributions, whereas those determined by a Type B evaluation are based on probability distributions, which—in the sense of Bayesian statistics—express a degree of belief. Thus, the two types of standard uncertainty are based on two different concepts of probability (see section 4.1 for details). The authors of the GUM were well aware of this fact, since they state " ... *that in both cases the distributions are models that are used to represent the state of our knowledge."* (GUM, 4.1.6, [5, 4]). With this remark, they obviously want to make clear that type A standard uncertainties should also be considered as parameters characterizing probability distributions that express a degree of belief.

When the uncertainty propagation law is used to calculate the combined standard uncertainty, it is important to take into account significant correlations. In certain cases this can lead to a considerable change in the value of the combined standard uncertainty. This is illustrated by a case that occurs in inter-laboratory comparisons.

Example 6.1 (Effect of correlations when using the E_n criterion)

In inter-laboratory comparisons, the so-called E_n criterion is used to verify the conformity of the measurement methods used by the participating laboratories. This criterion allows the assessment of the uncertainty of the difference between a measured value obtained by one laboratory and the weighted average of the measured values of all participating laboratories.

The calculations are based on the measured values x_i ($i = 1, \dots, n$) obtained by each laboratory and the associated standard uncertainties $u(x_i)$. First, the weighted mean

$$\bar{x} = \sum_{i=1}^{n} w_i x_i$$

and its associated standard uncertainty

$$u(\bar{x}) = \left(\sum_{i=1}^{n} \frac{1}{u^2(x_i)} \right)^{-1/2}$$

is calculated, where the weights are given by

$$w_i = \left(\frac{u(\bar{x})}{u(x_i)} \right)^2, \qquad i = 1, \dots, n.$$

Subsequently, the deviations

$$d_i = x_i - \bar{x}, \qquad i = 1, \dots, n,$$

and associated standard uncertainties

$$u(d_i) = \sqrt{u^2(x_i) + u^2(\bar{x}) - 2u(x_i, \bar{x})}, \qquad i = 1, \dots, n,$$

are calculated, where the covariance $u(x_i, \bar{x})$ describes the correlation between the measured value x_i and the weighted mean $\bar{x}$ of *all* measured values.

The individual covariances $u(x_i, \bar{x})$ are calculated, taking into account the fact that the quantities measured by each laboratory are stochastically independent. These calculations yield

$$u(x_i, \bar{x}) = \mathrm{Cov}\left(x_i, \sum_{k=1}^{n} w_k X_k \right) =$$

$$\sum_{k=1}^{n} w_k \mathrm{Cov}\,(X_i, X_k) = w_i \mathrm{Var}(X_i) = w_i u^2(x_i), \qquad i = 1, \dots, n.$$

It follows by inserting the equation for the weights w_i.

$$u(x_i, \bar{x}) = u^2(\bar{x}), \qquad i = 1, \dots, n.$$

Inserting this result into the equation for the calculation of the standard uncertainties $u(d_i)$ finally yields

$$u(d_i) = \sqrt{u^2(x_i) - u^2(\overline{x})}, \qquad i = 1, \dots, n.$$

This expression is well-defined because $u^2(x_i) > u^2(\overline{x})$ is always true. This assertion can be understood as follows: The sum of all weights w_i is one. All weights are positive. It follows that any weighting factor must be less than one, which finally gives the assertion.

Once the deviations d_i and their associated standard uncertainties have been determined, the numbers

$$E_n = \frac{d_i}{ku(d_i)} = \frac{x_i - \overline{x}}{k\sqrt{u^2(x_i) - u^2(\overline{x})}}$$

can be calculated for each laboratory, where k denotes the usual coverage factor according to the GUM (in most cases $k = 2$ is used).

For each participating laboratory, the number E_n is the ratio of the deviation from the weighted mean and its expanded uncertainty. If the calculation yields $|E_n| \le 1$ (i. e. if the weighted mean is just within the individual "error bar" given by the expanded uncertainty), the respective measurement x_i can be considered as compatible with the weighted mean [100].

Due to the strong positive correlation between the measured values x_i and their weighted mean, the uncertainty of the deviations of the individual values from this mean is much smaller in this example.

Once the value y of the measurand Y and its associated uncertainty have been determined, the evaluation of the measured data is complete. However, in certain cases, e. g. in industrial and commercial applications, where regulatory requirements apply, or where health and safety aspects have to be taken into account, it is usually required that the uncertainty is stated as an interval centred at the result of the measurement, encompassing a large fraction of the values that can reasonably be attributed to the measurand (GUM, 6.1.2 [5, 4]). For this purpose the so-called *expanded uncertainty* has been introduced in the GUM (GUM, 6.2 [5, 4]), which has to be calculated from the standard uncertainty according to the equation

$$U(y) = k\,u(y),$$

where k denotes the so-called *coverage factor*, which according to the GUM usually takes values between 2 and 3 (GUM, 6.3.1 [5, 4]). However, regardless of the underlying probability distribution of the output quantity, we always have $k \le 1/\sqrt{1 - p}$ (see example 4.80), i. e. in the case of $p = 95\,\%$ we obtain $k \le 4.47$.

The measurement result of a measurand Y is expressed by

$$y \pm U,$$

where y denotes the best estimate of the measurand Y and $(y - U, y + U)$ represents an interval that is expected to contain a large fraction of the values that could reasonably be attributed to the measurand (GUM, 6.2.1 [5, 4]). According to the GUM, the expanded uncertainty U determines only half of a symmetric interval around the best estimate y, encompassing the fraction p of the probability distribution characterized by that value and its associated standard uncertainty, where p denotes the coverage probability (or level of confidence) of that interval (GUM, 6.2.2 [5, 4]).

The expanded uncertainty of the GUM has similarities to the coverage interval discussed in section 5.8. The problem, however, is that in order to compute a coverage interval for a given coverage probability, it is actually necessary to know the underlying posterior probability density function of the measurand under consideration (the output quantity). However, since the methods of the GUM do not strictly follow BAYESian statistics, no posterior probability density is available after the evaluation of the measured data. If this were the case, a relationship could be established between the coverage probability p and the coverage factor k, and k could be uniquely determined.

The information provided by the best estimate y and its associated standard uncertainty $u(y)$ is not sufficient to specify a probability density function that adequately describes the available knowledge about the measurand under consideration. As a possible way out of this dilemma, the approximate method of the *effective degrees of freedom* is proposed in the GUM (GUM, G.4 [5, 4]). This method is based on publications by B. L. WELCH [101, 102, 103] and F. E. SATTERTHWAITE [104]. It assumes an approximate t distribution that matches the measured data as closely as possible. However, this method is not uncontroversial [105].

In practice, the method of effective degrees of freedom is rarely used, but in most cases it is simply assumed that the probability distribution function characterized by y and $u(y)$ is a normal distribution and that the effective degrees of freedom are large enough to justify this assumption as a sufficiently close approximation. In this case, a choice of $k = 2$ results in a coverage interval with a coverage probability (level of confidence) of about 95 %, while $k = 3$ would yield a coverage probability of approximately 99 %.

A From univariate to multivariate uncertainty

The content of this appendix is a revised version of an unpublished article,[1] written in 2010, i. e. prior to the publication of Suppl. 2 of the *Guide to the Expression of Uncertainty in Measurement* (GUM). Its purpose was to serve as a tutorial on multivariate uncertainty calculations.

A.1 Introduction

The ISO/IEC document *Guide to the Expression of Uncertainty in Measurement* (GUM) [5, 4] deals mainly with measurement models having only a single output quantity. Such a model is called here a *univariate model*. However, it is sometimes the case that more than one output quantity is required, all of which are related to a common set of input quantities. In such cases the models are called here *multivariate models*. In order to evaluate the measurement uncertainties associated with the estimated expectations of these output quantities, the uncertainty propagation as treated in the GUM requires an appropriate extension, as mentioned in clauses 3.1.7 and F.1.2.3 of the GUM, but only treated in more detail in example H.2 of this guide. Meanwhile, Suppl. 2 of the GUM [106] provides such an extension.

A.2 Univariate uncertainty calculations

In order to better understand the concept that leads to multivariate uncertainty calculations, a brief overview of univariate uncertainty calculations is given first, because as will become clear later, multivariate uncertainty calculations are only a generalization of univariate uncertainty calculations.

We start with the simplest case, where the output quantity Y depends on only one input quantity X. The model is given by the equation

$$Y = f(X), \tag{A.1}$$

i. e. the measurand Y is an arbitrary real function of the measurand X. The simplest case of such a functional relation is the linear equation

$$Y = b + cX, \tag{A.2}$$

---

1 The original version of this article is available on the preprint server of the Cornell University Library (arxiv.org): Michael P. Krystek, *From Univariate to Multivariate Uncertainty Calculation*, arXiv:1008.2700 [physics.data-an].

https://doi.org/10.1515/9783111453712-007

where b and c are known real constants, i. e. they are not part of the measurement task. If we take the expectation of this equation, we get

$$y = b + cx, \tag{A.3}$$

where $x = E[X]$ and $y = E[Y]$ denote the expectations of X and Y respectively. Recall now that the uncertainty $u(y)$ associated with the expectation y (i. e. with the value y of the measurand Y and *not* with the measurand itself, as is often mistakenly stated) is defined to be

$$u(y) = \sqrt{\text{Var}(y)}, \tag{A.4}$$

where the variance of Y is given by

$$\text{Var}(Y) = E\left[(Y - y)^2\right]. \tag{A.5}$$

Inserting the equations (A.2) and (A.3) into this equation gives

$$\text{Var}(Y) = E\left[c(X - x)\right]^2 = c^2 E\left[(X - x)^2\right] = c^2 \text{Var}(X). \tag{A.6}$$

Thus, the uncertainty $u(y)$ associated with the expectation y is given by

$$u(y) = c\,u(x),$$

as follows from equations (A.4) and (A.6), i. e. simply by multiplying the uncertainty $u(x)$ associated with the measured value x of the input quantity X by the given real constant c.

In the case of a linear model, we have been able to relate the uncertainty $u(x)$ associated with x to the uncertainty $u(y)$ associated with y. However, it is by no means guaranteed that such a simple propagation of the uncertainty is always possible. To see why this might be a problem, consider that the function f in the model equation (A.1) is non-linear. Taking the best estimate of this equation according to the recommendations of the GUM (see GUM, 4.1.4 [5]) gives $y = f(x)$, where x and y denote the expectations of X and Y respectively. If we now try to calculate the variance of Y as before according to the equation (A.5), we realize that we are not able to relate it to the variance of X in a simple way, i. e. the uncertainty propagation is not possible by the same means as in the linear case.

In order to facilitate the uncertainty calculation, we follow a suggestion of C. F. Gauss [66], who proposed to linearize the model equation $Y = f(X)$, i. e. to approximate the function f by its tangent at the expectation x. As follows from calculus, the equation of this tangent is given by

$$\frac{Y - y}{X - x} = \tan \alpha = \left(\frac{df(X)}{dX}\right)_{X=x},$$

where α denotes the angle of the tangent with the x-axis at the point x. This equation can also be written as

$$Y = y + c_X(X - x), \tag{A.7}$$

where the real constant

$$c_X = \left(\frac{\mathrm{d}f(X)}{\mathrm{d}X}\right)_{X=x}$$

is conventionally called the *sensitivity coefficient* of X. The sensitivity coefficient is a measure of how much the output quantity changes about its expectation y when the input quantity changes about its expectation x, provided that such a change is small enough to justify a linearization.

Inserting equation (A.7) into equation (A.5) and using equations (A.4) gives

$$u(y) = c_X u(x),$$

i. e. the uncertainties of X and Y are now proportional again, as in the linear case. However, this time we have to make sure that c_X is not zero, i. e. the tangent must not be parallel to the x-axis. Furthermore, a linearization is only possible if f is differentiable at x.

Usually the measurand Y depends on more than one input quantity. Therefore, assuming that there are n input quantities to consider, we have to replace the simple model equation (A.1) by the model equation

$$Y = f(X_1, \dots, X_n). \tag{A.8}$$

A simple example of this type of model is

$$Y = X_1 + X_2, \tag{A.9}$$

which is often used when a measured quantity X_1 is subject to a systematic measurement error X_2 and needs to be corrected accordingly. Taking the expectation of this model equation yields

$$y = x_1 + x_2, \tag{A.10}$$

where y denotes the expectation of Y and x_1 and x_2 denote the expectations of X_1 and X_2 respectively. From the equations (A.9) and (A.10), as well as the defining equation (A.5) of the variance, we obtain

$$\mathrm{Var}(Y) = \mathrm{E}\big[(X_1 - x_1) + (X_2 - x_2)\big]^2 = \mathrm{Var}(X_1) + \mathrm{Var}(X_2) + 2u(x_1, x_2),$$

where, according to the convention of the GUM

$$u(x_i, x_j) = \mathrm{Cov}(X_i, X_j) = \mathrm{E}\big[(X_i - x_i)(X_j - x_j)\big] \tag{A.11}$$

denotes the covariance associated with the expectations x_1 and x_2 (note that the covariance is *not* associated with the measurands themselves, as is often mistakenly stated, but with the respective expectations). Thus, from the defining equation (A.4) of the uncertainty we obtain

$$u(y) = \sqrt{u^2(x_1) + u^2(x_2) + 2u(x_1, x_2)}. \tag{A.12}$$

If the correlation coefficient

$$\rho(x_i, x_j) = \frac{u(x_i, x_j)}{u(x_i)u(x_j)}, \qquad -1 \le \rho(x_i, x_j) \le +1, \tag{A.13}$$

is introduced, the equation (A.12) can also be written as

$$u(y) = \sqrt{u^2(x_1) + u^2(x_2) + 2\rho(x_1, x_2)u(x_1)u(x_2)}. \tag{A.14}$$

This equation is particularly useful if the quantities X_1 and X_2 happen to be uncorrelated, i. e. if $\rho(x_1, x_2) = 0$ holds, in which case the equation (A.14) can be simplified to

$$u(y) = \sqrt{u^2(x_1) + u^2(x_2)},$$

or if they are totally correlated, i. e. $\rho(x_1, x_2) = \pm 1$ applies, in which case the equation (A.14) can be simplified to

$$u(y) = |u(x_1) \pm u(x_2)| .$$

Note that compared to the uncorrelated case, the uncertainty will be increased by a positive correlation and decreased by a negative correlation.

We now consider the case of a non-linear model with two input quantities given by the equation

$$Y = f(X_1, X_2). \tag{A.15}$$

Taking the best estimate of this equation according to the recommendations of the GUM (see GUM, 4.1.4 [5]) yields

$$y = f(x_1, x_2).$$

Since we already know from experience that a non-linear model is not suitable for simple uncertainty propagation, we linearize the model equation again. In this case we approximate the function $f(X_1, X_2)$ by the tangent plane at the point (x_1, x_2). From differential geometry we know that this tangent plane, if it exists, is given by

$$Y = y + (X_1 - x_1)\left(\frac{\partial f(X_1, X_2)}{\partial X_1}\right)_{(X_1, X_2)=(x_1, x_2)}$$

$$+ (X_2 - x_2)\left(\frac{\partial f(X_1, X_2)}{\partial X_2}\right)_{(X_1, X_2)=(x_1, x_2)}. \tag{A.16}$$

This equation can also be written as

$$Y = y + c_1(X_1 - x_1) + c_2(X_2 - x_2), \tag{A.17}$$

where the real constants

$$c_i = \left(\frac{\partial f(X_1, X_2)}{\partial X_i}\right)_{(X_1, X_2)=(x_1, x_2)}, \qquad i = 1, 2,$$

are again called sensitivity coefficients. The sensitivity coefficient c_i is a measure of how much the output quantity changes with respect to its expectation y when the input quantity X_i changes with respect to its expectation x_i, provided that such a change is small enough to justify a linearization.

Using the defining equations (A.5) and (A.11) of the variance and the covariance respectively, we obtain from equation (A.17) by a short calculation

$$\mathrm{Var}(Y) = \mathrm{E}\left[c_1(X_1 - x_1) + c_2(X_2 - x_2)\right]^2 =$$

$$c_1^2\mathrm{Var}(X_1) + c_2^2\mathrm{Var}(X_2) + 2c_1c_2u(x_1, x_2)$$

and thus, using the defining equation (A.4) of the uncertainty, finally

$$u(y) = \sqrt{c_1^2 u^2(x_1) + c_2^2 u^2(x_2) + 2c_1c_2u(x_1, x_2)},$$

or, again using the correlation coefficient given by equation (A.13),

$$u(y) = \sqrt{c_1^2 u^2(x_1) + c_2^2 u^2(x_2) + 2c_1c_2\rho(x_1, x_2)u(x_1)u(x_2)}. \tag{A.18}$$

If the quantities X_1 and X_2 are uncorrelated, i. e. if $\rho(x_1, x_2) = 0$ is valid, the equation (A.18) can be simplified to

$$u(y) = \sqrt{c_1^2 u^2(x_1) + c_2^2 u^2(x_2)}.$$

On the other hand, if $\rho(x_1, x_2) = \pm 1$ is valid, i. e. if the quantities X_1 and X_2 are totally correlated, the equation (A.18) can be simplified to

$$u(y) = |c_1 u(x_1) \pm c_2 u(x_2)|.$$

Note that in this case, compared to the uncorrelated case, the uncertainty is generally *not* increased by a positive correlation and decreased by a negative correlation, because the result depends also on the sensitivity coefficients c_1 and c_2, which are not necessarily positive.

We are now ready to consider the most general univariate model, as given by equation (A.8). Taking the best estimate of this equation according to the recommendations of the GUM (see GUM, 4.1.4 [5]) we get

$$y = f(x_1, \dots, x_n).$$

The linearization of equation (A.8) can be done analogously to the case given by the model equation (A.15), which led to equation (A.16). The generalization of the latter equation gives

$$Y = y + \sum_{i=1}^{n}(X_i - x_i)\left(\frac{\partial f(X_1, \dots, X_n)}{\partial X_i}\right)_{(X_1, \dots, X_n)=(x_1, \dots, x_n)}, \tag{A.19}$$

This equation can also be written as

$$Y = y + \sum_{i=1}^{n} c_i(X_i - x_i), \tag{A.20}$$

where the real constants

$$c_i = \left(\frac{\partial f(X_1, \ldots, X_n)}{\partial X_i}\right)_{(X_1,\ldots,X_n)=(x_1,\ldots,x_n)}, \qquad i = 1, \ldots, n, \tag{A.21}$$

denote again the sensitivity coefficients.

Using the defining equations (A.5) and (A.11) of the variance and covariance respectively, we obtain from equation (A.20) by a short calculation

$$\text{Var}(Y) = E\left[\sum_{i=1}^{n} c_i(X_i - x_i)\right]^2 = \sum_{i=1}^{n} c_i^2 \text{Var}(X_i) + 2\sum_{i=1}^{n-1}\sum_{j=i+1}^{n} c_i c_j u(x_i, x_j)$$

and thus, using the defining equation (A.4) of the uncertainty, finally

$$u(y) = \sqrt{\sum_{i=1}^{n} c_i^2 u^2(x_i) + 2\sum_{i=1}^{n-1}\sum_{j=i+1}^{n} c_i c_j u(x_i, x_j)}, \tag{A.22}$$

or using the correlation coefficient from equation (A.13),

$$u(y) = \sqrt{\sum_{i=1}^{n} c_i^2 u^2(x_i) + 2\sum_{i=1}^{n-1}\sum_{j=i+1}^{n} c_i c_j \rho(x_i, x_j) u(x_i) u(x_j)}. \tag{A.23}$$

The equations (A.22) and (A.23) are equivalent to the equations stated in section 5.2 of the GUM [5, 4] for correlated inputs. If the input quantities are not correlated, i.e. if $\rho(x_i, x_j) = 0$ for $i \neq j$, the equation (A.23) can be simplified to

$$u(y) = \sqrt{\sum_{i=1}^{n} c_i^2 u^2(x_i)},$$

which is equivalent to the equation given in clause 5.1 of the GUM [5, 4] for this particular case.

A.3 Multivariate uncertainty calculation

After a comprehensive review of univariate uncertainty calculations, which has shown where the formulae given in the GUM come from, we are now in a good position to deal with the multivariate case.

As mentioned in the introduction, the multivariate case is characterized by the fact that there is more than one output quantity, where all output quantities are related to a common set of input quantities. If we have n input quantities and m output quantities, where n and m are natural numbers that are not necessarily equal, the multivariate model is given by the system of equations

$$Y_i = f_i(X_1, \ldots, X_n), \qquad i = 1, \ldots, m, \tag{A.24}$$

where the (generally non-linear) real functions f_i may or may not be equal. For $m = 1$ we obviously return to the univariate case, which can be seen as a special case of multivariate calculations.

According to the recommendations of the GUM (see GUM, 4.1.4 [5]), the best estimates resulting from the system of equations (A.24) are given by

$$y_i = f_i(x_1, \ldots, x_n), \qquad i = 1, \ldots, m.$$

The linearization of equation (A.24) can be obtained by a generalization of equation (A.19), which gives

$$Y_i = y_i + \sum_{j=1}^{n} (X_j - x_j) \left(\frac{\partial f_i(X_1, \ldots, X_n)}{\partial X_j} \right)_{(X_1, \ldots, X_n) = (x_1, \ldots, x_n)}, \qquad i = 1, \ldots, m.$$

This system of equations can also be written as

$$Y_i = y_i + \sum_{j=1}^{n} c_{ij}(X_j - x_j), \qquad i = 1, \ldots, m. \tag{A.25}$$

where the real constants

$$c_{ij} = \left(\frac{\partial f_i(X_1, \ldots, X_n)}{\partial X_j} \right)_{(X_1, \ldots, X_n) = (x_1, \ldots, x_n)}, \qquad i = 1, \ldots, m, \quad j = 1, \ldots, n, \tag{A.26}$$

are called sensitivity coefficients, as in the univariate case. The sensitivity coefficient c_{ij} is a measure of how much the i-th output quantity Y_i changes with respect to its expected value y_i when the j-th input quantity X_j changes with respect to its expected value x_j, provided that such a change is small enough to justify a linearization.

Using the defining equations (A.5) and (A.11) for the variance and the covariance respectively, and the equation (A.4) for the definition of the uncertainty, we obtain by some simple calculations from the system of equations (A.25) the systems of equations

$$u^2(y_i) = \sum_{j=1}^{n} c_{ij}^2 u^2(x_j) + 2 \sum_{j=1}^{n-1} \sum_{k=j+1}^{n} c_{ij} c_{ik} u(x_j, x_k), \qquad i = 1, \ldots, m, \tag{A.27}$$

and

$$u(y_i, y_j) = \sum_{k=1}^{n} c_{ik}^2 u^2(x_k) + 2 \sum_{k=1}^{n-1} \sum_{\ell=k+1}^{n} c_{ik} c_{j\ell} u(x_k, x_\ell), \tag{A.28}$$

$$i = 1, \ldots, m, \quad j = i + 1, \ldots, m,$$

respectively, for the uncertainties $u(y_i)$ and the correlations $u(y_i, y_j)$ of the output quantities. Note that in the multivariate case, unlike the univariate case, we also obtain the covariances of the output quantities in addition to the variances.

Using the definition of the correlation coefficient according to equation (A.13), we can combine equations (A.27) and (A.28) to obtain the system of equations

$$\rho(y_i, y_j) = \sum_{k=1}^{n} \sum_{\ell=1}^{n} c_{ik} c_{j\ell} \rho(x_k, x_\ell) \frac{u(x_k)u(x_\ell)}{u(y_i)u(y_j)}, \qquad i, j = 1, \dots, m. \tag{A.29}$$

If the input quantities are uncorrelated, i. e. if $\rho(x_k, x_\ell) = 0$ holds for $i \neq j$, this system of equations can be simplified to

$$\rho(y_i, y_j) = \sum_{k=1}^{n} c_{ik} c_{jk} \frac{u^2(x_k)}{u(y_i)u(y_j)}, \qquad i, j = 1, \dots, m. \tag{A.30}$$

Thus, in the multivariate case, the output quantities are generally correlated, even if the input quantities are uncorrelated. This is due to the fact that all output quantities depend on a common set of input quantities, as can easily be shown. In particular, if we use the model equations

$$Y_i = f_i(X_i), \qquad i = 1, \dots, m,$$

i. e. if the output quantities do not depend on a common set of input quantities, but rather each of them depends only on a single input quantity, we verify using the equations (A.26) and (A.21) that

$$c_{ij} = c_i \delta_{ij}, \qquad i, j = 1, \dots, m,$$

is valid, where

$$\delta_{ij} = \begin{cases} 1 & \text{if } i = j \\ 0 & \text{otherwise} \end{cases}$$

denotes the so-called KRONECKER symbol. Thus, in this particular case, the system of equations (A.27) and (A.28) becomes

$$u(y_i) = c_i u(x_i), \qquad i = 1, \dots, m, \tag{A.31}$$

and

$$u(y_i, y_j) = c_i c_j u(x_i, x_j), \qquad i = 1, \dots, m, \quad j = i + 1, \dots, m.$$

Using the defining equation (A.13) of the correlation coefficient, the latter system of equations can also be written as

$$\rho(y_i, y_j)u(y_i)u(y_j) = c_i c_j \rho(x_i, x_j)u(x_i)u(x_j), \qquad i, j = 1, \dots, m. \tag{A.32}$$

If the input quantities happen to be uncorrelated, i. e. if $\rho(x_k, x_\ell) = \delta_{k\ell}$ holds, we obtain from equations (A.32) and (A.31)

$$\rho(y_i, y_j) = \delta_{ij}, \qquad i, j = 1, \dots, m,$$

i. e. in the special case considered, the output quantities are uncorrelated.

A.4 Matrix representation

The multivariate uncertainty calculation can be written more compactly and elegantly by using a matrix representation. To do this, we subsume the input and output quantities by the vectors $X = (X_1, \ldots, X_n)^\mathsf{T}$ and $Y = (Y_1, \ldots, Y_m)^\mathsf{T}$, respectively. If we additionally introduce a vector $f(X)$ with vector components $f_i = f_i(X_1, \ldots, X_n) = f_i(X)$ ($i = 1, \ldots, m$), we can rewrite the system of equations (A.24) as

$$Y = f(X). \tag{A.33}$$

Taking the best estimate of this equation according to the recommendations of the GUM (see GUM, 4.1.4 [5]) gives

$$y = f(x),$$

with the vectors $x = (x_1, \ldots, x_n)^\mathsf{T}$ and $y = (y_1, \ldots, y_m)^\mathsf{T}$ of the expectations of the input and output quantities respectively.

The linearization of the model (A.33) can be written as

$$Y = y + C(X - x), \tag{A.34}$$

where C denotes the sensitivity matrix with respect to the input quantities (in mathematics this matrix is called the JACOBI matrix of the system of equations), which has as matrix elements the sensitivity coefficients as given by the equations (A.26). Note that in general the sensitivity matrix C is not a square matrix, but has m rows and n columns. In the univariate case, i. e. for $m = 1$, the sensitivity matrix even degenerates into a row vector. The matrix equation (A.34) is equivalent to the system of equations (A.25), as can be easily verified.

We now introduce the variance-covariance matrix (sometimes also called the uncertainty matrix) of the input quantities

$$U_X = \begin{pmatrix} u^2(x_1) & u(x_1, x_2) & \cdots & u(x_1, x_{n-1}) & u(x_1, x_n) \\ u(x_2, x_1) & u^2(x_2) & \cdots & u(x_2, x_{n-1}) & u(x_2, x_n) \\ \vdots & \vdots & \ddots & \vdots & \vdots \\ u(x_{n-1}, x_1) & u(x_{n-1}, x_2) & \cdots & u^2(x_{n-1}) & u(x_{n-1}, x_n) \\ u(x_n, x_1) & u(x_n, x_2) & \cdots & u(x_n, x_{n-1}) & u^2(x_n) \end{pmatrix} \tag{A.35}$$

and the output quantities

$$U_Y = \begin{pmatrix} u^2(y_1) & u(y_1, y_2) & \cdots & u(y_1, y_{m-1}) & u(y_1, y_m) \\ u(y_2, y_1) & u^2(y_2) & \cdots & u(y_2, y_{m-1}) & u(y_2, y_m) \\ \vdots & \vdots & \ddots & \vdots & \vdots \\ u(y_{m-1}, y_1) & u(y_{m-1}, y_2) & \cdots & u^2(y_{m-1}) & u(y_{m-1}, y_m) \\ u(y_m, y_1) & u(y_m, y_2) & \cdots & u(y_m, y_{m-1}) & u^2(y_m) \end{pmatrix}. \tag{A.36}$$

Since $u(x_i, x_j) = u(x_j, x_i)$ and $u(y_i, y_j) = u(y_j, y_i)$ follows from the equation (A.11), the matrix $\mathbf{U}_X$ is a symmetric $(n \times n)$-matrix and the matrix $\mathbf{U}_Y$ is a symmetric $(m \times m)$-matrix. Moreover, it can be shown that both matrices are positive definite, i. e. they have only positive real eigenvalues.

Using the sensitivity matrix $\mathbf{C}$ and the variance-covariance matrices $\mathbf{U}_X$ and $\mathbf{U}_Y$ of the input and output quantities respectively, the system of equations (A.27) and (A.28) can be written as

$$\mathbf{U}_Y = \mathbf{C}\mathbf{U}_X\mathbf{C}^\mathsf{T}, \tag{A.37}$$

where $\mathbf{C}^\mathsf{T}$ is the transposed matrix of $\mathbf{C}$. The matrix equation (A.37) describes the uncertainty propagation for the general multivariate case. However, the univariate case is also included here as a special case, where $\mathbf{C}$ degenerates to a row vector, $\mathbf{C}^\mathsf{T}$ to a column vector and $\mathbf{U}_Y$ to a single real positive number, which is the variance of the single output quantity.

The concept of correlation coefficients can be introduced into the matrix representation in several ways. One possibility is to use uncertainty vectors, another is to use variance matrices. Since variance matrices can be developed in a more natural way from variance-covariance matrices, we are going to use them.

The variance matrix of the input quantities is formed as a diagonal matrix of the n variances, i. e.

$$\mathbf{V}_X = \mathrm{diag}\big(u^2(x_1), \dots, u^2(x_n)\big).$$

Thus, if the input quantities are uncorrelated, we have $\mathbf{V}_X = \mathbf{U}_X$. Now we introduce the correlation matrix $\mathbf{R}_X$ of the input quantities, which has the n^2 correlation coefficients $\rho(x_i, x_j)$ $(i, j = 1, \dots, n)$ as matrix elements. This matrix is a symmetric positive definite square matrix and is equal to the unit matrix $\mathbf{I}$ if the input quantities are all uncorrelated. The relation between the correlation matrix, the variance matrix and the variance-covariance matrix of the input quantities is given by the matrix equation

$$\mathbf{U}_X = \mathbf{V}_X^{1/2}\mathbf{R}_X\mathbf{V}_X^{1/2}, \tag{A.38}$$

where $\mathbf{V}_X^{1/2} = \mathrm{diag}\big(u(x_1), \dots, u(x_n)\big)$, i. e. a diagonal matrix with the uncertainties as matrix elements. Correspondingly, for the output quantities we get the matrix equation

$$\mathbf{U}_Y = \mathbf{V}_Y^{1/2}\mathbf{R}_Y\mathbf{V}_Y^{1/2}. \tag{A.39}$$

Inserting the equations (A.38) and (A.39) into the equation (A.37) gives

$$\mathbf{R}_Y = \mathbf{V}_Y^{-1/2}\mathbf{C}\mathbf{V}_X^{1/2}\mathbf{R}_X\mathbf{V}_X^{1/2}\mathbf{C}^\mathsf{T}\mathbf{V}_Y^{-1/2}. \tag{A.40}$$

This equation is the matrix representation of the system of equations (A.29).

In the case of uncorrelated inputs, i. e. if $\mathbf{R}_X = \mathbf{I}$ is valid, we obtain $\mathbf{U}_X = \mathbf{V}_X$ from equation (A.38). Inserting this result into equation (A.37) gives

$$\mathbf{U}_Y = \mathbf{C}\mathbf{V}_X\mathbf{C}^\mathsf{T}. \tag{A.41}$$

If we combine the equations (A.39) and (A.41) and solve for $\mathbf{R}_Y$, we get

$$\mathbf{R}_Y = \mathbf{V}_Y^{-1/2} \mathbf{C} \mathbf{V}_X \mathbf{C}^\mathsf{T} \mathbf{V}_Y^{-1/2}, \tag{A.42}$$

which is the matrix representation of the system of equations (A.30). Note that since $\mathbf{V}_X \neq \mathbf{I}$ generally $\mathbf{R}_Y \neq \mathbf{I}$ follows, i. e. the output quantities are always correlated, regardless of whether the input quantities are correlated or not, unless $\mathbf{C}$ is a diagonal matrix. If $\mathbf{C}$ is a diagonal matrix, it follows from equation (A.41) that $\mathbf{U}_Y$ is also a diagonal matrix and thus $\mathbf{U}_Y = \mathbf{V}_Y = \mathbf{C} \mathbf{V}_X \mathbf{C}^\mathsf{T}$ holds, which finally gives $\mathbf{R}_Y = \mathbf{I}$ from equation (A.42).

A.5 Generalization

So far we have assumed that the model is given by a system of equations such as (A.24) or (A.33), which is completely solved in terms of the output quantities. Such a model is called an explicit model. However, there are cases where, for some reason, the model equations are not fully solved with respect to the output quantities. Such a model is called an implicit model and is given by

$$\mathbf{F}(\mathbf{X}, \mathbf{Y}) = \mathbf{0} \tag{A.43}$$

in the matrix representation (it is sufficient to restrict our considerations to this representation only), where $\mathbf{X} = (X_1, \dots, X_n)^\mathsf{T}$ and $\mathbf{Y} = (Y_1, \dots, Y_m)^\mathsf{T}$ again denote the vectors of the input and output quantities, respectively, and $\mathbf{F}(\mathbf{X}, \mathbf{Y})$ denotes a vector with components $F_i = F_i(X_1, \dots, X_n, Y_1, \dots, Y_m) = F_i(\mathbf{X}, \mathbf{Y})$ ($i = 1, \dots, M$), and $\mathbf{0}$ denotes the zero vector of length M. Note that we must require $M \geq m$, i. e. there are at least as many equations as there are output quantities in the model, but $M > m$ is also allowed, which often, but not always, leads to the overdetermined least squares case. Thus, the vector equation (A.43) represents the most general model, and all other models, whether multivariate or univariate, are just special cases of this model. In fact, the equation (A.33) can be seen as a special case of the equation (A.43) with $\mathbf{F}(\mathbf{X}, \mathbf{Y}) = \mathbf{f}(\mathbf{X}) - \mathbf{Y}$ and $M = m$.

Taking the best estimate of equation (A.43) according to the recommendations of the GUM (see GUM, 4.1.4 [5]) gives

$$\mathbf{F}(\mathbf{x}, \mathbf{y}) = \mathbf{0}. \tag{A.44}$$

In order to obtain the expectations of the output quantities, this system of equations must then be solved for the vector $\mathbf{y}$, which can be done either algebraically, if possible, or numerically using a suitable method.

Using equation (A.44), the linearization of equation (A.43) at the point given by the expectations $\mathbf{x}$ and $\mathbf{y}$ of the input and output quantities gives

$$\mathbf{C}_Y(\mathbf{Y} - \mathbf{y}) + \mathbf{C}_X(\mathbf{X} - \mathbf{x}) = \mathbf{0}, \tag{A.45}$$

where $\mathbf{C}_X$ and $\mathbf{C}_Y$ denote the sensitivity matrices with respect to the input and output quantities. $\mathbf{C}_X$ is a $(M \times n)$-matrix with matrix elements

$$c_{X,ij} = \left(\frac{\partial F_i(\mathbf{X}, \mathbf{Y})}{\partial X_j}\right)_{(X=x,Y=y)} , \qquad i = 1, \dots, M, \quad j = 1, \dots, n,$$

and $\mathbf{C}_Y$ is a $(M \times m)$-matrix with matrix elements

$$c_{Y,ij} = \left(\frac{\partial F_i(\mathbf{X}, \mathbf{Y})}{\partial Y_j}\right)_{(X=x,Y=y)} , \qquad i = 1, \dots, M, \quad j = 1, \dots, m.$$

Recall that M denotes the number of model equations, n the number of input quantities and m the number of output quantities.

For well-posed problems, $\mathbf{C}_Y^\mathsf{T}\mathbf{C}_Y$ is a symmetric and positive definite $(m \times m)$-matrix. Then the inverse matrix $(\mathbf{C}_Y^\mathsf{T}\mathbf{C}_Y)^{-1}$ exists, and we can write the equation (A.45) as

$$\mathbf{Y} - \mathbf{y} = -\mathbf{C}(\mathbf{X} - \mathbf{x}), \tag{A.46}$$

with the combined sensitivity matrix

$$\mathbf{C} = (\mathbf{C}_Y^\mathsf{T}\mathbf{C}_Y)^{-1}\mathbf{C}_Y^\mathsf{T}\mathbf{C}_X, \tag{A.47}$$

which is a $(m \times n)$-matrix. From equation (A.46), we deduce

$$(\mathbf{Y} - \mathbf{y})(\mathbf{Y} - \mathbf{y})^\mathsf{T} = \mathbf{C}(\mathbf{X} - \mathbf{x})(\mathbf{X} - \mathbf{x})^\mathsf{T}\mathbf{C}^\mathsf{T}, \tag{A.48}$$

Note that both $(\mathbf{X} - \mathbf{x})(\mathbf{X} - \mathbf{x})^\mathsf{T}$ and $(\mathbf{Y} - \mathbf{y})(\mathbf{Y} - \mathbf{y})^\mathsf{T}$ are matrices constructed as dyadic products of two vectors.

Taking the expectation of equation (A.48) we get

$$\mathbf{U}_Y = \mathbf{C}\mathbf{U}_X\mathbf{C}^\mathsf{T}, \tag{A.49}$$

because $\mathbf{U}_X = \mathsf{E}\left[(\mathbf{X} - \mathbf{x})(\mathbf{X} - \mathbf{x})^\mathsf{T}\right]$ and $\mathbf{U}_Y = \mathsf{E}\left[(\mathbf{Y} - \mathbf{y})(\mathbf{Y} - \mathbf{y})^\mathsf{T}\right]$ is valid, as can be easily verified from the equations (A.35) and (A.36) using the definitions (A.5) and (A.11) of the variance and the covariance respectively.

Since equation (A.49) is formally the same as equation (A.37), all what was said about correlations in the previous section also applies here, if only the sensitivity matrix $\mathbf{C}$ in equations (A.40) to (A.42) and in the accompanying text is taken to mean the matrix product as given in equation (A.47).

A.6 Coverage regions

Before discussing the concept of a coverage region, we recall the meaning of the so-called expanded uncertainty according to the GUM. If there is only one measurand, say

Y, the complete measurement result is given by the best estimate y and the associated standard uncertainty $u(y)$. However, for some reason most people prefer to report the measurement result as $y \pm U(y)$, with $U(y) = ku(y)$, where $U(y)$ is called the *expanded uncertainty* and k the *coverage factor* according to the GUM, although this does not add any new information.

The expanded uncertainty is defined in the GUM as a *quantity defining an interval around the result of a measurement that can be expected to encompass a large fraction of the distribution of values that could reasonably be attributed to the measurand* (GUM, 2. 3.5 [5]), where *the choice of the factor k, usually in the range 2 to 3, is based on the coverage probability or confidence level required of the interval* (GUM, 3.3.7 [5]). The implication of this definition is that the probability of finding the values of the measurand Y within a symmetrical interval $(y - U(y), y + U(y))$ centred on the best estimate y is the coverage probability p, i. e.

$$\mathsf{P}\left(|Y - y| < U(y)\right) = \mathsf{P}\left(|Y - y| < ku(y)\right) = p. \tag{A.50}$$

Note that this equation is usually interpreted as a formula to calculate k for a given probability p, provided the probability density function of the measurand Y is known, i. e. k cannot be chosen arbitrarily.

If there are m measurands Y_i $(i = 1, \dots, m)$, we neither can use the equation (A.50), nor can we simply use $\mathsf{P}\left(|Y_1 - y_1| < U(y_1), \dots, |Y_m - y_m| < U(y_m)\right) = p$ instead, because then we have not taken into account possible correlations, i. e. we have violated the principle of including *all* available information.

In order to generalize the equation (A.50), we note that we can equivalently write

$$\mathsf{P}\left(\frac{(Y - y)^2}{u^2(y)} < k^2\right) = p. \tag{A.51}$$

Now a generalization becomes straightforward, giving

$$\mathsf{P}\left((\mathbf{Y} - \mathbf{y})^{\mathsf{T}} \mathbf{U}_Y^{-1} (\mathbf{Y} - \mathbf{y}) < k^2\right) = p. \tag{A.52}$$

It should be clear that the equation (A.51) is only a special case of the equation (A.52). Thus, the coverage region for the multivariate case is given by

$$(\mathbf{Y} - \mathbf{y})^{\mathsf{T}} \mathbf{U}_Y^{-1} (\mathbf{Y} - \mathbf{y}) < k^2,$$

which is the mathematical representation of an ellipsoid in a space of the dimension determined by the number of measurands represented by the vector $\mathbf{Y}$ and this ellipsoid is centred on a point given as coordinates by the best estimates of the output quantities. Thus, in the bivariate case, the coverage region is an ellipse, and in the univariate case it degenerates to an interval. Note that equation (A.52) must be used to calculate the coverage factor k for a given probability p, provided that the joint probability function for $\mathbf{Y}$ is known.

To conclude the discussion on coverage areas, we will add one more remark. If we combine the equations (A.46) and (A.49), we get the remarkable result

$$(Y - y)^{\mathsf{T}} U_Y^{-1} (Y - y) = (X - x)^{\mathsf{T}} U_X^{-1} (X - x),$$

i. e. the input and output ranges belong to the same coverage factor. However, this does *not* mean that the coverage probabilities of the input and output coverage regions are necessarily the same. This can only be the case, of course, if the joint probability functions were the same, which cannot generally be assumed.

If we require the coverage probabilities of the input and output coverage regions to be the same, we must satisfy the condition

$$P\left((Y - y)^{\mathsf{T}} U_Y^{-1} (Y - y) < k_Y^2\right) = P\left((X - x)^{\mathsf{T}} U_X^{-1} (X - x) < k_X^2\right),$$

where $k_Y \neq k_X$ is generally valid. This shows why it is important to understand that the coverage factor cannot be chosen arbitrarily.

A.7 Summary of multivariate calculations

For ease of reference, the details of the multivariate uncertainty calculations are summarized in this section. These calculations consist of the following steps
1. We choose a mathematical model to describe our measurement task

$$F(X, Y) = 0. \tag{A.53}$$

This model is generally a system of M equations, not necessarily linear, relating n input quantities, represented by a vector X, to m output quantities, represented by a vector Y.
2. We combine our measurement results in a vector x and the associated variances and covariances in the $(n \times n)$-matrix U_X. Alternatively, we can also represent the standard uncertainties by a diagonal matrix

$$V_X^{1/2} = \mathrm{diag}\big(u(x_1), \dots, u(x_n)\big)$$

and the correlation coefficients by a $(n \times n)$-matrix R_X with diagonal elements all equal to one. In this case, we obtain the variance-covariance matrix U_X from the matrix equation

$$U_X = V_X^{1/2} R_X V_X^{1/2}. \tag{A.54}$$

Note that the meaning of this matrix equation is simply that we have to multiply each row and column of the correlation matrix R_Y by the respective standard uncertainty to obtain the variance-covariance matrix U_X.
3. We solve the system of equations

$$F(x, y) = 0 \tag{A.55}$$

for the best estimates $y = (y_1, \ldots, y_m)^\mathsf{T}$ of the measurands $Y = (Y_1, \ldots, Y_m)^\mathsf{T}$ (output quantities). For a non-linear system of equations, this may require the use of an appropriate numerical method if the solution cannot be found analytically. Note that the system of equations (A.55) is exactly the same as the system of equations (A.53); we just have to replace the vectors X and Y by the vectors x and y respectively.

4. To obtain the $(M \times n)$-matrix C_X and the $(M \times m)$-matrix C_Y, we calculate the partial derivatives

$$c_{X,ij} = \left(\frac{\partial F_i(X, Y)}{\partial X_j} \right)_{(X=x, Y=y)} , \qquad i = 1, \ldots, M, \quad j = 1, \ldots, n, \qquad \text{(A.56)}$$

and

$$c_{Y,ij} = \left(\frac{\partial F_i(X, Y)}{\partial Y_j} \right)_{(X=x, Y=y)} , \qquad i = 1, \ldots, M, \quad j = 1, \ldots, m. \qquad \text{(A.57)}$$

These derivatives should preferably be calculated analytically. However, in some cases this is either impossible or inconvenient, in which case we can use a suitable method of numerical, or better automatic differentiation (for details of this technique see e. g. [107]).

Note that for the explicit model, i. e. if $F(X, Y) = F(X) - Y$ happens to be valid, we have $M = m$ and $C_Y = -I$, where I denotes the $(m \times m)$ identity matrix.

5. The next step is to calculate the sensitivity matrix C using the matrix equation

$$C = (C_Y^\mathsf{T} C_Y)^{-1} C_Y^\mathsf{T} C_X . \qquad \text{(A.58)}$$

Note that for the common case $M = m$ the matrix C_Y is square, and the matrix equation (A.58) can be simplified to

$$C = C_Y^{-1} C_X . \qquad \text{(A.59)}$$

For an explicit model this equation can even be simplified to $C = -C_X$.

6. The final step is to compute the variance-covariance matrix U_Y of the output quantities from the matrix equation

$$U_Y = C U_X C^\mathsf{T} . \qquad \text{(A.60)}$$

7. If the correlation matrix R_Y of the output quantities is required, we can use the matrix equation

$$R_Y = V_Y^{-1/2} U_Y V_Y^{-1/2} . \qquad \text{(A.61)}$$

The meaning of this equation is that we have to divide each row and column of the variance-covariance matrix U_Y by the square root of the respective diagonal element to obtain the correlation matrix R_Y.

A.8 Examples

In order to illustrate the procedure outlined in the previous section and to facilitate a better understanding of multivariate uncertainty calculations, some examples are given in the following.

Example A.1 (Resistance and reactance)

This example is motivated by the example given in annex H.2 of the GUM. The model in this case is explicit.

The resistance (here denoted by Y_1) and the reactance (here denoted by Y_2) of a circuit element are determined by measuring the amplitude (here denoted by X_1) of a sinusoidally alternating potential difference across its terminals, the amplitude (here denoted by X_2) of the alternating current flowing through it, and the phase shift angle (here denoted by X_3) of the alternating potential difference relative to the alternating current. In this case, the model is given by the system of equations

$$\frac{X_1}{X_2} \cos X_3 - Y_1 = 0 \tag{A.62}$$

$$\frac{X_1}{X_2} \sin X_3 - Y_2 = 0 \tag{A.63}$$

Taking the data published in annex H.2 of the GUM of the GUM, we have the input data vector $x = (4.9990\,\text{V}, 19.6610\,\text{mA}, 1.04446\,\text{rad})^\mathsf{T}$ and the standard uncertainties $u(x_1) = 0.0032\,\text{V}$, $u(x_2) = 0.0095\,\text{A}$ and $u(x_3) = 0.00075\,\text{rad}$ respectively. In addition, the correlation matrix

$$R_X = \begin{pmatrix} 1 & -0.36 & 0.86 \\ -0.36 & 1 & -0.65 \\ 0.86 & -0.65 & 1 \end{pmatrix}$$

is given. Using equation (A.54) we therefore obtain the variance-covariance matrix

$$U_X = 10^{-6} \begin{pmatrix} 10.24\,\text{V}^2 & -10.944\,\text{V}\cdot\text{mA} & 2.064\,\text{V} \\ -10.944\,\text{V}\cdot\text{mA} & 90.25\,(\text{mA})^2 & -4.63125\,\text{mA} \\ 2.064\,\text{V} & -4.63125\,\text{mA} & 0.5625 \end{pmatrix}.$$

The best estimates of the output quantities can be calculated from the equations

$$y_1 = \frac{x_1}{x_2} \cos x_3,$$

$$y_2 = \frac{x_1}{x_2} \sin x_3,$$

which gives $y = (127.732\,\Omega, 219.847\,\Omega)^\mathsf{T}$. This result agrees with the result given in annex H.2 of the GUM.

Applying the equation (A.56), we obtain from the equations (A.62) and (A.63)

$$
C_X = \begin{pmatrix} \dfrac{y_1}{x_1} & -\dfrac{y_1}{x_2} & -y_2 \\[2ex] \dfrac{y_2}{x_1} & -\dfrac{y_2}{x_2} & y_1 \end{pmatrix}
$$

or

$$
C_X = \begin{pmatrix} 25.5515\,\text{A}^{-1} & -6496.7280\,\text{V}\cdot\text{A}^{-2} & -219.8465\,\text{V}\cdot\text{A}^{-1} \\ 43.9781\,\text{A}^{-1} & -11181.8581\,\text{V}\cdot\text{A}^{-2} & 127.7322\,\text{V}\cdot\text{A}^{-1} \end{pmatrix}.
$$

Since we have an explicit model, this immediately leads to the sensitivity matrix

$$
C = \begin{pmatrix} -25.5515\,\text{A}^{-1} & 6496.7280\,\text{V}\cdot\text{A}^{-2} & 219.8465\,\text{V}\cdot\text{A}^{-1} \\ -43.9781\,\text{A}^{-1} & 11181.8581\,\text{V}\cdot\text{A}^{-2} & -127.7322\,\text{V}\cdot\text{A}^{-1} \end{pmatrix}.
$$

Applying the equation (A.60) we get

$$
U_Y = C U_X C^{\mathsf{T}} = \begin{pmatrix} 0.004897\,\Omega^2 & -0.012240\,\Omega^2 \\ -0.012240\,\Omega^2 & 0.087448\,\Omega^2 \end{pmatrix},
$$

i.e. the standard uncertainty of the resistance y_1 is $u(y_1) = 0.070\,\Omega$ and of the reactance y_2 it is $u(y_2) = 0.296\,\Omega$. These results differ only slightly from those given in annex H.2 of the GUM.

Using the equation (A.61) we finally obtain the correlation matrix

$$
R_Y = \begin{pmatrix} 1 & -0.592 \\ -0.592 & 1 \end{pmatrix},
$$

i.e. the correlation coefficient $r(y_1, y_2) = -0.592$ is not the same as that given in annex H.2 of the GUM, but it is close.

The slight differences between our results and those given in annex H.2 of the GUM are probably due to a limited numerical accuracy of the calculations leading to the results given in the GUM. Since we have checked our calculations with a computer algebra system, we are confident that our results are correct.

Example A.2 (Circle parameters)

It is well known that a circle in a plane is completely determined by three points in that plane. This example will show how the parameters of the circle, i.e. the co-ordinates of its centre and its radius, are affected by the uncertainty of the point co-ordinates.

If the co-ordinates of the three points are denoted by (X_1, X_2), (X_3, X_4), (X_5, X_6) respectively, the centre co-ordinates of the circle by (Y_1, Y_2) and its radius by Y_3, we have the model equations

$$
\begin{aligned}
(X_1 - Y_1)^2 + (X_2 - Y_2)^2 - Y_3^2 &= 0, \\
(X_3 - Y_1)^2 + (X_4 - Y_2)^2 - Y_3^2 &= 0, \\
(X_5 - Y_1)^2 + (X_6 - Y_2)^2 - Y_3^2 &= 0.
\end{aligned} \tag{A.64}
$$

We assume that the vector $x = (-1, 0, 0, 1, 1, 0)^\mathsf{T}$ is given and that the data are uncorrelated, i. e. the correlation matrix of the input data is equal to the identity matrix and the variance-covariance matrix is a diagonal matrix. We assume that the latter matrix is given by $U_X = \sigma^2 I$ with an arbitrary real constant σ representing the standard uncertainty of the input data.

In order to calculate the best estimates of the output quantities, we have to solve the system of equations

$$(x_1 - y_1)^2 + (x_2 - y_2)^2 - y_3^2 = 0 ,$$
$$(x_3 - y_1)^2 + (x_4 - y_2)^2 - y_3^2 = 0 ,$$
$$(x_5 - y_1)^2 + (x_6 - y_2)^2 - y_3^2 = 0 ,$$

for y_1, y_2 and y_3. After some algebraic manipulation we get the formulae

$$y_1 = \frac{(x_4 - x_6)x_1^2 + (x_6 - x_2)x_3^2 + (x_2 - x_4)x_5^2}{2[(x_4 - x_6)x_1 + (x_6 - x_2)x_3 + (x_2 - x_4)x_5]} -$$

$$\frac{(x_4 - x_6)(x_6 - x_2)(x_2 - x_4)}{2[(x_4 - x_6)x_1 + (x_6 - x_2)x_3 + (x_2 - x_4)x_5]} ,$$

$$y_2 = \frac{(x_3 - x_5)x_2^2 + (x_5 - x_1)x_4^2 + (x_1 - x_3)x_6^2}{2[(x_3 - x_5)x_2 + (x_5 - x_1)x_4 + (x_1 - x_3)x_6]} -$$

$$\frac{(x_3 - x_5)(x_5 - x_1)(x_1 - x_3)}{2[(x_3 - x_5)x_2 + (x_5 - x_1)x_4 + (x_1 - x_3)x_6]} ,$$

$$y_3 = \sqrt{(x_1 - y_1)^2 + (x_2 - y_2)^2} .$$

Substituting the input data into these formulae gives $y = (0, 0, 1)^\mathsf{T}$ for the best estimates of the output quantities, i. e. the result is a unit circle centred at the origin of the co-ordinate system.

Applying the equations (A.56) and (A.57), we obtain from the equations (A.64)

$$C_X = 2 \begin{pmatrix} x_1 - y_1 & x_2 - y_2 & 0 & 0 & 0 & 0 \\ 0 & 0 & x_3 - y_1 & x_4 - y_2 & 0 & 0 \\ 0 & 0 & 0 & 0 & x_5 - y_1 & x_6 - y_2 \end{pmatrix}$$

and

$$C_Y = -2 \begin{pmatrix} x_1 - y_1 & x_2 - y_2 & y_3 \\ x_3 - y_1 & x_4 - y_2 & y_3 \\ x_5 - y_1 & x_6 - y_2 & y_3 \end{pmatrix} ,$$

or by inserting the input and output data

$$C_X = 2 \begin{pmatrix} -1 & 0 & 0 & 0 & 0 & 0 \\ 0 & 0 & 0 & 1 & 0 & 0 \\ 0 & 0 & 0 & 0 & 1 & 0 \end{pmatrix} \tag{A.65}$$

and

$$C_Y = 2 \begin{pmatrix} 1 & 0 & -1 \\ 0 & -1 & -1 \\ -1 & 0 & -1 \end{pmatrix}. \tag{A.66}$$

Since we have the case $M = m$ here, we can use the equation (A.59). Inserting the results of (A.65) and (A.66) into this matrix equation gives the sensitivity matrix

$$C = \frac{1}{2} \begin{pmatrix} -1 & 0 & 0 & 0 & -1 & 0 \\ -1 & 0 & 0 & -2 & 1 & 0 \\ 1 & 0 & 0 & 0 & -1 & 0 \end{pmatrix}.$$

Applying the equation (A.60) we obtain

$$U_Y = \frac{\sigma^2}{2} \begin{pmatrix} 1 & 0 & 0 \\ 0 & 3 & -1 \\ 0 & -1 & 1 \end{pmatrix},$$

i. e. we find the standard uncertainties of the centre co-ordinates $u(y_1) = \sigma/\sqrt{2}$ and $u(y_2) = \sigma\sqrt{3/2}$, and the standard uncertainty of the radius $u(y_3) = \sigma/\sqrt{2}$.
Using the equation (A.61) we finally obtain the correlation matrix

$$R_Y = \begin{pmatrix} 1 & 0 & 0 \\ 0 & 1 & -\dfrac{1}{\sqrt{3}} \\ 0 & -\dfrac{1}{\sqrt{3}} & 1 \end{pmatrix},$$

i. e. the centre co-ordinates are not correlated with each other and the radius is only correlated with one of the centre co-ordinates with a correlation coefficient of $-1/\sqrt{3}$. This result was expected due to the symmetry of the problem.

Example A.3 (Alignment)
Given two line segments of equal length in a plane, there exists a transformation, consisting of a rotation and a translation, which makes it possible to align the two line segments in such a way that the respective end points of the line segments coincide (we must specify which points must be paired in order to obtain a unique solution). This example shows how the parameters of the alignment transformation are affected by the uncertainties of the co-ordinates of the end points of the two line segments.

If we denote the co-ordinates of the end points of the two line segments by (X_1, X_2), (X_3, X_4) and (X_5, X_6), (X_7, X_8) respectively, the angle of rotation by Y_1 and the components of translation by Y_2 and Y_3, the model is given by the system of equations

$$X_5 - X_1 \cos Y_1 + X_2 \sin Y_1 - Y_2 = 0,$$
$$X_6 - X_1 \sin Y_1 - X_2 \sin Y_1 - Y_3 = 0,$$
$$X_7 - X_3 \sin Y_1 + X_4 \sin Y_1 - Y_2 = 0,$$
$$X_8 - X_3 \sin Y_1 - X_4 \cos Y_1 - Y_3 = 0.$$

(A.67)

This model implicitly contains the requirement that the point with co-ordinates (X_1, X_2) must be transformed into the point with co-ordinates (X_5, X_6) (this ensures the uniqueness of the transformation). Note that although we have more equations as output quantities here, we do not have an overdetermined system of equations, i. e. we are not dealing with a least squares fit.

We assume that the vector $x = (0, 0, 1, 0, 0, 1, 0, 2)^\mathsf{T}$ is given, and the data are uncorrelated, i. e. the correlation matrix of the input data is equal to the identity matrix and the variance-covariance matrix is a diagonal matrix. We assume that the latter matrix is given by $\mathbf{U}_X = \sigma^2 \mathbf{I}$ with an arbitrary real constant σ representing the standard uncertainty of the input data.

In order to calculate the best estimates of the output quantities, we need to solve the system of equations

$$x_5 - x_1 \cos y_1 + x_2 \sin y_1 - y_2 = 0,$$
$$x_6 - x_1 \sin y_1 - x_2 \cos y_1 - y_3 = 0,$$
$$x_7 - x_3 \sin y_1 + x_4 \sin y_1 - y_2 = 0,$$
$$x_8 - x_3 \sin y_1 - x_4 \cos y_1 - y_3 = 0,$$

for y_1, y_2 and y_3. After some algebraic manipulation we get the formulae

$$y_1 = \arctan \frac{(x_2 - x_4)(x_5 - x_7) - (x_1 - x_3)(x_6 - x_8)}{(x_1 - x_3)(x_5 - x_7) + (x_2 - x_4)(x_6 - x_8)},$$

$$y_2 = \frac{x_5 + x_7}{2} - \frac{x_1 + x_3}{2} \cos y_1 - \frac{x_2 + x_4}{2} \sin y_1,$$

$$y_3 = \frac{x_6 + x_8}{2} + \frac{x_1 + x_3}{2} \sin y_1 - \frac{x_2 + x_4}{2} \cos y_1.$$

Substituting the input data into these formulae gives $y = (\pi/2, 0, 1)^\mathsf{T}$ for the best estimates of the output quantities, i. e. the line segments can be aligned by a counter-clockwise rotation with an angle of 90° around the origin of the co-ordinate system and a translation of one unit along its y-axis.

Applying the equations (A.56) and (A.57), we obtain from the equations (A.67)

$$C_X = \begin{pmatrix} -\cos y_1 & \sin y_1 & 0 & 0 & 1 & 0 & 0 & 0 & 0 \\ -\sin y_1 & -\cos y_1 & 0 & 0 & 0 & 0 & 1 & 0 & 0 \\ 0 & 0 & -\cos y_1 & \sin y_1 & 0 & 0 & 1 & 0 \\ 0 & 0 & -\sin y_1 & -\cos y_1 & 0 & 0 & 0 & 1 \end{pmatrix}$$

and

$$C_Y = \begin{pmatrix} x_1 \sin y_1 + x_2 \cos y_1 & -1 & 0 \\ -x_1 \sin y_1 + x_2 \sin y_1 & 0 & -1 \\ x_3 \sin y_1 + x_4 \cos y_1 & -1 & 0 \\ -x_3 \cos y_1 + x_4 \sin y_1 & 0 & -1 \end{pmatrix},$$

or by inserting the input and output data

$$C_X = \begin{pmatrix} 0 & 1 & 0 & 0 & 1 & 0 & 0 & 0 \\ -1 & 0 & 0 & 0 & 0 & 1 & 0 & 0 \\ 0 & 0 & 0 & 1 & 0 & 0 & 1 & 0 \\ 0 & 0 & -1 & 0 & 0 & 0 & 0 & 1 \end{pmatrix} \tag{A.68}$$

and

$$C_Y = \begin{pmatrix} 0 & -1 & 0 \\ 0 & 0 & -1 \\ 1 & -1 & 0 \\ 0 & 0 & -1 \end{pmatrix}. \tag{A.69}$$

Since we have the case $M > m$ here, we need to use the full equation (A.58). Inserting the results (A.68) and (A.69) into this matrix equation gives the sensitivity matrix

$$C = \frac{1}{2} \begin{pmatrix} 0 & -2 & 0 & 2 & -2 & 0 & 2 & 0 \\ 0 & -2 & 0 & 0 & -2 & 0 & 0 & 0 \\ 1 & 0 & 1 & 0 & 0 & -1 & 0 & -1 \end{pmatrix}$$

Applying the equation (A.60) we get

$$U_Y = \sigma^2 \begin{pmatrix} 4 & 2 & 0 \\ 2 & 2 & 0 \\ 0 & 0 & 1 \end{pmatrix},$$

i. e. the uncertainty of the angle of rotation is $u(y_1) = 2\sigma$ and the uncertainties of the components of translation are $u(y_2) = \sqrt{2}\,\sigma$ and $u(y_3) = \sigma$ respectively.

Using the equation (A.61) we finally obtain the correlation matrix

$$R_Y = \begin{pmatrix} 1 & \dfrac{1}{\sqrt{2}} & 0 \\ \dfrac{1}{\sqrt{2}} & 1 & 0 \\ 0 & 0 & 1 \end{pmatrix},$$

i. e. the components of the translation are not correlated, but the angle of rotation is correlated with the translation in x-direction.

B Dealing with systematic measurement errors

The content of this appendix is a revised version of an unpublished article,[1] written in 2010. Its purpose was to serve as a tutorial on the application of BAYESian methods to systematic errors.

B.1 Introduction

The *Guide to the Expression of Uncertainty in Measurement* (GUM) [5, 4] was published in 1995, i. e. about thirty years ago. Thus, the evaluation of measurement uncertainty should be well established today. Unfortunately, the GUM has little to offer with regard to the consideration of systematic effects.

The recommendation given in the GUM is to correct for known systematic effects, but unfortunately the reader is not given any detailed information on what to do if he is unable (or unwilling) to apply this rule. This particular problem (the problem of a known bias) has been dealt with for a long time, although this simple solution, which is easy to apply, does not seem to be widely known and has already led to some misunderstandings in the past.

However, there are other systematic effects, such as variations in ambient temperature, cosine error in length measurement, noisy signals and others, which must be taken into account in practical problems of measurement evaluation. Although most scientists are aware of the need to take these effects into account, they often do not know how to incorporate them into their particular uncertainty evaluation. The problem seems to be that there is no simple method (no cookbook recipe). The main question is how to use the knowledge of the systematic effects to obtain the expected value of the associated systematic measurement error and the associated standard uncertainty.

The purpose of this appendix is to show how systematic measurement errors can be treated on the basis of BAYES probability theory, on which most of the rules contained in the GUM are based, without explicitly mentioning it.

We will begin with some preliminary remarks on some essential terminology. Then a brief overview is given of useful methods and tools, such as the product rule of probability theory, BAYES's theorem, the principle of maximum entropy, and the marginalization equation. An outline of a method for dealing with systematic measurement bias is then presented. Finally, some simple examples of practical interest are given to demonstrate the applicability of the proposed method.

---

1 The original version of this article is available on the preprint server of the Cornell University Library (arxiv.org): Michael Krystek, *Bayesian theory of systematic measurement deviations*, arXiv:1009.0942 [physics.data-an].

https://doi.org/10.1515/9783111453712-008

B.2 Preliminary remarks

According to the BAYESian theory of measurement uncertainty of K. WEISE and W. WÖGER [98], the probability density function (pdf) is the mathematical representation of the state of knowledge about the values of the measured quantities of interest. All values of interest concerning these quantities can be obtained directly from their respective pdfs.

Each pdf can be characterized by two essential values. If we denote the measured quantity by X and its pdf by $p(X)$, we can calculate the value

$$x = E[X] = \int_{-\infty}^{+\infty} X p(X) dX,$$

which is called the expected value (or just expectation) of X. The other value of interest is the variance of the pdf, given by

$$\text{Var}(X) = E\left[(X - x)^2\right]. \tag{B.1}$$

This value characterizes the dispersion of the pdf around the expectation.

The pdf also contains the information about the value to be considered as the best estimate of the true value of the measured quantity. Today's international agreements follow a proposal by C. F. GAUSS [66], that the best estimate to be reported must be determined as that value around which the dispersion of the pdf is smallest, i. e. around which the pdf is most concentrated.

If we take an arbitrary value ξ, then the pdf dispersion around this value is given by

$$E\left[(X - \xi)^2\right] = \text{Var}(X) + (x - \xi)^2. \tag{B.2}$$

Obviously this dispersion has a minimum at $\xi = x$. So the dispersion at this particular value is equal to the variance of the pdf.

The measurement uncertainty associated with any value ξ is defined as the square root of the dispersion, i. e.

$$u(\xi) = \sqrt{E\left[(X - \xi)^2\right]}. \tag{B.3}$$

Therefore, the smallest possible uncertainty $u(x)$ is associated with the best estimate x as stated in the GUM. This smallest possible uncertainty is called the standard uncertainty.

In general, instead of the quantity of interest X, the quantity

$$Y = X + X_{\text{syst}} \tag{B.4}$$

is measured, where X_{syst} is called the systematic measurement error. Taking the expectation of the equation (B.4) and solving it for the value x gives

$$x = y - x_{\text{syst}}. \tag{B.5}$$

Thus, in order to obtain the best estimate x of the quantity of interest X, we must correct our measured value for the systematic error by subtracting the best estimate x_{syst} of that error.

The standard measurement uncertainty associated with x is

$$u(x) = \sqrt{u^2(y) + u^2(x_{\mathrm{syst}})}. \tag{B.6}$$

Note that even for the case $x_{\mathrm{syst}} = 0$ we still have to take into account the uncertainty of the systematic error according to equation (B.6), because we always have $u(x_{\mathrm{syst}}) \neq 0$, unless x_{syst} is perfectly known which can never be justified.

If two quantities X and Y are (fully or partially) correlated, we can introduce their covariance

$$\mathrm{Cov}(XY) = \mathrm{E}\left[(X - x)(Y - y)\right],$$

which is a generalization of the variance given by equation (B.1).

B.3 Ignoring an existing systematic error

The recommended procedure according to the GUM is to correct for a known systematic measurement error. However, sometimes this is not done for various reasons. In this case, instead of the best expectation x according to (B.5), the uncorrected value x^* is used as the best expectation of the quantity X, ignoring our knowledge of systematic effects. The standard measurement uncertainty associated with x^* certainly cannot be $u(x)$, because this uncertainty is associated with x and not with $x^* \neq x$. We have to use another uncertainty $u(x^*)$ instead.

By combining the equations (B.1), (B.2) and (B.3) with $\xi = x^*$ we get

$$\mathrm{E}\left[(X - x^*)^2\right] = u^2(x) + (x - x^*)^2. \tag{B.7}$$

But our assumption was that we did not need to correct for any systematic effect, i. e. we actually used the model $X = X^*$. Under this assumption, the left side of the equation (B.7), which is equal to $u^2(x^*)$, becomes the variance that must be associated with x^*. Since we have not corrected for the known systematic measurement error, we still have $x_{\mathrm{syst}} = x - x^*$. So we now get from equation (B.7)

$$u(x^*) = \sqrt{u^2(x) + x_{\mathrm{syst}}^2}.$$

It turns out that $u(x^*) \geq u(x)$, where the equality sign is only valid for $x_{\mathrm{syst}} = 0$. This was to be expected, of course, since ignoring known information should always increase the uncertainty.

We have only given a brief outline here, concerning the case where the correction for systematic effects is ignored. For further details see [108, 109].

B.4 BAYES' theorem and marginalization

Since we will be using the product rule, BAYES's theorem, and marginalization in the following, we will give a brief introduction to this topic here. More details can be found in various textbooks on BAYESian probability theory.

Suppose we have two random variables X and Y, and their joint pdf is given by $p(X, Y)$. Then, according to the product rule of probability theory, we have

$$p(X, Y) = p(X|Y)p(Y), \tag{B.8}$$

where $p(X|Y)$ is the conditional pdf of X given Y, and $p(Y)$ is the pdf of Y. The product rule can be generalized to more than two quantities by recursively applying the equation (B.8). If X and Y are stochastically independent, $p(X|Y) = p(X)$ holds and the equation (B.8) changes to $p(X, Y) = p(X)p(Y)$, i. e. the joint pdf of independent quantities is the product of the individual pdfs.

Since $p(Y, X) = p(X, Y)$ is valid, we obtain from the equation (B.8), by exchanging X and Y,

$$p(X|Y) = \frac{p(Y|X)p(X)}{p(Y)}, \tag{B.9}$$

where $p(X|Y)$ and $p(Y)$ are defined as before, $p(Y|X)$ is the conditional pdf of Y given X, and $p(X)$ is the pdf of X. If we now replace Y by a constant value y, e. g. a measured value, the equation (B.9) can be written as

$$p(X|y) = Cp(y|X)p(X), \tag{B.10}$$

where $p(y) = 1/C$ has been used, with a normalization constant C, because $p(y)$ is constant if y is constant. The equation (B.10) is a possible representation of BAYES's theorem. The function $p(X|y)$ is called the posterior pdf of the quantity X, given the value y. It represents the state of knowledge after y has been obtained by a measurement. The function $p(X)$ is called the prior pdf of the quantity X. It represents the state of knowledge before a measurement is taken. The function $p(y|X)$ is proportional to the so-called likelihood function. This function is related to the probability of measuring the value y if the quantity X were known.

If more than one quantity is given and more than one value has been measured, e. g. the quantities X_i ($i = 1, 2, \dots$) and the values y_k ($k = 1, 2, \dots$), Bayes' theorem can be generalized to

$$p(X_1, X_2, \dots | y_1, y_2, \dots) = Cp(y_1, y_2, \dots | X_1, X_2, \dots)p(X_1, X_2, \dots). \tag{B.11}$$

Here C is again a normalization constant, but different from the one used in equation (B.10). The likelihood function now represents the probability of measuring the values $y_1, y_2, \dots$ if the quantities $X_1, X_2, \dots$ were known.

If the joint pdf $p(X, Y)$ of two quantities X and Y is known, but we are only interested in the pdf of the quantity X, we can use the marginalization equation

$$p(X) = \int_{-\infty}^{+\infty} p(X, Y) \, dY.$$

The marginalization equation can also be generalized:

$$p(X) = \int_{-\infty}^{+\infty} \cdots \int_{-\infty}^{+\infty} p(X, Y_1, Y_2, \ldots, Y_n) \, dY_1 dY_2 \cdots dY_n.$$

Marginalization is a powerful tool that allows us to remove so-called nuisance quantities, i. e. quantities that need to be included in our calculation but are of no interest to our final result.

B.5 The principle of maximum entropy

Suppose we have some prior knowledge about a quantity X, i. e. the prior pdf $p(X)$ is known. We also know some values y_k ($k = 1, 2, \ldots$). If we knew the likelihood function $p(y_1, y_2, \ldots | X)$, we could use the BAYES theorem (B.11) to get the posterior pdf $p(X | y_1, y_2, \ldots)$. But in many cases the question is how to get the likelihood function. Following a suggestion from E. T. JAYNES [83, 84, 85], we can find this function by applying the principle of maximum entropy (PME).[2]

In order to apply the principle of maximum entropy, we have to maximize the relative entropy[3] of the posterior pdf with respect to the prior pdf, which serves as an invariant measure function, i. e. in our case the functional

$$S = -\int_{-\infty}^{+\infty} p(X) p(y_1, y_2, \ldots | X) \log p(y_1, y_2, \ldots | X) \, dX.$$

Since the posterior pdf has to be normalized, we have to obey the constraint

$$\int_{-\infty}^{+\infty} p(X | y_1, y_2, \ldots) \, dX = 1. \tag{B.12}$$

---

2 The principle of maximum entropy is often used only to obtain a prior pdf. In fact, E. T. JAYNES himself strongly advocated this approach under the assumption that the likelihood is already known. However, the principle of maximum entropy is not restricted to this particular application. We use it here differently, assuming that the prior pdf is known from the mathematical model of the measurement. This approach is justified by a theorem of B. O. KOOPMAN [110] and E. J. G. PITMAN [111], which states that the necessary and sufficient condition for a sampling distribution is that it has the general form of a maximum entropy distribution.

3 The relative entropy [112] is a generalization of SHANNON's information entropy [86].

If additional information is known, we can use further constraints, e. g.

$$\int\limits_{-\infty}^{+\infty} g_k(X)p(X|y_1, y_2, \ldots)\mathrm{d}X = g_k(x), \qquad k = 1, 2, \ldots, n. \tag{B.13}$$

Given these constraints, we must therefore maximize the expression

$$J = -\int\limits_{-\infty}^{+\infty} p(X)p(y_1, y_2, \ldots |X)\log p(y_1, y_2, \ldots |X)\mathrm{d}X$$

$$+ \sum_{k=0}^{n} \lambda_k \left(g_k(x) - \int\limits_{-\infty}^{+\infty} g_k(X)p(X)p(y_1, y_2, \ldots |X)\mathrm{d}X \right)$$

by the variation of the function $p(y_1, y_2, \ldots |X)$, where λ_k ($k = 0, 1, 2, \ldots$) are the so-called LAGRANGE parameters of the variational problem. The minimum of J is given by $\delta J = 0$, where

$$\delta J = -\int\limits_{-\infty}^{+\infty} p(X)(1 + \log p(y_1, y_2, \ldots |X))\delta p(y_1, y_2, \ldots |X)\mathrm{d}X -$$

$$\sum_{k=0}^{n} \lambda_k \int\limits_{-\infty}^{+\infty} p(X)g_k(X)\delta p(y_1, y_2, \ldots |X)\mathrm{d}X.$$

The result of the variation is

$$p(y_1, y_2, \ldots |X) = \exp\left(-1 - \sum_{k=0}^{n} \lambda_k g_k(X)\right).$$

If we insert this result into the equation (B.11), we finally get

$$p(X|y_1, y_2, \ldots) = C \exp\left(-\sum_{k=1}^{n} \lambda_k g_k(X)\right) p(X), \tag{B.14}$$

where $C = \exp(-1 - \lambda_0)$ has been used. The normalization constant C can be obtained by inserting (B.14) into (B.12) and the LAGRANGE parameters λ_k ($k = 1, 2, \ldots$) can be calculated by inserting (B.14) into (B.13).

We demonstrate the use of the PME with two simple examples:

Example B.1 (Uniform distribution)

In the first example we assume that the only knowledge we have about a quantity X is that it can take any value within the interval $(x_{\min}, x_{\max})$. This can be expressed by the pdf

$$p(X) \propto \Theta(X - x_{min})\Theta(x_{max} - X), \tag{B.15}$$

where

$$\Theta(x) = \begin{cases} 0 & \text{if } x < 0 \\ 1 & \text{if } x \geq 0 \end{cases}$$

denotes the HEAVISIDE step function.

Apart from the normalization requirement we have no other constraints, i.e. all LAGRANGE parameters λ_k ($k = 1, 2, \dots$) must be set to zero in (B.14). Thus, with (B.15) we get from (B.14)

$$p(X|x_{min}, x_{max}) = C\,\Theta(X - x_{min})\Theta(x_{max} - X).$$

The normalization constant is obtained by inserting this result into (B.12), which finally gives

$$p(X|x_{min}, x_{max}) = \frac{1}{x_{max} - x_{min}}\Theta(X - x_{min})\Theta(x_{max} - X),$$

i.e. we get a uniform pdf that assigns the same probability to each value of the quantity X within the interval (x_{min}, x_{max}). This is a reasonable result based on our limited knowledge of the quantity X.

Example B.2 (Exponential distribution)

In the second example we assume that we know about a quantity X that it cannot take negative values and that we know only one such value x. Our knowledge can be expressed by the prior

$$p(X) \propto \Theta(X) \tag{B.16}$$

and by the reasonable assumption that x is the expectation of X, i.e.

$$x = \int_{-\infty}^{+\infty} X p(X|x)\mathrm{d}X. \tag{B.17}$$

In addition to the normalization requirement, we now have one constraint. So the LAGRANGE parameter λ_1 is not zero, while all other LAGRANGE parameters are zero. Comparing (B.13) and (B.17) we see that we have $g_1(X) = X$ and therefore $g_1(x) = x$. Thus, from (B.14) and (B.16) we get

$$p(X|x) = C\exp(-\lambda_1 X)\Theta(X). \tag{B.18}$$

Substituting this result into (B.12) gives

$$C\int_0^\infty \exp(-\lambda_1 X)\,\mathrm{d}X = \frac{C}{\lambda_1} = 1$$

and inserting it into (B.17)

$$C \int_0^\infty X \exp\left(-\lambda_1 X\right) \mathrm{d}X = \frac{C}{\lambda_1^2} = x \,,$$

i. e. we have $C = \lambda_1 = 1/x$. With this result we obtain from (B.18)

$$p(X|x) = \frac{1}{x} \exp\left(-\frac{X}{x}\right) \Theta(X) \,.$$

This is the pdf of an exponential distribution. It allows us to calculate the standard uncertainty associated with the value x. We have

$$E[X^2] = \int_0^\infty X^2 \, p(X|x) \mathrm{d}X = 2x^2 \,.$$

Thus, from the definition of the standard uncertainty we obtain $u(x) = x$.

B.6 A procedure to handle systematic errors

Having discussed some of the basic ideas of the BAYESian theory of measurement uncertainty, we will now outline a procedure for dealing with systematic measurement errors.

To start with, we first need a mathematical model of the systematic measurement error. If X_{syst} denotes this quantity, and we know that it depends on some other quantities $X_1, X_2, \dots$ as well as on some known values $y_1, y_2, \dots$, we can write a model equation

$$X_{\mathrm{syst}} = f(X_1, X_2, \dots; y_1, y_2, \dots) \,, \tag{B.19}$$

summarizing the relation between the systematic measurement error X_{syst} and the quantities and values that influence it.

The systematic measurement error X_{syst} and all quantities $X_1, X_2, \dots$ appearing in equation (B.19) must be treated as random quantities. The values $y_1, y_2, \dots$ are assumed to be given. Thus, we have a joint pdf $p(X_{\mathrm{syst}}, X_1, X_2, \dots | y_1, y_2, \dots)$ of all considered stochastic quantities $X_1, X_2, \dots$, given the values $y_1, y_2, \dots$.

Using the product rule, we can split the joint pdf $p(X_{\mathrm{syst}}, X_1, X_2, \dots | y_1, y_2, \dots)$ to obtain a product of pdfs that can be further processed, i. e.

$$p(X_{\mathrm{syst}}, X_1, X_2, \dots | y_1, y_2, \dots) =$$
$$p(X_{\mathrm{syst}}|X_1, X_2, \dots, y_1, y_2, \dots) p(X_1, X_2, \dots | y_1, y_2, \dots) \,, \tag{B.20}$$

where $p(X_{\mathrm{syst}}|X_1, X_2, \dots, y_1, y_2, \dots)$ is the conditional pdf of X_{syst} given the quantities $X_1, X_2, \dots$ and the values $y_1, y_2, \dots$, and $p(X_1, X_2, \dots | y_1, y_2, \dots)$ is the pdf of the quantities $X_1, X_2, \dots$ given the values $y_1, y_2, \dots$.

The conditional pdf $p(X_{\text{syst}}|X_1, X_2, \dots, y_1, y_2, \dots)$ is obtained from the model equation (B.19) as follows

$$p(X_{\text{syst}}|X_1, X_2, \dots, y_1, y_2, \dots) = \delta[X_{\text{syst}} - f(X_1, X_2, \dots; y_1, y_2, \dots)],$$

where $\delta(\cdot)$ denotes DIRAC's δ function, defined by

$$\int\limits_{-\infty}^{+\infty} f(x)\,\delta(x - y)\,\mathrm{d}x = f(y).$$

In order to proceed with the second factor $p(X_1, X_2, \dots|y_1, y_2, \dots)$ on the right-hand side of equation (B.20), we can use the product rule of probability theory recursively until we get pdfs that we can specify, using all available knowledge about the quantities involved. In some cases, it may be helpful to use the PME to get the pdf under consideration. Sometimes it might also be useful to use the marginalization equation to get rid of nuisance quantities. The whole procedure can be complicated and usually requires some experience. Unfortunately, we cannot give more details here, as each case may have its own peculiarities.

As a final result of the procedure outlined above, we hopefully have a closed formula for the pdf $p(X_{\text{syst}}, X_1, X_2, \dots|y_1, y_2, \dots)$, which can now be used to calculate the expectations

$$E(X_{\text{syst}}) = \int\limits_{-\infty}^{+\infty} \cdots \int\limits_{-\infty}^{+\infty} X_{\text{syst}}\, p(X_{\text{syst}}, X_1, X_2, \dots|y_1, y_2, \dots)\,\mathrm{d}X_{\text{syst}}\,\mathrm{d}X_1\,\mathrm{d}X_2 \dots \tag{B.21}$$

and

$$E(X_{\text{syst}}^2) = \int\limits_{-\infty}^{+\infty} \cdots \int\limits_{-\infty}^{+\infty} X_{\text{syst}}^2\, p(X_{\text{syst}}, X_1, X_2, \dots|y_1, y_2, \dots)\,\mathrm{d}X_{\text{syst}}\,\mathrm{d}X_1\,\mathrm{d}X_2 \dots. \tag{B.22}$$

These two values can then be used to obtain the value x_{syst} of the systematic measurement error and its associated uncertainty $u(x_{\text{syst}})$.

We should note here that there are cases where we cannot evaluate the integrals (B.21) and (B.22) in closed analytical form. Indeed, this will happen in many practical cases. Then we can use some suitable approximations, such as the so-called LAPLACE approximation, to treat the problem still analytically, if such an approximation can be justified, or we may have to use numerical methods to evaluate the integrals. Since these integrals are usually of high dimensionality, Monte Carlo methods may be appropriate. For details of such methods see e. g. Suppl. 1 of the GUM [54, 55] and [113], which deal with Monte Carlo methods applied to measurement uncertainty calculations. Numerical methods do not even require the pdf to be specified as a closed formula, but only a suitable algorithm for its calculation, thus providing the greatest possible flexibility for a numerical approach to the problem.

In the following sections, some examples of the calculation of the values of the systematic measurement errors and the associated uncertainties are given to demonstrate the application of the procedure outlined in this section. The examples have been chosen so that they are both simple enough to allow all the formulae involved to be evaluated analytically, and are also of some practical interest.

B.7 The influence of temperature on length

It is well known that changes in environmental temperature are one of the major sources of systematic measurement error in length measurement. We will now examine this effect in detail.

Let us assume that we have a rod of a known material that has been calibrated for its length in the past. From the calibration certificate, we know the expectation of the length of this rod, denoted by ℓ_0, at a given temperature, denoted by ϑ_0, which is usually the reference temperature of 20 °C standardized in ISO 1, as well as the standard measurement uncertainty associated with this length, denoted by $u(\ell_0)$. We want to repeat the measurement of the length at the specified temperature. Let the quantity associated with this length be denoted by L_0. We try to keep the environmental temperature as constant as possible at the specified value, but we cannot be sure that the temperature is ϑ_0, as we require, but rather T. If the latter is the case, we will not measure the length L_0, but the length

$$L = L_0\big[1 + \alpha(T - \vartheta_0)\big], \tag{B.23}$$

where α denotes the known coefficient of thermal expansion of the material. Since we are not sure of the value of the temperature, we have to assume a systematic measurement error

$$X_{\text{syst}} = L - L_0. \tag{B.24}$$

Combining equations (B.23) and (B.24) gives the model equation

$$X_{\text{syst}} = L_0\alpha(T - \vartheta_0). \tag{B.25}$$

We are interested in the expectation x_{syst} of this systematic measurement error and its associated standard uncertainty $u(x_{\text{syst}})$.

The values of α and ϑ_0 are assumed to be known, but the quantities X_{syst}, L_0 and T must be treated as random quantities. Suppose we have used a special temperature recording device (e. g. a minimum-maximum thermometer) during our length measurement, which gives us only the minimum and maximum temperatures $\vartheta_{\min}$ and $\vartheta_{\max}$ respectively. Thus, all we have is the information that T is somewhere in the interval $(\vartheta_{\min}, \vartheta_{\max})$ and that it can take any value there.

Let $p(X_{\text{syst}}, L_0, T | \alpha, \vartheta_0, \vartheta_{\min}, \vartheta_{\max})$ denote the joint pdf of the quantities X_{syst}, L_0 and T, given that the coefficient of thermal expansion α and the temperatures ϑ_0, $\vartheta_{\min}$ and

$\vartheta_{\max}$ are known. Using the product rule of probability theory, we get

$$p(X_{\text{syst}}, L_0, T | \alpha, \vartheta_0, \vartheta_{\min}, \vartheta_{\max}) = p(X_{\text{syst}} | L_0, T, \alpha, \vartheta_0) p(L_0, T | \vartheta_{\min}, \vartheta_{\max}), \quad (B.26)$$

where $p(X_{\text{syst}} | L_0, T, \alpha, \vartheta_0)$ denotes the conditional pdf of X_{syst}, given that L_0 and T are known in addition to α and ϑ_0, and $p(L_0, T | \vartheta_{\min}, \vartheta_{\max})$ denotes the joint pdf of L_0 and T, given that $\vartheta_{\min}$ and $\vartheta_{\max}$ are known. Note that both X_{syst} and L_0 do not depend on $\vartheta_{\min}$ and $\vartheta_{\max}$, but T does. On the other hand, L_0 and T do not depend on α and ϑ_0. Consequently, the corresponding arguments of the respective pdf have been omitted in the equation (B.26).

Applying the product rule of probability theory again, we get

$$p(L_0, T | \vartheta_{\min}, \vartheta_{\max}) = p(L_0) p(T | \vartheta_{\min}, \vartheta_{\max}), \quad (B.27)$$

where $p(T | \vartheta_{\min}, \vartheta_{\max})$ denotes the conditional pdf of T, given $\vartheta_{\min}$ and $\vartheta_{\max}$, and $p(L_0)$ denotes the pdf of L_0 which is independent of T, $\vartheta_{\min}$ and $\vartheta_{\max}$.

The conditional pdf $p(X_{\text{syst}} | L_0, T, \alpha, \vartheta_0)$ can be derived from the model equation (B.25) as

$$p(X_{\text{syst}} | L_0, T, \alpha, \vartheta_0) = \delta[X_{\text{syst}} - \alpha(T - \vartheta_0)], \quad (B.28)$$

where $\delta(\cdot)$ is Dirac's δ function.

The pdf $p(T | \vartheta_{\min}, \vartheta_{\max})$ can be obtained by applying the PME. Since our only prior knowledge is that T is somewhere in the interval $(\vartheta_{\min}, \vartheta_{\max})$, we get the uniform (rectangular) distribution

$$p(T | \vartheta_{\min}, \vartheta_{\max}) = \frac{\Theta(T - \vartheta_{\min}) \Theta(\vartheta_{\max} - T)}{\vartheta_{\max} - \vartheta_{\min}}, \quad (B.29)$$

where $H(\cdot)$ denotes the HEAVISIDE step function. This is a reasonable result because, based on our limited knowledge, any temperature within the interval $(\vartheta_{\min}, \vartheta_{\max})$ should be equally likely.

Combining the equations (B.26), (B.27), (B.28) and (B.29) we finally get

$$p(X_{\text{syst}}, L_0, T | \alpha, \vartheta_0, \vartheta_{\min}, \vartheta_{\max}) =$$
$$\delta[X_{\text{syst}} - L_0 \alpha(T - \vartheta_0)] \frac{\Theta(T - \vartheta_{\min}) \Theta(\vartheta_{\max} - T)}{\vartheta_{\max} - \vartheta_{\min}} p(L_0). \quad (B.30)$$

We can now calculate the expectations

$$E[X_{\text{syst}}] = \int_{-\infty}^{+\infty} \int_{-\infty}^{+\infty} \int_{-\infty}^{+\infty} X_{\text{syst}} \, p(X_{\text{syst}}, L_0, T | \alpha, \vartheta_0, \vartheta_{\min}, \vartheta_{\max}) dX_{\text{syst}} dL_0 dT$$

and

$$E[X_{\text{syst}}^2] = \int_{-\infty}^{+\infty} \int_{-\infty}^{+\infty} \int_{-\infty}^{+\infty} X_{\text{syst}}^2 \, p(X_{\text{syst}}, L_0, T | \alpha, \vartheta_0, \vartheta_{\min}, \vartheta_{\max}) dX_{\text{syst}} dL_0 dT.$$

Using equation (B.30), the definitions of DIRAC's δ function and HEAVISIDE's step function, integration with respect to X_{syst} gives

$$E(X_{\text{syst}}) = \frac{\alpha}{\vartheta_{\text{max}} - \vartheta_{\text{min}}} \int_{-\infty}^{+\infty} \int_{\vartheta_{\text{min}}}^{\vartheta_{\text{max}}} L_0(T - \vartheta_0)p(L_0)dL_0dT$$

and

$$E(X_{\text{syst}}^2) = \frac{\alpha^2}{\vartheta_{\text{max}} - \vartheta_{\text{min}}} \int_{-\infty}^{+\infty} \int_{\vartheta_{\text{min}}}^{\vartheta_{\text{max}}} L_0^2(T - \vartheta_0)^2 p(L_0)dL_0dT,$$

and finally the integration with respect to L_0 and T

$$E[X_{\text{syst}}] = \alpha\ell_0 \left(\frac{\vartheta_{\text{max}} + \vartheta_{\text{min}}}{2} - \vartheta_0 \right) \tag{B.31}$$

and

$$E[X_{\text{syst}}^2] = \alpha^2 \left[\left(\frac{\vartheta_{\text{max}} + \vartheta_{\text{min}}}{2} - \vartheta_0 \right)^2 + \frac{1}{12}(\vartheta_{\text{max}} - \vartheta_{\text{min}})^2 \right] E[L_0^2], \tag{B.32}$$

with

$$\ell_0 = E[L_0] = \int_{-\infty}^{+\infty} L_0\, p(L_0)dL_0$$

and

$$E[L_0^2] = \int_{-\infty}^{+\infty} L_0^2\, p(L_0)dL_0.$$

From the equation (B.31) we obtain directly

$$X_{\text{syst}} = \alpha\ell_0 \left(\frac{\vartheta_{\text{max}} + \vartheta_{\text{min}}}{2} - \vartheta_0 \right)$$

and using the definition of the standard uncertainty, we get from the equations (B.31) and (B.32)

$$u(x_{\text{syst}}) = \alpha\sqrt{\left(\frac{\vartheta_{\text{max}} + \vartheta_{\text{min}}}{2} - \vartheta_0 \right)^2 u^2(\ell_0) + \frac{(\vartheta_{\text{max}} - \vartheta_{\text{min}})^2}{12}(\ell_0^2 + u^2(\ell_0))}. \tag{B.33}$$

Sometimes the formula

$$u(x_{\text{syst}}) = \alpha\ell_0 \frac{\vartheta_{\text{max}} - \vartheta_{\text{min}}}{2\sqrt{3}}$$

can be found in the literature, which is a good approximation of the equation (B.33), provided that $u(\ell_0) \ll \ell_0$ is valid.

B.8 The cosine error in length measurement

When the distance between two parallel planes is to be determined, the measurement must be made perpendicular to these planes. Any deviation from the perpendicularity requirement leads to a systematic measurement error known as the cosine error. Here we will show how to deal with this error.

Suppose we want to measure the distance D between two parallel planes. If Φ is the angle by which the direction of measurement deviates from perpendicularity, then the quantity

$$L = \frac{D}{\cos \Phi}$$ (B.34)

instead of the correct distance D. It follows that we must expect a systematic measurement error

$$X_{\text{syst}} = L - D.$$ (B.35)

Combining equations (B.34) and (B.35) gives the model equation

$$X_{\text{syst}} = L(1 - \cos \Phi),$$

which, using a trigonometric identity, can also be written as

$$X_{\text{syst}} = 2L \sin^2 \frac{\Phi}{2}.$$ (B.36)

We are interested in knowing the expectation x_{syst} of this systematic measurement error and its associated standard uncertainty $u(x_{\text{syst}})$.

We assume that the value φ_{m} is known and that the angle Φ can take any value in the interval $(-\varphi_{\text{m}}, \varphi_{\text{m}})$. This is the only information that is available.

The three quantities X_{syst}, L and Φ must be treated as random quantities. If φ_{m} is known, X_{syst}, L and Φ have a joint pdf $p(X_{\text{syst}}, L, \Phi|\varphi_{\text{m}})$, which can be written using the product rule of probability theory as

$$p(X_{\text{syst}}, L, \Phi|\varphi_{\text{m}}) = p(X_{\text{syst}}|L, \Phi, \varphi_{\text{m}})p(L, \Phi|\varphi_{\text{m}}),$$ (B.37)

where $p(X_{\text{syst}}|L, \Phi, \varphi_{\text{m}})$ denotes the conditional pdf of the systematic measurement error X_{syst}, if the length L and the angle Φ are given besides φ_{m}, and $p(L, \Phi|\varphi_{\text{m}})$ denotes the joint pdf of L and Φ if φ_{m} is known. Since L does not depend on φ_{m}, we have

$$p(L, \Phi|\varphi_{\text{m}}) = p(L)p(\Phi|\varphi_{\text{m}}).$$ (B.38)

The conditional pdf $p(X_{\text{syst}}|L, \Phi, \varphi_{\text{m}})$ can be immediately derived from the model equation (B.36) to be

$$p(X_{\text{syst}}|L, \Phi, \varphi_{\text{m}}) = \delta\left(X_{\text{syst}} - 2L \sin^2 \frac{\Phi}{2}\right),$$ (B.39)

where $\delta(\cdot)$ denotes DIRAC's δ function.

If we assume, as stated above, that the angle Φ can take any value within the interval $(-\varphi_m, \varphi_m)$, the pdf $p(\Phi|\varphi_m)$ can be obtained by applying the PME, which gives the uniform (rectangular) distribution

$$p(\Phi|\varphi_m) = \frac{\Theta(\Phi + \varphi_m)\Theta(\varphi_m - \Phi)}{2\varphi_m}, \tag{B.40}$$

where $H(\cdot)$ denotes the HEAVISIDE step function.

Combining equations (B.37), (B.38), (B.39) and (B.40) gives

$$p(X_{\text{syst}}, L, \Phi|\varphi_m) = \frac{1}{2\varphi_m}\delta\left(X_{\text{syst}} - 2L\sin^2\frac{\Phi}{2}\right)\Theta(\Phi + \varphi_m)\Theta(\varphi_m - \Phi)p(L). \tag{B.41}$$

As we will see later, the pdf $p(L)$ does not need to be specified in detail. We only need to require that this function can be normalized in the usual way, in order to ensure that $p(X_{\text{syst}}, L, \Phi|\varphi_m)$ is normalized.

We can now calculate the expectations

$$E[X_{\text{syst}}] = \int_{-\infty}^{+\infty}\int_{-\infty}^{+\infty}\int_{-\infty}^{+\infty} X_{\text{syst}}\, p(X_{\text{syst}}, L, \Phi|\varphi_m)\, dX_{\text{syst}}\, dL\, d\Phi,$$

and

$$E[X_{\text{syst}}^2] = \int_{-\infty}^{+\infty}\int_{-\infty}^{+\infty}\int_{-\infty}^{+\infty} X_{\text{syst}}^2\, p(X_{\text{syst}}, L, \Phi|\varphi_m)\, dX_{\text{syst}}\, dL\, d\Phi,$$

respectively. Using equation (B.41), the definitions of DIRAC's δ-function and of the HEAVISIDE step-function, integration with respect to X_{syst} and L gives

$$E[X_{\text{syst}}] = \frac{\ell}{\varphi_m}\int_{-\varphi_m}^{+\varphi_m}\sin^2\frac{\Phi}{2}\, d\Phi,$$

and

$$E[X_{\text{syst}}^2] = \frac{2}{\varphi_m}E[L^2]\int_{-\varphi_m}^{+\varphi_m}\sin^4\frac{\Phi}{2}\, d\Phi,$$

with

$$\ell = \int_{-\infty}^{+\infty} L\, p(L)dL$$

and

$$E[L^2] = \int_{-\infty}^{+\infty} L^2\, p(L)dL,$$

and the integration with respect to Φ finally

$$E(X_{\text{syst}}) = \ell\left(1 - \frac{\sin\varphi_m}{\varphi_m}\right) \tag{B.42}$$

and

$$E[X_{\text{syst}}^2] = \left(\frac{3}{2} - 2\frac{\sin\varphi_m}{\varphi_m} + \frac{\sin 2\varphi_m}{4\varphi_m}\right)E[L^2]. \tag{B.43}$$

From equation (B.42), we get directly

$$x_{\text{syst}} = \ell\left(1 - \frac{\sin\varphi_m}{\varphi_m}\right), \tag{B.44}$$

and using the definition of the standard uncertainty we obtain from equations (B.42) and (B.43)

$$u(x_{\text{syst}}) = \sqrt{\left(\frac{3}{2} - 2\frac{\sin\varphi_m}{\varphi_m} + \frac{\sin 2\varphi_m}{4\varphi_m}\right)u^2(\ell) + \ell^2\left[\frac{1}{2} + \frac{\sin 2\varphi_m}{4\varphi_m} - \left(\frac{\sin\varphi_m}{\varphi_m}\right)^2\right]}. \tag{B.45}$$

In practice, the angle φ_m can usually be assumed to be very small. In this case we can expand the right-hand sides of equations (B.44) and (B.45) into a TAYLOR series with respect to φ_m. Taking into account terms up to the second order, we obtain the approximations

$$x_{\text{syst}} \approx \frac{\ell}{6}\varphi_m^2, \tag{B.46}$$

and

$$u(x_{\text{syst}}) \approx \varphi_m^2\sqrt{\frac{1}{5}\left(\frac{u^2(\ell)}{4} + \frac{\ell^2}{9}\right)}. \tag{B.47}$$

This result shows that the cosine error is a so-called second order systematic effect, since there is no linear term in φ_m.

In the case that $u(\ell) \ll \ell$ is valid, equation (B.47) can be approximated even further, yielding

$$u(x_{\text{syst}}) \approx \frac{\ell}{3\sqrt{5}}\varphi_m^2. \tag{B.48}$$

The final approximations given by the expressions (B.46) and (B.48) are identical to the results given in section F.2.4.4 of the GUM under the assumption of a rectangular distribution for the angle.

B.9 The influence of form deviations on the distance

If the distance between two spheres, i. e. the distance between their respective centres, is to be determined, the influence of the shape deviations of the spheres must be taken into account as a systematic measurement error. We will now show how this can be done.

It is well known from geometry that the distance between two spheres is equal to the sum of their radii, i. e.

$$D = R_1 + R_2 \tag{B.49}$$

where the distance is denoted by D and the radii of the two spheres are denoted by R_1 and R_2 respectively.

In practice, however, we cannot assume that the spheres are geometrically perfect objects. Usually we have to expect form deviations. In order to simplify the following calculations, we will assume, without loss of generality, that one of the two spheres, say the first one, has no form deviations, while the other does. We must therefore assume that the second sphere has a radius of R_2^* instead of R_2, which gives a distance of

$$D^* = R_1 + R_2^* \tag{B.50}$$

instead of the distance D. This introduces a systematic measurement error

$$X_{\text{syst}} = D^* - D.$$

Combining equations (B.49) and (B.50) gives the model equation

$$X_{\text{syst}} = R_2^* - R_2. \tag{B.51}$$

We are interested in the expectation x_{syst} of this systematic measurement error and its associated standard uncertainty $u(x_{\text{syst}})$.

In order to introduce the form deviations of the sphere into our calculations, we can enclose the real surface between two concentric ideal spheres with radii R_{min} and R_{max} respectively. It is assumed that these spheres are chosen so that the difference $R_{\text{max}} - R_{\text{min}}$ is a minimum (minimum zone fit). Thus, the radius R_2^* of the sphere is only known to lie somewhere in the interval $(R_{\text{min}}, R_{\text{max}})$. But since we have to determine the boundaries of this interval from our measured data, we have to associate uncertainties with these boundaries, i. e. we have a situation of uncertain boundaries.

In practice, it is more convenient to replace the quantities R_{min} and R_{max} by the form deviation

$$F = R_{\text{max}} - R_{\text{min}}$$

and the mean radius

$$R_{\text{m}} = \frac{R_{\text{max}} + R_{\text{min}}}{2},$$

from which

$$R_{\text{min}} = R_{\text{m}} - \frac{F}{2} \tag{B.52}$$

and

$$R_{\text{max}} = R_{\text{m}} + \frac{F}{2} \tag{B.53}$$

can be obtained.

All five quantities X_{syst}, R_2^*, R_2, R_{m} and F must be treated as random quantities. Let $p(X_{\text{syst}}, R_2^*, R_2, R_{\text{m}}, F)$ denote their joint pdf. We can recursively use the product rule of probability theory to split this pdf, which gives us

$$p(X_{\text{syst}}, R_2^*, R_2, R_{\text{m}}, F) = p(X_{\text{syst}}|R_2^*, R_2, F, R_{\text{m}})p(R_2^*|R_2, R_{\text{m}}, F)p(R_2, R_{\text{m}}, F), \quad \text{(B.54)}$$

where $p(X_{\text{syst}}|R_2^*, R_2, R_{\text{m}}, F)$ denotes the conditional pdf of X_{syst}, given that R_2^*, R_2, R_{m} and F are known, $p(R_2^*|R_2, R_{\text{m}}, F)$ denotes the joint pdf of R_2^*, given R_2, R_{m} and F, and $p(R_2, R_{\text{m}}, F)$ denotes the joint pdf of R_2, R_{m} and F.

The conditional pdf $p(X_{\text{syst}}|R_2^*, R_2, R_{\text{m}}, F)$ can be deduced directly from the model equation (B.51) to be

$$p(X_{\text{syst}}|R_2^*, R_2, R_{\text{m}}, F) = \delta(X_{\text{syst}} - R_2^* + R_2). \quad \text{(B.55)}$$

If the mean radius R_{m} and the form deviation F are given, then the boundaries R_{min} and R_{max} are also given, because they are related to R_{m} and F by equations (B.52) and (B.53). Thus, we have a situation that allows us to get the conditional pdf $p(R_2^*|R_2, R_{\text{m}}, F)$ of the radius R_2^* by applying the PME, which gives the uniform (rectangular) distribution

$$p(R_2^*|R_2, R_{\text{m}}, F) = \frac{1}{F} \Theta\left(R_2^* - R_{\text{m}} + \frac{F}{2}\right) \Theta\left(R_{\text{m}} + \frac{F}{2} - R_2^*\right), \quad \text{(B.56)}$$

where $H(\cdot)$ denotes the HEAVISIDE step function.

Combining the equations (B.54), (B.55) and (B.56) we get

$$p(X_{\text{syst}}, R_2^*, R_2, R_{\text{m}}, F) = \frac{1}{F}\delta(X_{\text{syst}} - R_2^* + R_2) \times$$

$$\Theta\left(R_2^* - R_{\text{m}} + \frac{F}{2}\right) \Theta\left(R_{\text{m}} + \frac{F}{2} - R_2^*\right) p(R_2, R_{\text{m}}, F). \quad \text{(B.57)}$$

We do not specify the pdf $p(R_2, R_{\text{m}}, F)$ here, because it later turns out that this function may be arbitrary. It is sufficient to assume that it can be normalized as usual to ensure that $p(X_{\text{syst}}, R_2^*, R_2, R_{\text{m}}, F)$ is also normalized.

We can now calculate the expectations

$$E(X_{\text{syst}}) = \int_{-\infty}^{+\infty}\int_{-\infty}^{+\infty}\int_{-\infty}^{+\infty}\int_{-\infty}^{+\infty}\int_{-\infty}^{+\infty} X_{\text{syst}} p(X_{\text{syst}}, R_2^*, R_2, R_{\text{m}}, F)dX_{\text{syst}}dR_2^*dR_2dR_{\text{m}}dF,$$

and

$$E(X_{\text{syst}}^2) = \int_{-\infty}^{+\infty}\int_{-\infty}^{+\infty}\int_{-\infty}^{+\infty}\int_{-\infty}^{+\infty}\int_{-\infty}^{+\infty} X_{\text{syst}}^2 p(X_{\text{syst}}, R_2^*, R_2, R_{\text{m}}, F)dX_{\text{syst}}dR_2^*dR_2dR_{\text{m}}dF.$$

Using equation (B.57), the definitions of DIRAC's δ-function and the HEAVISIDE step function, integration with respect to X_{syst} gives

$$E[X_{\text{syst}}] = \int\limits_{-\infty}^{+\infty} \int\limits_{-\infty}^{+\infty} \int\limits_{-\infty}^{+\infty} \int\limits_{R_{\text{m}}-F/2}^{R_{\text{m}}+F/2} \frac{1}{F}(R_2^* - R_2)p(R_2, R_{\text{m}}, F)\mathrm{d}R_2^* \mathrm{d}R_2 \mathrm{d}R_{\text{m}} \mathrm{d}F,$$

and

$$E[X_{\text{syst}}^2] = \int\limits_{-\infty}^{+\infty} \int\limits_{-\infty}^{+\infty} \int\limits_{-\infty}^{+\infty} \int\limits_{R_{\text{m}}-F/2}^{R_{\text{m}}+F/2} \frac{1}{F}(R_2^* - R_2)^2 p(R_2, R_{\text{m}}, F)\mathrm{d}R_2^* \mathrm{d}R_2 \mathrm{d}R_{\text{m}} \mathrm{d}F.$$

Further integration with respect to R_2^* gives

$$E[X_{\text{syst}}] = \int\limits_{-\infty}^{+\infty} \int\limits_{-\infty}^{+\infty} \int\limits_{-\infty}^{+\infty} (R_{\text{m}} - R_2)p(R_2, R_{\text{m}}, F)\mathrm{d}R_2 \mathrm{d}R_{\text{m}} \mathrm{d}F,$$

and

$$E[X_{\text{syst}}^2] = \int\limits_{-\infty}^{+\infty} \int\limits_{-\infty}^{+\infty} \int\limits_{-\infty}^{+\infty} \left[(R_{\text{m}} - R_2)^2 + \frac{F^2}{12}\right] p(R_2, R_{\text{m}}, F)\mathrm{d}R_2 \mathrm{d}R_{\text{m}} \mathrm{d}F.$$

By performing the last three integrations, using marginalization where necessary, we obtain

$$E[X_{\text{syst}}] = r_{\text{m}} - r_2, \tag{B.58}$$

and

$$E[X_{\text{syst}}^2] = E[R_{\text{m}}^2] + E[R_2^2] - 2E[R_{\text{m}}R_2] + \frac{E[F^2]}{12}, \tag{B.59}$$

or with

$$r_{\text{m}} = \int\limits_{-\infty}^{+\infty} \int\limits_{-\infty}^{+\infty} \int\limits_{-\infty}^{+\infty} R_{\text{m}}\, p(R_2, R_{\text{m}}, F)\mathrm{d}R_2 \mathrm{d}R_{\text{m}} \mathrm{d}F,$$

$$r_2 = \int\limits_{-\infty}^{+\infty} \int\limits_{-\infty}^{+\infty} \int\limits_{-\infty}^{+\infty} R_2\, p(R_2, R_{\text{m}}, F)\mathrm{d}R_2 \mathrm{d}R_{\text{m}} \mathrm{d}F,$$

$$E[R_{\text{m}}^2] = \int\limits_{-\infty}^{+\infty} \int\limits_{-\infty}^{+\infty} \int\limits_{-\infty}^{+\infty} R_{\text{m}}^2\, p(R_2, R_{\text{m}}, F)\mathrm{d}R_2 \mathrm{d}R_{\text{m}} \mathrm{d}F,$$

$$E[R_2^2] = \int\limits_{-\infty}^{+\infty} \int\limits_{-\infty}^{+\infty} \int\limits_{-\infty}^{+\infty} R_2^2\, p(R_2, R_{\text{m}}, F)\mathrm{d}R_2 \mathrm{d}R_{\text{m}} \mathrm{d}F,$$

$$E[R_m R_2] = \int\limits_{-\infty}^{+\infty} \int\limits_{-\infty}^{+\infty} \int\limits_{-\infty}^{+\infty} R_m R_2 \, p(R_2, R_m, F) dR_2 dR_m dF ,$$

and

$$E[F^2] = \int\limits_{-\infty}^{+\infty} \int\limits_{-\infty}^{+\infty} \int\limits_{-\infty}^{+\infty} F^2 \, p(R_2, R_m, F) dR_2 dR_m dF .$$

From equation (B.58) we get directly

$$x_{syst} = r_m - r_2 , \tag{B.60}$$

and using the definition of uncertainty and covariance, we obtain from equations (B.58) and (B.59)

$$u(x_{syst}) = \sqrt{u^2(r_m) + u^2(r_2) - 2\mathrm{Cov}(R_m R_2) + \frac{f^2 + u^2(f)}{12}} , \tag{B.61}$$

with

$$f = E[F] = \int\limits_{-\infty}^{+\infty} \int\limits_{-\infty}^{+\infty} \int\limits_{-\infty}^{+\infty} F \, p(R_2, R_m, F) dR_2 dR_m dF .$$

Normally the quantities R_m and R_2 are different, because R_2 will in most cases be obtained from a least squares fit of the sphere, while R_m is the mean radius of a minimum zone sphere fit. However, if $R_2 = R_m$ is valid, equations (B.60) and (B.61) can be simplified to

$$x_{syst} = 0$$

and

$$u(x_{syst}) = \sqrt{\frac{f^2 + u^2(f)}{12}} , \tag{B.62}$$

i. e. the systematic measurement error becomes zero. But note that the standard uncertainty $u(x_{syst})$ is still not zero as long as the form deviation is known to be present, even if we assume $f = 0$ to be valid.

If the form deviation is large and the standard uncertainty of the form deviation is negligible compared to the form deviation itself, i. e. if $u(f) \ll f$ is valid, equation (B.62) can be approximated by

$$u(x_{syst}) \approx \frac{f}{2\sqrt{3}} .$$

In this case, the uncertainty of the systematic error is mainly determined by the estimated form deviation.

In our calculations we have assumed that only one of the two spheres has form deviations while the other is perfect. In practice, however, both spheres have form deviations and each of them, of course, gives a contribution to the systematic measurement error according to the formulae (B.60) and (B.61). Obtaining the formulae in this case is a straightforward generalization, which can be easily carried out, and is therefore not shown here.

B.10 Noise as a systematic error

A signal is often corrupted by noise, which can be treated as a systematic measurement error.[4] To keep the calculations simple, we will only consider the case of a constant signal corrupted by noise. The noise amplitude varies rapidly during the measurement period.

Let us assume that we are only able to detect the maximum and minimum amplitude values x_{max} and x_{min} respectively, because the measuring instrument is not able to resolve individual amplitude values.

In order to describe the observed fluctuations, we use the simple noise model

$$X_{syst} = A \sin \Phi, \tag{B.63}$$

where A denotes the amplitude of the noise and Φ denotes the phase, which can take any value in the interval $[0, 2\pi)$. We want to know the expectation x_{syst} of this systematic error and its associated standard uncertainty $u(x_{syst})$.

Since only the values x_{max} and x_{min} are known, we have to treat the quantities X_{syst}, A and Φ as random. Their joint pdf is denoted by $p(X_{syst}, A, \Phi | x_{max}, x_{min})$, given the values x_{max} and x_{min}, which can be written by recursive application of the product rule of probability theory as

$$p(X_{syst}, A, \Phi | x_{max}, x_{min}) = p(X_{syst} | A, \Phi, x_{max}, x_{min}) p(A | x_{max}, x_{min}) p(\Phi), \tag{B.64}$$

where $p(X_{syst} | A, \Phi, x_{max}, x_{min})$ is the conditional pdf of the systematic measurement error, given that the amplitude A and the phase Φ, as well as the amplitude values x_{max} and x_{min} are known, $p(A | x_{max}, x_{min})$ is the conditional pdf of A, given that the values x_{max} and x_{min} are known, and $p(\Phi)$ the pdf of the phase Φ, taking into account the latter two cases that the amplitude A and the phase Φ are stochastically independent and, furthermore, that the latter quantity is independent of the values x_{max} and x_{min}.

The conditional pdf $p(X_{syst} | \Phi, x_{max}, x_{min})$ can be immediately derived from the model equation (B.63) as follows

$$p(X_{syst} | \Phi, x_{max}, x_{min}) = \delta \left(X_{syst} - A \sin \Phi \right), \tag{B.65}$$

where $\delta(\cdot)$ denotes DIRAC's δ-function.

---

4 Noise is normally considered to be a random error. However, if we have a model that describes how the noise is generated, we can consider noise to be a systematic error. This view is certainly controversial, but the difference between a random error and a systematic error is that the latter varies predictably (VIM, 2.17 [3]), whereas the former does not (VIM, 2.19 [3]). Therefore, it is reasonable to consider the noise as a systematic error if there is a model to predict it.

The pdf $p(A|x_{max}, x_{min})$ can be obtained by applying the PME. Since our prior knowledge is only that A is in the interval (x_{min}, x_{max}), we get the uniform (rectangular) distribution function

$$p(A|x_{max}, x_{min}) = \frac{\Theta(A - x_{min})\Theta(x_{max} - A)}{x_{max} - x_{min}}, \tag{B.66}$$

where $\Theta(\cdot)$ denotes the HEAVISIDE step function.

Since we have assumed that the phase Φ can take any value within the interval $(0, 2\pi)$, the pdf $p(\Phi)$ can be obtained by applying the PME, giving the uniform (rectangular) distribution function

$$p(\Phi) = \frac{\Theta(\Phi)\Theta(2\pi - \Phi)}{2\pi}, \tag{B.67}$$

where $\Theta(\cdot)$ again denotes the HEAVISIDE step function.

Combining the equations (B.64), (B.65), (B.66) and (B.67) gives

$$p(X_{syst}, A, \Phi|x_{max}, x_{min}) =$$

$$\frac{\delta(X_{syst} - A \sin \Phi)\Theta(A - x_{min})\Theta(x_{max} - A)\Theta(\Phi)\Theta(2\pi - \Phi)}{2\pi(x_{max} - x_{min})}. \tag{B.68}$$

We can now calculate the expectations

$$E[X_{syst}] = \int_{-\infty}^{+\infty}\int_{-\infty}^{+\infty}\int_{-\infty}^{+\infty} X_{syst}\,p(X_{syst}, A, \Phi|x_{max}, x_{min})dX_{syst}dAd\Phi \tag{B.69}$$

and

$$E[X_{syst}^2] = \int_{-\infty}^{+\infty}\int_{-\infty}^{+\infty}\int_{-\infty}^{+\infty} X_{syst}^2\,p(X_{syst}, A, \Phi|x_{max}, x_{min})dX_{syst}dAd\Phi. \tag{B.70}$$

Using equation (B.68), the definitions of DIRAC's δ-function and the HEAVISIDE step function, the integration with respect to X_{syst} and A gives

$$E[X_{syst}] = \frac{x_{max} + x_{min}}{4\pi} \int_0^{2\pi} \sin \Phi d\Phi$$

and

$$E[X_{syst}^2] = \frac{(x_{max} - x_{min})^2 + 3x_{max}x_{min}}{6\pi} \int_0^{2\pi} \sin^2 \Phi d\Phi,$$

and the integration with respect to Φ finally

$$E[X_{syst}] = 0 \tag{B.71}$$

and

$$E[X_{syst}^2] = \frac{(x_{max} - x_{min})^2}{6} + \frac{1}{2} x_{max} x_{min} \,. \tag{B.72}$$

From the equation (B.71) we get directly

$$x_{syst} = 0$$

and from equations (B.71) and (B.72), using the definition of the standard uncertainty, we get

$$u(x_{syst}) = \sqrt{\frac{(x_{max} - x_{min})^2}{6} + \frac{1}{2} x_{max} x_{min}} \,, \tag{B.73}$$

i. e. we do not need to correct for a systematic effect due to noise, but we must expect an uncertainty due to it.

To the best of the author's knowledge, the result (B.73) is not yet mentioned in the literature. Instead, another result is often given for the noise corruption problem. This is obtained if the amplitude A is not treated as a stochastic quantity, but is assumed to be a constant with the value

$$A = \frac{x_{max} - x_{min}}{2} \,.$$

With this assumption we have

$$p(A|x_{max}, x_{min}) = \delta\left(A - \frac{x_{max} - x_{min}}{2}\right)$$

instead of the equation (B.66), where $\delta(\cdot)$ denotes DIRAC's δ-function, which gives

$$p(X_{syst}, A, \Phi | x_{max}, x_{min}) =$$

$$\frac{1}{2\pi} \delta(X_{syst} - A \sin \Phi) \delta\left(A - \frac{x_{max} - x_{min}}{2}\right) \Theta(\Phi) \Theta(2\pi - \Phi)$$

instead of the pdf given by equation (B.68). Using this pdf in the equations (B.69) and (B.70), the definitions of Dirac's δ-function, and the HEAVISIDE step function, the integration with respect to X_{syst} and A now gives

$$E[X_{syst}] = \frac{x_{max} - x_{min}}{4\pi} \int_0^{2\pi} \sin \Phi d\Phi$$

and

$$E[X_{syst}^2] = \frac{(x_{max} - x_{min})^2}{8\pi} \int_0^{2\pi} \sin^2 \Phi d\Phi \,,$$

and the integration with respect to Φ finally

$$E[X_{syst}] = 0 \tag{B.74}$$

and

$$E[X_{syst}^2] = \frac{(x_{max} - x_{min})^2}{8} .$$ (B.75)

From the equation (B.74) we get directly

$$x_{syst} = 0$$ (B.76)

and using the equations (B.74) and (B.75), as well as the definition of the standard uncertainty we get

$$u(x_{syst}) = \frac{x_{max} - x_{min}}{2\sqrt{2}} = \frac{A}{\sqrt{2}} ,$$ (B.77)

i. e. it turns out again that we do not need to correct for the systematic effect of the noise, but we still have to expect an uncertainty due to it, which is now equal to the effective amplitude value. This is the result often found in the literature.

Therefore, if the amplitude of a sinusoidal fluctuation is constant and only its phase varies randomly, we obtain a standard uncertainty equal to the effective amplitude value. In fact, the assumption of a sinusoidal fluctuation is not even necessary in such a case. We only need the weaker assumption that the fluctuations are an ergodic random process (for more details on ergodicity see e. g. [114]). In this case, we can replace the expectations by the averages over the measurement time T, provided that T is sufficiently long. Thus, if the observed ergodic fluctuation is given by the function $x(t)$ as a function of time t, we have

$$E[X_{syst}] = \overline{x(t)} = \frac{1}{T} \int_0^T x(t)dt$$ (B.78)

and

$$E[X_{syst}^2] = \overline{x^2(t)} = \frac{1}{T} \int_0^T x^2(t)dt ,$$ (B.79)

where $\overline{x(t)}$ is the time average of the fluctuating signal $x(t)$ and $x_{eff}^2 = \overline{x^2(t)}$ is the square of its effective amplitude value.

From the equation (B.78) we therefore get

$$x_{syst} = \overline{x(t)}$$ (B.80)

and using the definition of uncertainty from equations (B.78) and (B.79) we obtain

$$u(x_{syst}) = \sqrt{x_{eff}^2 - \overline{x(t)}^2} .$$ (B.81)

These two equations give the general result for a systematic effect caused by noise, if that noise can be assumed to be an ergodic random process. In the case of a sinusoidal process with a constant amplitude $(x_{max} - x_{min})/2$ and a random phase, we again obtain the equations (B.76) and (B.77).

Note that a sinusoidal fluctuation with random phase and random amplitude is not ergodic. In this case the formulae (B.80) and (B.81) are not applicable, as we have already shown above for the example of a uniform pdf for the amplitude fluctuations.

If the noise cannot be assumed to be ergodic, and we cannot specify the pdf of the amplitude fluctuations, we can still specify the standard uncertainty of the systematic measurement error, provided that we know at least the expectation a of the amplitude and the associated standard uncertainty $u(a)$. In this case we use the joint pdf

$$p(X_{syst}, A, \Phi | a, u(a)) = \frac{1}{2\pi} \delta(X_{syst} - A \sin \Phi) \Theta(\Phi) \Theta(2\pi - \Phi) p(A | a, u(a))$$

instead of equation (B.68). Inserting this pdf into equations (B.69) and (B.70) and integrating with respect to X_{syst} and Φ yields

$$E[X_{syst}] = 0 \tag{B.82}$$

and

$$E[X_{syst}^2] = \frac{1}{2} \int_{-\infty}^{+\infty} A^2 p(A | a, u(a)) dA = \frac{E[A^2]}{2}. \tag{B.83}$$

Thus, from equation (B.82) we again get $x_{syst} = 0$, and from equations (B.82) and (B.83), using the definition of the standard uncertainty, we get

$$u(x_{syst}) = \sqrt{\frac{a^2 + u^2(a)}{2}}. \tag{B.84}$$

This result holds for any pdf underlying A. It can be shown that if A has a uniform pdf, the result (B.73) is consistent with equation (B.84). If A varies according to a GAUSSian pdf, i. e. in the case of a so-called GAUSSian noise with mean $a = 0$ and standard deviation $u(a) = \sigma$, we obtain from equation (B.84) the result

$$u(x_{syst}) = \frac{\sigma}{\sqrt{2}}.$$

This formula shows that for a GAUSSian noise, the standard deviation of the noise amplitude plays the role of the root-mean-square value.

C Bayesian linking of key comparisons

The content of this appendix is a revised version of an unpublished article,[1] written in 2015. A Bayesian approach to the problem of linking the results of key comparison measurements is presented. The mathematical treatment is based on Bayesian statistics. This robust statistical analysis provides expressions and standard uncertainties for the key comparison reference value (KCRV) and the degree of equivalence (DOE), as well as a conformity check, without assuming the priority of the comparison contributions. In addition to deriving the mathematical formulae to be used for this type of "distributed linkage", a synthetic and a real linkage example are presented and possible applications of this linkage method are discussed.

C.1 Introduction

The CIPM Mutual Recognition Arrangement (MRA), which came into force in October 1999 and has since been signed by 97 National Metrology Institutes (NMIs), 4 international organizations and 149 designated institutes worldwide, describes the principles and requirements for the mutual recognition of national measurement standards and of calibration and measurement certificates issued by NMIs. In addition, the CIPM MRA covers a further 150 institutes designated by the signatory bodies. The MRA, chapter 3 states:

> ... the technical basis of this arrangement is the set of results obtained in the course of time through key comparisons carried out by the Consultative Committees of the CIPM, the BIPM, and the regional metrology organizations (RMOs), and published by the BIPM and maintained in the key comparison database. Key comparisons carried out by Consultative Committees or the BIPM are referred to as CIPM key comparisons; key comparisons carried out by regional metrology organizations are referred to as RMO key comparisons; RMO key comparisons must be linked to the corresponding CIPM key comparisons by means of joint participants. The degree of equivalence derived from an RMO key comparison has the same status as that derived from a CIPM key comparison.

To further support mutual confidence in the validity of the calibration and measurement certificates issued by the participating institutes, so-called supplementary comparisons are additionally carried out. The metrology community has followed these documented

---

1 The original version of this article is available on the preprint server of the Cornell University Library (arxiv.org): M. Krystek, H. Bosse, *A Bayesian approach to the linking of key comparisons*, arXiv:1501.07134v1 [stat.AP].

https://doi.org/10.1515/9783111453712-009

requirements to regularly perform international comparison measurements, the results of which are continuously registered and made available to the public in the BIPM Key Comparison Database (KCDB).

The issue of linking by joint participants is already addressed in the MRA, but without prescribing a specific mathematical linking procedure. Various approaches to robust statistical linking of the results of comparisons, e. g. by transferring a key comparison reference value (KCRV) from one comparison to another, have been discussed and proposed in the literature and have also been applied to the analysis of comparison results [115, 116, 117, 118].

Most of the proposed approaches for robust statistical linking follow the basic assumption of the MRA, namely that the results of an RMO key comparison must be linked to the results of a CIPM key comparison. The CIPM key comparison in this concept is generally regarded as a primary comparison, both with respect to the chronological order and with respect to the participating laboratories, because

> ... participation in a CIPM key comparison is open to laboratories having the highest technical competence and experience, normally the member laboratories of the appropriate Consultative Committee.[2]

In this sense, the linkage of CIPM key comparisons and RMO key comparisons can be considered as a hierarchical process. However, [118] also briefly discussed approaches to linking comparisons by global minimization using generalized least squares methods [119], which can be seen as examples of non-hierarchical linking methods.

In the field of dimensional metrology, under the responsibility of the CIPM Consultative Committee for Length (CCL), international comparisons were started immediately after the signing of the MRA. The results of the first CIPM key comparison on gauge blocks, for example, were already published in 2002 [120].

Based on the experience gained from international comparisons in the field of dimensional metrology, the CCL Working Group on Dimensional Metrology (CCL-WGDM[3]) has published a document describing the characteristics of dimensional comparisons and also proposing an alternative comparison scheme, the so-called CCL-RMO key comparison scheme. The background and details of the proposed comparison scheme are not repeated here, but some key statements from that document are quoted:

- The CCL-RMO key comparisons follow the idea of several comparisons mutually linked together, without the necessity of a top level comparison delivering a KCRV.
- The classical scheme of CCL and RMO key comparisons can be considered to be a special case of the more general CCL-RMO scheme.

---

2 See section 6.1 of the CIPM Mutual Recognition Arrangement (MRA).
3 In June 2009, the structure of the CCL Working Groups changed: MRA-related work is the responsibility of the WG-MRA, while linking issues are the responsibility of the Task Group on Linking (TG-L).

-   In terms of linking laboratories within and across regions by calculation of their respective degrees of equivalence, the classical (hierarchical) and the CCL-RMO comparison scheme can be regarded to be equivalent. The linking of the comparisons is guaranteed by common participation of selected laboratories.

In [121] a generalized formalism for linking a CIPM key comparison with similar regional key comparisons was presented. As an example, this procedure was applied to link the results of a regional (SIM) gauge block comparison [122] with the results of the first CIPM gauge block key comparison [120]. In the procedure applied, a so-called linking invariant parameter was calculated on the basis of the results of the linking laboratories, i. e. those laboratories which participated in both comparisons, in this case using different transfer standards.

This proposed method was discussed and well accepted in the CCL-WGDM, but it was also criticized that it still required a comparison to be designated as primary. This condition is not fully consistent with the linking procedure to be applied for analysing comparisons according to the proposed CCL-RMO key comparison scheme. It was argued that an adapted "distributed linking" procedure should be developed, which should provide a statistically rigorous linking taking into account all available information of the comparisons to be linked, but without the need to designate one comparison as primary.

In the following, a Bayesian approach for the implementation of such a "distributed linking" is presented and an example for the linking of two gauge block comparisons is discussed, referring to the data published in [120],[122].

C.2 The scenario

Consider the following scenario: A group of laboratories (group A) has participated in a key comparison of a travelling standard A and another group of laboratories (group B) has participated in a key comparison of a travelling standard B. The task of each group was to measure a single measurand of the respective travelling standard (e. g. the length of a gauge block). In order to link the results of the two key comparisons, a specific group of laboratories (group C) participated in both key comparisons, i. e. the laboratories of this group were members of both group A and group B and thus measured both travelling standards. All laboratories have reported the respective measurement results and the associated standard uncertainties. In addition, the laboratories of group C have reported the covariances associated with the measurement results of the two different travelling standards. It should also be assumed that the standards were stable, i. e. that they did not change their properties over time during the key comparisons.

In order to simplify the organization of the data, we assume that the laboratories are labelled by assigning a unique natural number to each of them, starting with one in ascending order, regardless of their membership in one of the groups. In the following, we will use the labels as indices for the respective data. It should be understood that although we use an ordered set of indices, it is completely arbitrary which index is assigned to which laboratory. Any permutation of the indices used would also work.

Once the indices have been assigned, we can refer to the index sets of the laboratories rather than the groups. Those laboratories which e. g. measured only the travelling standard A are in the index set $\Im_A \setminus \Im_B$.

C.3 The information available

Once we have assigned labels (indices) to the laboratories, we can summarize the available data in a clear way simply by using the same indices for the laboratories and the data. In the following, we will denote the measurement results of the travelling standard A obtained by the laboratories of group A by $x_{A,i}$ ($i \in \Im_A$) and the standard uncertainties associated with these results by $u(x_{A,i})$ ($i \in \Im_A$). Similarly, we will denote by $x_{B,i}$ ($i \in \Im_B$) the measurement results obtained by the laboratories of group B with respect to the travelling standard B, and the standard uncertainties associated with these results by $u(x_{B,i})$ ($i \in \Im_B$). It remains only to consider the covariances associated with the results of the measurements of the two travelling standards, as reported by the laboratories of group C. These covariances will be denoted in the following by $u(x_{A,i}, x_{B,i})$ ($i \in (\Im_A \cap \Im_B)$). Equivalently, we can use the correlation coefficients defined by

$$r_{AB,i} = \frac{u(x_{A,i}, x_{B,i})}{u(x_{A,i})u(x_{B,i})}, \qquad i \in (\Im_A \cap \Im_B),$$

instead of the covariances. We will often use the correlation coefficients as abbreviations in the following calculations.

Under the condition that the two travelling standards under consideration have been stable during the key comparison, the assumption is justified that all measurement results are the experimental realization of only two different quantities, i. e. the measurands of the travelling standards A and B, denoted in the following by Y_A and Y_B, respectively. Furthermore, we assume that the measurement results have already been appropriately corrected for possible systematic deviations and that the uncertainties associated with the corrections have been included in the combined standard uncertainties associated with the corrected results.

Note that we use capital letters to denote the measurands and *not* the respective measured value of a measurand, which we will denote with lower case letters throughout this appendix. However, we will not differentiate between a measurand and its value (an ideal value to be estimated, sometimes called *true value*) and use the same letter for both, since the respective meaning follows from the context.

C.4 Establishing key comparison reference values

Preliminary remarks

Using the ordinary arithmetic mean of the measured values to calculate a Key Comparison Reference Value (KCRV) cannot be justified because it does not take into account any of the known uncertainty values and thus does not make use of all the available information as required by the *Guide to the Expression of Uncertainty in Measurement* (GUM) [5]. We will therefore apply the Bayesian theory of measurement uncertainty [98], which is entirely based on Bayesian statistics and the principle of maximum entropy (PME). The idea of this theory is first to establish a joint probability density function (pdf) of the measurands under consideration, taking into account the data obtained by measurement or otherwise, together with their associated uncertainties, as well as physical relations and prior information about the measurands. This joint pdf expresses the state of incomplete knowledge about the measurands and is subsequently used to calculate the expectations of the measurands and their associated covariance matrix.

The joint probability density function

Since the measurements of all laboratories participating in the key comparison are assumed to be independent, we can assign probability density functions to the data of each laboratory individually. These pdfs belong to only two different types, because the data types of groups A and B, except for the data of group C, are similar and will therefore lead to the same type of pdf. However, the data type of group C is dissimilar to the data types of groups A and B, excluding the data of group C, and therefore we must expect to get a different type of pdf for this group.

In order to assign pdfs[4] to the data, we follow a suggestion by Jaynes [83], [84], [85] to use the principle of maximum entropy (PME) for this purpose. The PME requires maximizing the relative information entropy [86], [112] by a suitable pdf under the constraints imposed on the pdf by the normalization condition and the data. We make the usual assumption that the measured data and the associated variances and covariances can be equated with the corresponding expectations.

The PME yields in the present circumstances

$$p\left(x_{A,i}, u(x_{A,i}) \mid Y_A\right) = \frac{1}{u(x_{A,i})\sqrt{2\pi}} \exp\left(-\frac{(Y_A - x_{A,i})^2}{2u^2(x_{A,i})}\right), \qquad i \in \mathfrak{I}_A \setminus \mathfrak{I}_B, \quad \text{(C.1)}$$

---

4 We will use the abbreviation
$$p(X|Y) = p_{X|Y}(x|y)$$
for the conditional pdf to simplify the notation of the formulae.

$$p(x_{\mathrm{B},i}, u(x_{\mathrm{B},i}) \mid Y_{\mathrm{B}}) = \frac{1}{u(x_{\mathrm{B},i})\sqrt{2\pi}} \exp\left(-\frac{(Y_{\mathrm{B}} - x_{\mathrm{B},i})^2}{2u^2(x_{\mathrm{B},i})}\right), \qquad i \in \mathfrak{I}_{\mathrm{B}} \setminus \mathfrak{I}_{\mathrm{A}},$$

and

$$p(x_{\mathrm{A},i}, x_{\mathrm{B},i}, u(x_{\mathrm{A},i}), u(x_{\mathrm{B},i}), r_{\mathrm{AB},i} \mid Y_{\mathrm{A}}, Y_{\mathrm{B}}) =$$

$$\frac{1}{2\pi\, u(x_{\mathrm{A},i})u(x_{\mathrm{B},i})\sqrt{1 - r_{\mathrm{AB},i}^2}} \exp\left(\frac{r_{\mathrm{AB},i}(Y_{\mathrm{A}} - x_{\mathrm{A},i})(Y_{\mathrm{B}} - x_{\mathrm{B},i})}{(1 - r_{\mathrm{AB},i}^2)u(x_{\mathrm{A},i})u(x_{\mathrm{B},i})}\right) \tag{C.2}$$

$$\times \exp\left(-\frac{1}{2(1 - r_{\mathrm{AB},i}^2)}\left[\frac{(Y_{\mathrm{A}} - x_{\mathrm{A},i})^2}{u^2(x_{\mathrm{A},i})} + \frac{(Y_{\mathrm{B}} - x_{\mathrm{B},i})^2}{u^2(x_{\mathrm{B},i})}\right]\right), \qquad i \in \mathfrak{I}_{\mathrm{A}} \cap \mathfrak{I}_{\mathrm{B}}.$$

These pdfs can be regarded as sampling distributions and can be used to construct the likelihood function in the usual way. Note that the values $x_{\mathrm{A},i}$ ($i \in \mathfrak{I}_{\mathrm{A}}$) and $x_{\mathrm{B},i}$ ($i \in \mathfrak{I}_{\mathrm{B}}$), as well as their associated uncertainties $u(x_{\mathrm{A},i})$ and $u(x_{\mathrm{B},i})$ respectively are given data, while Y_{A} and Y_{B} are unknown quantities.

The prior pdf $p(Y_{\mathrm{A}}, Y_{\mathrm{B}})$ would allow us to express our knowledge about the quantities Y_{A} and Y_{B}. However, to be conservative, we use Jeffreys' prior here [36], i. e. we assign a constant to this prior pdf. If we knew more about Y_{A} and Y_{B}, we could change the prior accordingly.

Using Bayes' theorem we obtain the posterior pdf

$$p(Y_{\mathrm{A}}, Y_{\mathrm{B}} \mid \mathfrak{D}) = C e^{-\chi^2/2},$$

where the set $\mathfrak{D}$ represents all known data,

$$\chi^2 = \sum_{i \in (\mathfrak{I}_{\mathrm{A}} \setminus \mathfrak{I}_{\mathrm{B}})} \frac{(Y_{\mathrm{A}} - x_{\mathrm{A},i})^2}{u^2(x_{\mathrm{A},i})} + \sum_{i \in (\mathfrak{I}_{\mathrm{B}} \setminus \mathfrak{I}_{\mathrm{A}})} \frac{(Y_{\mathrm{B}} - x_{\mathrm{B},i})^2}{u^2(x_{\mathrm{B},i})}$$

$$+ \sum_{i \in (\mathfrak{I}_{\mathrm{A}} \cap \mathfrak{I}_{\mathrm{B}})} \frac{1}{1 - r_{\mathrm{AB},i}^2}\left[\frac{(Y_{\mathrm{A}} - x_{\mathrm{A},i})^2}{u^2(x_{\mathrm{A},i})} + \frac{(Y_{\mathrm{B}} - x_{\mathrm{B},i})^2}{u^2(x_{\mathrm{B},i})}\right] \tag{C.3}$$

$$- \sum_{i \in (\mathfrak{I}_{\mathrm{A}} \cap \mathfrak{I}_{\mathrm{B}})} \frac{2r_{\mathrm{AB},i}}{1 - r_{\mathrm{AB},i}^2} \frac{(Y_{\mathrm{A}} - x_{\mathrm{A},i})(Y_{\mathrm{B}} - x_{\mathrm{B},i})}{u(x_{\mathrm{A},i})u(x_{\mathrm{B},i})}$$

and C denotes a normalization constant. This is the posterior pdf of the quantities Y_{A} and Y_{B} given the data $\mathfrak{D}$.

After some algebraic transformations, equation (C.3) can be written as

$$\chi^2 = \frac{1}{1 - \tilde{r}_{\mathrm{AB}}^2}\left(\frac{(Y_{\mathrm{A}} - \hat{y}_{\mathrm{A}})^2}{u^2(\hat{y}_{\mathrm{A}})} - 2\tilde{r}_{\mathrm{AB}}\frac{(Y_{\mathrm{A}} - \hat{y}_{\mathrm{A}})(Y_{\mathrm{B}} - \hat{y}_{\mathrm{B}})}{u(\hat{y}_{\mathrm{A}})u(\hat{y}_{\mathrm{B}})} + \frac{(Y_{\mathrm{B}} - \hat{y}_{\mathrm{B}})^2}{u^2(\hat{y}_{\mathrm{B}})}\right) + q^2 \tag{C.4}$$

with

$$\hat{y}_A = \frac{bs_1 + cs_2}{ab - c^2}$$

$$\hat{y}_B = \frac{cs_1 + as_2}{ab - c^2}$$

$$u(\hat{y}_A) = \sqrt{\frac{b}{ab - c^2}}$$

$$u(\hat{y}_B) = \sqrt{\frac{a}{ab - c^2}}$$

$$\tilde{r}_{AB} = \frac{c}{\sqrt{ab}}$$

$$
\begin{aligned}
q^2 = & \sum_{i \in (\Im_A \setminus \Im_B)} \frac{(x_{A,i} - \hat{y}_A)^2}{u^2(x_{A,i})} + \sum_{i \in (\Im_B \setminus \Im_A)} \frac{(x_{B,i} - \hat{y}_B)^2}{u^2(x_{B,i})} \\
& + \sum_{i \in (\Im_A \cap \Im_B)} \frac{1}{1 - r_{AB,i}^2} \left(\frac{(x_{A,i} - \hat{y}_A)^2}{u^2(x_{A,i})} + \frac{(x_{B,i} - \hat{y}_B)^2}{u^2(x_{B,i})} \right) \\
& - \sum_{i \in (\Im_A \cap \Im_B)} \frac{2 r_{AB,i}}{1 - r_{AB,i}^2} \frac{(x_{A,i} - \hat{y}_A)(x_{B,i} - \hat{y}_B)}{u(x_{A,i}) u(x_{B,i})}
\end{aligned}
\tag{C.5}
$$

and the abbreviations

$$a = \sum_{i \in (\Im_A \setminus \Im_B)} \frac{1}{u^2(x_{A,i})} + \sum_{i \in (\Im_A \cap \Im_B)} \frac{1}{(1 - r_{AB,i}^2) u^2(x_{A,i})}$$

$$b = \sum_{i \in (\Im_B \setminus \Im_A)} \frac{1}{u^2(x_{B,i})} + \sum_{i \in (\Im_A \cap \Im_B)} \frac{1}{(1 - r_{AB,i}^2) u^2(x_{B,i})}$$

$$c = \sum_{i \in (\Im_A \cap \Im_B)} \frac{r_{AB,i}}{(1 - r_{AB,i}^2) u(x_{A,i}) u(x_{B,i})}$$

$$s_1 = \sum_{i \in (\Im_A \setminus \Im_B)} \frac{x_{A,i}}{u^2(x_{A,i})} + \sum_{i \in (\Im_A \cap \Im_B)} \frac{1}{1 - r_{AB,i}^2} \left(\frac{x_{A,i}}{u^2(x_{A,i})} - \frac{r_{AB,i} x_{B,i}}{u(x_{A,i}) u(x_{B,i})} \right)$$

$$s_2 = \sum_{i \in (\Im_B \setminus \Im_A)} \frac{x_{B,i}}{u^2(x_{B,i})} + \sum_{i \in (\Im_A \cap \Im_B)} \frac{1}{1 - r_{AB,i}^2} \left(\frac{x_{B,i}}{u^2(x_{B,i})} - \frac{r_{AB,i} x_{A,i}}{u(x_{A,i}) u(x_{B,i})} \right)$$

Note that the auxiliary quantities a, b, c, s_1, s_2, and thus the quantities $\hat{y}_A$, $\hat{y}_B$, $u(\hat{y}_A)$, $u(\hat{y}_B)$, $\tilde{r}_{AB}$ depend only on the data obtained by the measurement. Thus, q^2 does not depend on the quantities Y_A and Y_B and can therefore be absorbed by the normalization constant. Therefore, after renormalization, we finally get

$$p(Y_A, Y_B | \mathfrak{D}) = \frac{1}{2\pi u(\hat{y}_A) u(\hat{y}_B) \sqrt{1 - \tilde{r}_{AB}^2}}$$

$$\times \exp\left[-\frac{1}{2(1 - \tilde{r}_{AB}^2)} \left(\frac{(Y_A - \hat{y}_A)^2}{u^2(\hat{y}_A)} - 2\tilde{r}_{AB} \frac{(Y_A - \hat{y}_A)(Y_B - \hat{y}_B)}{u(\hat{y}_A) u(\hat{y}_B)} + \frac{(Y_B - \hat{y}_B)^2}{u^2(\hat{y}_B)} \right) \right],$$

i. e. the posterior pdf of the quantities Y_A and Y_B is a bivariate Gaussian pdf.

The key comparison reference values

Since the posterior pdf of the quantities Y_A and Y_B is a bivariate Gaussian pdf, it follows that

$$\mathsf{E}\,[Y_A] = \hat{y}_A \qquad \mathsf{E}\,[Y_B] = \hat{y}_B$$

are the expectations of the quantities Y_A and Y_B, respectively, and

$$\mathbf{U}_Y = \begin{pmatrix} u^2(\hat{y}_A) & \tilde{r}_{AB} u(\hat{y}_A) u(\hat{y}_B) \\ \tilde{r}_{AB} u(\hat{y}_A) u(\hat{y}_B) & u^2(\hat{y}_B) \end{pmatrix}$$

their associated variance-covariance matrix. We now identify the values $\hat{y}_A$ and $\hat{y}_B$ as the key comparison reference values and the matrix $\mathbf{U}_Y$ as their associated variance-covariance matrix.

For convenience, we summarize the results, using the covariances rather than the correlation coefficients.

$$a = \sum_{i \in (\mathfrak{I}_A \setminus \mathfrak{I}_B)} \frac{1}{u^2(x_{A,i})} + \sum_{i \in (\mathfrak{I}_A \cap \mathfrak{I}_B)} \frac{u^2(x_{B,i})}{u^2(x_{A,i}) u^2(x_{B,i}) - u^2(x_{A,i}, x_{B,i})}, \qquad \text{(C.6)}$$

$$b = \sum_{i \in (\mathfrak{I}_B \setminus \mathfrak{I}_A)} \frac{1}{u^2(x_{B,i})} + \sum_{i \in (\mathfrak{I}_A \cap \mathfrak{I}_B)} \frac{u^2(x_{A,i})}{u^2(x_{A,i}) u^2(x_{B,i}) - u^2(x_{A,i}, x_{B,i})},$$

$$c = \sum_{i \in (\mathfrak{I}_A \cap \mathfrak{I}_B)} \frac{u(x_{A,i}, x_{B,i})}{u^2(x_{A,i}) u^2(x_{B,i}) - u^2(x_{A,i}, x_{B,i})},$$

$$s_1 = \sum_{i \in (\mathfrak{I}_A \setminus \mathfrak{I}_B)} \frac{x_{A,i}}{u^2(x_{A,i})} + \sum_{i \in (\mathfrak{I}_A \cap \mathfrak{I}_B)} \frac{u^2(x_{B,i}) x_{A,i} - u(x_{A,i}, x_{B,i}) x_{B,i}}{u^2(x_{A,i}) u^2(x_{B,i}) - u^2(x_{A,i}, x_{B,i})},$$

$$s_2 = \sum_{i \in (\mathfrak{I}_B \setminus \mathfrak{I}_A)} \frac{x_{B,i}}{u^2(x_{B,i})} + \sum_{i \in (\mathfrak{I}_A \cap \mathfrak{I}_B)} \frac{u^2(x_{A,i}) x_{B,i} - u(x_{A,i}, x_{B,i}) x_{A,i}}{u^2(x_{A,i}) u^2(x_{B,i}) - u^2(x_{A,i}, x_{B,i})}. \qquad \text{(C.7)}$$

Key comparison reference values (KCRVs):

$$\hat{y}_A = \frac{bs_1 + cs_2}{ab - c^2},$$ (C.8)

$$\hat{y}_B = \frac{cs_1 + as_2}{ab - c^2}.$$ (C.9)

Standard uncertainties of the KCRVs:

$$u(\hat{y}_A) = \sqrt{\frac{b}{ab - c^2}},$$ (C.10)

$$u(\hat{y}_B) = \sqrt{\frac{a}{ab - c^2}}.$$ (C.11)

Covariance of the KCRVs:

$$u(\hat{y}_A, \hat{y}_B) = \frac{c}{ab - c^2}.$$ (C.12)

C.5 Degrees of equivalence

The degrees of equivalence (DOE) are defined as

$$d_{A,i} = x_{A,i} - \hat{y}_A, \qquad i \in \mathfrak{I}_A,$$

and

$$d_{B,i} = x_{B,i} - \hat{y}_B, \qquad i \in \mathfrak{I}_B.$$

These values represent the deviation of each laboratory from the respective KCRV.
  In accordance with the rules of variance propagation, we obtain

$$u(d_{A,i}) = \sqrt{u^2(x_{A,i}) - u^2(\hat{y}_A)}, \qquad i \in \mathfrak{I}_A,$$ (C.13)

and

$$u(d_{B,i}) = \sqrt{u^2(x_{B,i}) - u^2(\hat{y}_B)}, \qquad i \in \mathfrak{I}_B,$$ (C.14)

for the respective uncertainties. The negative sign under the square root is due to the correlations $u(\hat{y}_A, x_{B,k}) = u^2(\hat{y}_A)$ and $u(\hat{y}_B, x_{B,k}) = u^2(\hat{y}_B)$ as shown in the section attached at the end of this appendix.

C.6 Conformity tests

Once the evaluation procedure has been completed, we need to check that the results obtained are consistent with the data supplied. This can be done in the following way. We notice that χ^2 is represented on the one hand by the equation (C.3) and on the other hand by the equation (C.4). Therefore, we can calculate the expectation $E[\chi^2]$ in two different ways, which results in

$$E[\chi^2] = N = 2 + q^2,$$

with

$$N = \text{card } \mathfrak{I}_A + \text{card } \mathfrak{I}_B$$

and q^2 given by equation (C.5). Thus, we obtain

$$q^2 = N - 2. \tag{C.15}$$

Note that the right-hand side of this equation can be regarded as the number of degrees of freedom, because it is simply the total number of measured data minus the number of estimated quantities.

However, the condition (C.15) cannot be expected to be strictly satisfied, but $N - 2$ is the most likely value of q^2, because if the data are indeed statistically distributed according to the model and the given measurement uncertainties, then equation the (C.15) follows. A significant deviation from this result is a signal that either the model is wrong or the data are suspect. Assuming that the model is correct, a result $q^2 > N - 2$ leads to the conclusion that there is a strong possibility that either some of the measurement uncertainties have been underestimated or that the corresponding measurement results contain uncorrected (or unknown) systematic deviations. Thus,

$$q^2 \leq N - 2$$

can be taken as an indication of the conformity of the estimated results with the data.

C.7 Examples

In order to demonstrate the application of the proposed method, we apply it to two examples. Our first example is synthetic. The data have been generated by simulated sampling from Gaussian distributions as given by the equations (C.1) to (C.2), using the parameters[5] $Y_A = 110$, $\sigma_A = 20$, $Y_B = 120$, $\sigma_B = 50$, $\rho = 0.5$ and a sample size of $n = 50$. Subsequently, the sampled data were evaluated using the usual formulae for sample mean, sample variance and sample covariance. The resulting data and their evaluation are shown in Tab. C.1 and Tab. C.2.

As can be seen, the data pass the conformity test. However, it turns out that the KCRVs are slightly larger than those used in the simulation. This is due to the fact that the correlation coefficients estimated from the sampled data tend to be too large compared to the correlation coefficient used for the simulation. This illustrates the effect of small sample sizes.

Tab. C.1: Data of the synthetic example obtained from the sampled data by applying the usual formulae for sample mean, sample variance and sample covariance.

Institute	$x_{A,i}$	$u(x_{A,i})$	$x_{B,i}$	$u(x_{B,i})$	$r_{AB,i}$
LAB-01	113.4	2.9			
LAB-02	112.1	2.8			
LAB-03	113.0	2.5			
LAB-04	110.6	2.6			
LAB-05	109.4	2.4			
LAB-06	107.0	2.6			
LAB-07	104.7	2.8			
LAB-08	109.0	2.6			
LAB-09	111.0	2.4	120.1	6.5	0.8
LAB-10	109.4	2.8	117.3	7.3	0.8
LAB-11	111.1	2.8	125.0	6.4	0.8
LAB-12	115.3	2.4	135.7	6.7	0.7
LAB-13			129.7	6.1	
LAB-14			129.1	7.5	
LAB-15			125.0	7.1	
LAB-16			123.6	6.6	
LAB-17			123.0	6.9	

As a second example, data from the CCL-K1 [120] and SIM.L-K1 [122] gauge block comparisons were evaluated. The metrology institutes CENAM, NIST and NRC participated in both comparisons and thus acted as linking laboratories. For convenience, the results and their associated standard uncertainties, as reported in the respective publications, are repeated in Tab. C.3. Covariance values were unfortunately not reported, although systematic deviations in the measurement of gauge blocks in each of the linking laboratories using the same measurement system will inevitably cause correlations, even if a correction for these systematic effects is applied, as is usually good practice. Thus, following the recommendation of the GUM, in the absence of this information,[6] all covariances were assumed to be zero, knowing that this is certainly not true.

The degrees of equivalence for the deviation from the nominal values and their associated standard uncertainties, as well as the key comparison reference values and their associated standard uncertainties for groups A and B respectively, obtained by

---

5 σ and ρ denote the standard deviation and the correlation coefficient of a Gaussian distribution, respectively.

6 It should be noted that this recommendation is consistent with the *modus operandi* of Bayesian theory, where only *known* information is taken into account.

Tab. C.2: Degrees of equivalence for the deviation from the nominal values and their associated standard uncertainties, as well as the key comparison reference values and their associated standard uncertainties for the groups A and B, respectively, obtained by the application of the proposed linking method for the synthetic example.

Institute	$d_{A,i}$	$u(d_{A,i})$	$d_{B,i}$	$u(d_{B,i})$
LAB-01	2.491	2.815		
LAB-02	1.191	2.712		
LAB-03	2.091	2.401		
LAB-04	−0.309	2.505		
LAB-05	−1.509	2.296		
LAB-06	−3.909	2.505		
LAB-07	−6.209	2.712		
LAB-08	−1.909	2.505		
LAB-09	0.091	2.296	−3.779	6.196
LAB-10	−1.509	2.712	−6.579	7.030
LAB-11	0.191	2.712	1.121	6.091
LAB-12	4.391	2.296	11.821	6.405
LAB-13			5.821	5.775
LAB-14			5.221	7.238
LAB-15			1.121	6.822
LAB-16			−0.279	6.300
LAB-17			−0.879	6.614

$$\hat{y}_A = 110.909, \quad u(\hat{y}_A) = 0.698$$
$$\hat{y}_B = 123.879, \quad u(\hat{y}_B) = 1.966$$
$$q^2/(N-2) = 0.89$$

applying the linking method proposed here, are given in Tab. C.4 on page 346 for the original data. However, the data do not pass the conformity test.

Assuming that the evaluation model is correct and that the respective measurement results have been corrected for systematic deviations, the test result leads to the conclusion that there is a strong possibility that some of the measurement uncertainties have been underestimated. Examining the uncertainty values given in Tab. C.3, it is obvious that "... *INMETRO1 submitted very optimistic uncertainty claims relative to conventional capabilities*" [122] (the corresponding value is printed in bold in the table). Since the gauge blocks at INMETRO were measured by two completely different instruments, involving different staff (the data denoted by INMETRO1 represent results from a research grade instrument, while those denoted by INMETRO2 represent results from the routine gauge block calibration service offered by INMETRO) [122], it is reasonable to question the very low uncertainty reported for INMETRO1. Therefore, this uncertainty value has been increased here from 4.0 nm to 11.2 nm, which is the minimum value for the data to

Tab. C.3: Results for the deviations from the nominal length and their associated standard uncertainties as taken from the CCL-K1 [120] and SIM.L-K1 [122] gauge block comparison reports (steel gauge block, nominal length 100 mm).

Institute	$x_{A,i}$/nm	$u(x_{A,i})$/nm	$x_{B,i}$/nm	$u(x_{B,i})$/nm
METAS	−96.0	13.0		
NPL	−140.0	33.0		
BNM-LNE	−110.0	16.0		
KRISS	−104.3	20.6		
NRLM	−89.4	16.3		
VNIIM	−104.0	15.0		
CSIRO	−114.0	16.0		
NIM	−90.0	10.3		
NIST	−117.0	17.9	−100.0	18.0
CENAM	−119.0	18.7	−93.0	23.0
NRC	−126.0	24.0	−124.0	26.0
INMETRO1			−98.0	4.0
INMETRO2			−68.0	29.0
INTI			−104.0	21.0
CEM			−148.0	17.0

pass the conformity test. The corresponding results obtained with this change are given in Tab. C.5 on page 347.

A comparison of the results given in Tab. C.4 with those given in Tab. C.5 shows that the reference value for group B and its associated standard uncertainty have changed significantly. This strongly emphasizes that it is imperative to carry out a *rigorous* conformity test. For example, the E_n criterion did not reveal any problem with the INMETRO1 data during the SIM.l-K1 comparison. In fact, the values $|E_{95}|$ as well as $|E|$ for the evaluation of the 100 mm steel gauge block data were even comparable to those of the NRC [122]. The usual χ^2-criterion at a 5 % confidence level also fails to indicate any problem with the data.

C.8 Uncorrelated measurement results

Here we consider the possibility that the measurement results of the linking laboratories (group C) have to be treated as if they were uncorrelated, because the respective covariances have not been reported.[7] In this case we have to set all covariances $u(x_{A,i}, x_{B,i})$

---

7 The author gratefully acknowledges the valuable comments of A. BALSAMO from INRIM (Italy) and the suggestion to add the information given in this section.

Tab. C.4: Degrees of equivalence for the deviation from the nominal values and their associated standard uncertainties, as well as the key comparison reference values and their associated standard uncertainties for the groups A and B respectively, obtained by applying the proposed linking method (results obtained for the original data in Tab. C.3).

Institute	$d_{A,i}/nm$	$u(d_{A,i})/nm$	$d_{B,i}/nm$	$u(d_{B,i})/nm$
METAS	7.6	12.1		
NPL	−36.4	32.6		
BNM-LNE	−6.4	15.2		
KRISS	−0.7	20.0		
NRLM	14.2	15.6		
VNIIM	−0.4	14.2		
CSIRO	−10.4	15.2		
NIM	13.6	9.1		
NIST	−13.4	17.2	0.5	17.6
CENAM	−15.4	18.1	7.5	22.7
NRC	−22.4	23.5	−23.5	25.7
INMETRO1			2.5	1.7
INMETRO2			32.5	28.8
INTI			−3.5	20.7
CEM			−47.5	16.6

$$\hat{y}_A = -103.6, \quad u(\hat{y}_A) = 4.9 \text{ nm}$$
$$\hat{y}_B = -100.5, \quad u(\hat{y}_B) = 3.6 \text{ nm}$$
$$q^2/(N-2) = 1.07 \quad \text{(conformity test failed)}$$

$(i \in (\mathfrak{I}_A \cap \mathfrak{I}_B))$ equal to zero in the equations (C.6) to (C.7). This yields

$$a = \sum_{i \in \mathfrak{I}_A} \frac{1}{u^2(x_{A,i})},$$

$$b = \sum_{i \in \mathfrak{I}_B} \frac{1}{u^2(x_{B,i})},$$

$$c = 0,$$

$$s_1 = \sum_{i \in \mathfrak{I}_A} \frac{x_{A,i}}{u^2(x_{A,i})},$$

$$s_2 = \sum_{i \in \mathfrak{I}_B} \frac{x_{B,i}}{u^2(x_{B,i})}.$$

Thus, we obtain from the equations (C.8) and (C.9) for the key comparison reference values

$$\hat{y}_A = \frac{s_1}{a} \quad \text{and} \quad \hat{y}_B = \frac{s_2}{b},$$

Tab. C.5: Degrees of equivalence for the deviation from the nominal values and their associated standard uncertainties, as well as the main reference values for comparison and their associated standard uncertainties for the groups A and B, respectively, obtained by applying the proposed linking method after increasing the uncertainty of INMETRO1 from 4.0 nm to 11.2 nm.

Institute	$d_{A,i}$/nm	$u(d_{A,i})$/nm	$d_{B,i}$/nm	$u(d_{B,i})$/nm
METAS	7.6	12.1		
NPL	−36.4	32.6		
BNM-LNE	−6.4	15.2		
KRISS	−0.7	20.0		
NRLM	14.2	15.6		
VNIIM	−0.4	14.2		
CSIRO	−10.4	15.2		
NIM	13.6	9.1		
NIST	−13.4	17.2	6.7	16.6
CENAM	−15.4	18.1	13.7	22.0
NRC	−22.4	23.5	−17.3	25.1
INMETRO1			8.7	8.9
INMETRO2			38.7	28.2
INTI			2.7	19.9
CEM			−41.3	15.6

$$\hat{y}_A = -103.6, \quad u(\hat{y}_A) = 4.9 \text{ nm}$$
$$\hat{y}_B = -106.7, \quad u(\hat{y}_B) = 6.8 \text{ nm}$$
$$q^2/(N-2) = 1.00$$

and from the equations (C.10) and (C.11) for their associated standard uncertainties

$$u(\hat{y}_A) = \frac{1}{\sqrt{a}} \quad \text{and} \quad u(\hat{y}_B) = \frac{1}{\sqrt{b}},$$

i. e. the usual weighted mean results for the KCRVs separately for each group participating in the key comparisons. The results of the groups are independent of each other as indicated by the covariance of the KCRVs, $u(\hat{y}_A, \hat{y}_B) = 0$, as obtained from the equation (C.12). This demonstrates that the linking procedure essentially depends on the covariances reported by the linking laboratories.

C.9 Conclusion

In this appendix we have derived formulae for the linking parameters Key Comparison Reference Value (KCRV) and Degree of Equivalence (DOE) for the analysis of two comparisons based on Bayesian statistics, i. e. taking into account all known information, but without the need to choose one comparison as primary. The applicability of the linking approach was demonstrated using an example from gauge block metrology.

The proposed approach can be described as a type of "distributed linking" that has been sought within the CCL working groups, particularly in the field of dimensional metrology. This "distributed linking" method can be used, for example, to compare the results of two comparisons started in parallel with different transfer standards, where it does not seem reasonable to choose one comparison as primary (defining a KCRV) and the other as secondary (linking its comparison results to the KCRV of the first). Such comparisons with more than one loop were organized when the number of interested participants became rather large, as for example in the completed line scale comparison EUROMET.L-K7 with 31 participants [123].

In a strict Bayesian sense, the use of the "distributed linking" approach takes into account all known information from both comparisons to determine the linking parameters, i. e. all measurement results and their uncertainties. On the other hand, when using the hierarchical linking approaches, it could be argued that not all information is taken into account because, for example, the results of an RMO key comparison are not used to reflect or re-analyse the results of a CIPM key comparison, even though the level of knowledge for both comparisons is increased by the double involvement of the linking laboratories. On the other hand, the results of a subsequent RMO comparison linked to a key comparison by the "distributed linking" approach described would also change the KCRV of the (already completed) key comparison. We see that this would pose new challenges for the operation of the KCDB.

The described approach can be extended to link more than two comparisons. Additional knowledge such as e. g. results from previous measurements can also be considered by choosing an appropriate prior pdf.

Calculation of covariances

The relations $u(\hat{y}_A, x_{B,k}) = u^2(\hat{y}_A)$ and $u(\hat{y}_B, x_{B,k}) = u^2(\hat{y}_B)$, respectively, have implicitly been used in the equations (C.13) and (C.14). In order to show that they are valid, we introduce the estimators

$$\hat{Y}_A = \frac{bS_1 + cS_2}{ab - c^2}$$

and

$$\hat{Y}_B = \frac{cS_1 + aS_2}{ab - c^2},$$

where

$$S_1 = \sum_{i \in (\mathfrak{I}_A \setminus \mathfrak{I}_B)} \frac{X_{A,i}}{u^2(x_{A,i})} + \sum_{i \in (\mathfrak{I}_A \cap \mathfrak{I}_B)} \frac{u^2(x_{B,i})X_{A,i} - u(x_{A,i}, x_{B,i})X_{B,i}}{u^2(x_{A,i})u^2(x_{B,i}) - u^2(x_{A,i}, x_{B,i})}$$

and

$$S_2 = \sum_{i \in (\mathfrak{I}_B \setminus \mathfrak{I}_A)} \frac{X_{B,i}}{u^2(x_{B,i})} + \sum_{i \in (\mathfrak{I}_A \cap \mathfrak{I}_B)} \frac{u^2(x_{A,i})X_{B,i} - u(x_{A,i}, x_{B,i})X_{A,i}}{u^2(x_{A,i})u^2(x_{B,i}) - u^2(x_{A,i}, X_{B,i})}.$$

are functions of the random variables $X_{A,i}$ ($i \in \mathfrak{I}_A$) and $X_{B,i}$ ($i \in \mathfrak{I}_B$). We assume the measured values $x_{A,i}$ ($i \in \mathfrak{I}_A$) and $x_{B,i}$ ($i \in \mathfrak{I}_B$) to be a realization of these random variables.

Since $E[X_{A,i}] = Y_A$, $E[X_{B,i}] = Y_B$, $Var[X_{A,i}] = u^2(x_{A,i})$, $Var[X_{B,i}] = u^2(x_{B,i})$ and $Cov[X_{A,i}, X_{B,i}] = u(x_{A,i}, x_{B,i})$ ($i \in \mathfrak{I}_A$, $i \in \mathfrak{I}_B$) it can be verified, that

$$E[\hat{S}_1] = aY_A - cY_B, \qquad E[\hat{S}_2] = bY_B - cY_A,$$

as well as

$$Var[\hat{S}_1] = a, \qquad Var[\hat{S}_2] = b, \qquad Cov[\hat{S}_1, \hat{S}_2] = -c$$

is valid. Using these results, we obtain

$$E[\hat{Y}_A] = \hat{y}_A, \qquad E[\hat{Y}_B] = \hat{y}_B,$$

$$Var[\hat{Y}_A] = u(\hat{y}_A), \qquad Var[\hat{Y}_B] = u(\hat{y}_B), \qquad Cov[\hat{Y}_A, \hat{Y}_B] = u(\hat{y}_A, \hat{y}_B),$$

i. e. the estimators $\hat{Y}_A$ and $\hat{Y}_B$ yield the correct results as given by equations (C.8) to (C.12).

We observe that

$$Cov[\hat{S}_1, X_{A,k}] = Cov[\hat{S}_2, X_{B,k}] = 1$$

and

$$Cov[\hat{S}_1, X_{B,k}] = Cov[\hat{S}_2, X_{A,k}] = 0.$$

Therefore,

$$u(\hat{y}_A, x_{A,k}) = Cov[\hat{Y}_A, X_{A,k}] = u^2(\hat{y}_A)$$

and

$$u(\hat{y}_B, x_{B,k}) = Cov[\hat{Y}_B, X_{B,k}] = u^2(\hat{y}_B).$$

That had to be proven.

List of Definitions

2.1 Quantity — VIM —— 8
2.2 Quantity —— 8
2.3 Quantity value —— 9
2.4 Quantity value — VIM —— 9
2.5 Measurement unit —— 11
2.6 Numerical quantity value —— 12
2.7 Quantity equation —— 13
2.8 Measurement —— 14
2.9 Measurand —— 14
2.10 Measurement result —— 15
2.11 Measured value —— 15
2.12 True quantity value —— 19
2.13 Measurement principle —— 21
2.14 Measurement method —— 21
2.15 Measurement procedure —— 23
2.16 Measurement accuracy —— 23
2.17 Measurement trueness —— 24
2.18 Measurement precision —— 25
2.19 Repeatability condition of measurement —— 27
2.20 Intermediate precision condition of measurement —— 27
2.21 Reproducibility condition of measurement —— 27
2.22 Resolution —— 27
2.23 Measurement error —— 29
2.24 Reference quantity value —— 29
2.25 Random measurement error —— 30
2.26 Systematic measurement error —— 31
2.27 Correction —— 31
2.28 Measurement uncertainty —— 34
2.29 Definitional uncertainty —— 36

3.1 Linear model —— 46
3.2 Non-linear model —— 47
3.3 Implicit model —— 48
3.4 Directed graph —— 74
3.5 Named graph —— 75
3.6 Path —— 75
3.7 Cycle —— 76
3.8 Directed acyclic graph —— 76
3.9 Predecessor, successor, isolated vertex —— 76
3.10 Connected graph —— 76
3.11 Input degree, output degree —— 76
3.12 Named, coloured vertex graph —— 77
3.13 Named, marked vertex graph —— 77
3.14 Dependency graph —— 78
3.15 Edge-weighted graph —— 85

https://doi.org/10.1515/9783111453712-010

4.1 Classical probability —— **91**
4.2 Random experiment —— **96**
4.3 Relative frequency —— **96**
4.4 Frequentist probability —— **99**
4.5 Subjective probability —— **103**
4.6 Sample space —— **106**
4.7 Proposition —— **106**
4.8 Event —— **108**
4.9 Elementary event —— **109**
4.10 Impossible event —— **109**
4.11 Certain event —— **109**
4.12 Mutually exclusive events —— **111**
4.13 Set difference —— **111**
4.14 Disjoint partition —— **111**
4.15 σ-algebra —— **113**
4.16 Measure axioms —— **116**
4.17 Mathematical probability —— **116**
4.18 Axioms of A. N. KOLMOGOROV —— **117**
4.19 Conditional probability —— **121**
4.20 Axioms of conditional probability —— **122**
4.21 Disjoint partition —— **127**
4.22 Stochastic independence —— **133**
4.23 Total stochastic independence —— **135**
4.24 Random quantity —— **136**
4.25 Discrete random quantity —— **138**
4.26 Continuous random quantity —— **138**
4.27 Probability distribution function —— **141**
4.28 Conditional probability distribution function —— **149**
4.29 Probability density function —— **151**
4.30 Conditional probability density function —— **155**
4.31 Expectation of a discrete random quantity —— **160**
4.32 Expectation of a continuous random quantity —— **162**
4.33 Expectation of a transformed random quantity —— **165**
4.34 Variance of a random quantity —— **168**
4.35 Standard deviation of a random quantity —— **168**
4.36 Multivariate random quantity —— **174**
4.37 Multivariate probability distribution function —— **176**
4.38 Multivariate probability density function —— **179**
4.39 Expectation of a multivariate random quantity —— **195**
4.40 Expectation of a function of a multivariate quantity —— **195**
4.41 Covariance of random quantities —— **200**
4.42 Uncorrelated random quantities —— **200**
4.43 Correlation coefficient —— **201**

5.1 Population —— **211**
5.2 Quantitative characteristic —— **211**
5.3 Population parameter —— **212**
5.4 Sample vector —— **212**
5.5 Statistic —— **213**

5.6   Ordinary statistics —— **214**
5.7   Order statistics —— **215**
5.8   Estimator —— **215**
5.9   Unbiased estimator —— **217**
5.10   Asymptotically unbiased estimator —— **218**
5.11   Efficient estimator —— **219**
5.12   FISHER information matrix —— **219**
5.13   Generalized moment of a probability density function —— **221**
5.14   Sample moment —— **222**
5.15   Moment estimator —— **222**
5.16   Likelihood function —— **225**
5.17   Number of degrees of freedom —— **239**
5.18   BAYESIAN estimator —— **242**
5.19   Expectation of the BAYESIAN estimator —— **243**
5.20   Confidence interval —— **255**
5.21   Coverage interval —— **261**
5.22   Probabilistic symmetric coverage interval —— **266**
5.23   Credible region —— **270**

6.1   Result of a direct measurement —— **277**
6.2   Propagation of measurement uncertainties —— **279**

List of Propositions

3.1 Solvability of an implicit model equation —— 49
3.2 Linearization of a non-linear function —— 58
3.3 Linearization of an explicit model function —— 62
3.4 Linearization of an implicit model function —— 65
3.5 Quadratic approximation of an explicit model function —— 72

4.1 Probability of the impossible event —— 117
4.2 Product rule —— 122
4.3 Conditional probability of the impossible event —— 123
4.4 Conditional probability of the complementary event —— 123
4.5 Conditional probability of conjoint events —— 124
4.6 Monotonicity relation of conditional probability —— 125
4.7 Boundedness of the conditional probability —— 125
4.8 Addition theorem of conditional probability —— 126
4.9 Multiplication theorem of conditional probability —— 126
4.10 Law of total probability —— 127
4.11 Theorem of BAYES-LAPLACE, version 1 —— 129
4.12 Theorem of BAYES-LAPLACE, version 2 —— 129
4.13 Stochastic independence —— 133
4.14 Stochastic independence of complementary events —— 134
4.15 Stochastic dependence of disjoint events —— 134
4.16 Properties of a probability distribution function —— 146
4.17 Rules of probability distribution functions —— 148
4.18 Conditions of a probability density function —— 153
4.19 Transformation of a probability density function —— 157
4.20 Linearity of expectations —— 165
4.21 Monotonicity of expectations —— 166
4.22 Properties of the variance and the standard deviation —— 170
4.23 Distribution function of independent quantities —— 176
4.24 Properties of a multivariate probability distribution —— 178
4.25 Conditions of a multivariate probability density —— 181
4.26 Density function of independent random quantities —— 182
4.27 Transformation of a multivariate probability density —— 183
4.28 Marginalization of probability distribution functions —— 186
4.29 Marginalization of probability density functions —— 187
4.30 Convolution of probability density functions —— 189
4.31 Linearity of the expectation —— 196
4.32 Expectation of a product of independent quantities —— 197
4.33 Properties of the covariance —— 200
4.34 Properties of the correlation coefficient —— 201
4.35 Correlation and stochastic dependence —— 203
4.36 Variance of a linear combination of random quantities —— 204
4.37 Upper bound of the probability of a function value —— 206
4.38 MARKOV's inequality —— 207
4.39 CHEBYSHEV's inequality —— 208
4.40 Central limit theorem —— 209

https://doi.org/10.1515/9783111453712-011

List of Examples

3.1 SHOCKLEY diode model —— **42**
3.2 Resistive voltage drop in an electric power line —— **44**
3.3 Ceiling of an aeroplane —— **44**
3.4 Mass of a mixture of liquids —— **45**
3.5 RICHMANN's mixture rule —— **45**
3.6 Area of a circular disk —— **46**
3.7 Volume of a straight circular cylinder —— **46**
3.8 Mass density of a liquid —— **46**
3.9 Electric power loss in a wire —— **47**
3.10 Van der Waals equation —— **47**
3.11 Semicircle in a plane —— **49**
3.12 Electric power loss in a wire; using submodels —— **50**
3.13 Derivation of RICHMANN's mixture rule —— **51**
3.19 Continuation of example 3.18 —— **66**
3.21 Cosine error in length measurement —— **70**
3.22 Continuation of example 3.21 —— **72**
3.23 OHM's law —— **79**
3.24 Resistance of an incandescent light bulb —— **80**
3.25 Electric power loss in a wire —— **81**
3.26 Continuation of example 3.25 —— **85**
3.27 Continuation of example 3.25 —— **86**

4.1 Rolling an ideally symmetric die —— **90**
4.2 French game of dice in the 17th century —— **90**
4.3 The game of dice of the CHEVALIER DE MÉRÉ —— **91**
4.4 Throwing a fair coin —— **92**
4.5 Drawing a card from a deck of cards —— **92**
4.6 Sums of the number of dots when rolling two dice —— **92**
4.7 BERTRAND's paradox —— **93**
4.8 Multiple rolling of an ideally symmetric die —— **95**
4.9 Probability of rain —— **100**
4.10 Insufficient knowledge —— **101**
4.11 Additional knowledge obtained by an experiment —— **102**
4.12 Shipwreck —— **103**
4.13 Influence of changing information on probability —— **103**
4.14 Sample spaces —— **107**
4.15 Some events when rolling a single die —— **108**
4.16 Power sets of the sample space —— **112**
4.17 σ-algebras —— **114**
4.18 Drawing cards —— **118**
4.19 Two rolls with an ideally symmetric die —— **119**
4.20 Random experiment —— **120**
4.21 Probability of complementary events —— **124**
4.22 Drawing coloured objects —— **124**
4.23 A shipwrecked person on an island —— **125**
4.24 Flipping a fair coin two times —— **126**
4.25 Disjoint partition applied to a game of dice —— **127**

https://doi.org/10.1515/9783111453712-012

4.26 Reliability of products —— 127
4.27 Surveys of tuberculosis —— 130
4.28 Lung cancer caused by smoking —— 132
4.29 Stochastic dependence in roulette —— 134
4.30 Flipping two distinguishable fair coins —— 135
4.31 Rolling an ideally symmetric die —— 136
4.32 Flipping a fair coin several times —— 137
4.33 Rolling a ball —— 138
4.34 Life expectancy of incandescent lamps —— 139
4.35 Sums of the number of dots when rolling two dice —— 140
4.36 Sums of the number of dots when rolling two dice —— 142
4.37 Breakage of a thin wire under strain —— 145
4.38 Continuation of example 4.36 —— 148
4.39 Continuation of example 4.37 —— 149
4.40 Probability of failure of a system —— 149
4.41 Continuation of example 4.37 —— 151
4.42 Continuation of example 4.36 —— 155
4.43 Linear transformation of a continuous quantity —— 157
4.44 Quadratic transformation of a continuous quantity —— 158
4.45 Logarithmic transformation of a continuous quantity —— 159
4.46 Expectation, J. BERNOULLI [25] —— 160
4.47 Rolling an ideally symmetric die —— 160
4.48 Continuation of example 4.36 —— 160
4.49 A six turns up at the k-th roll of a die —— 161
4.50 Continuation of example 4.37 —— 163
4.51 Expectation of a quadratic function —— 164
4.52 Continuation of example 4.48 —— 170
4.53 Continuation of example 4.50 —— 170
4.54 Transformation of temperature scales —— 171
4.55 Standardized and reduced random quantities —— 171
4.56 Position of an object in space —— 173
4.57 Complex valued random quantities —— 173
4.58 Sums of the number of dots when rolling two dice —— 174
4.59 Measurement of a current-voltage characteristic —— 175
4.60 Measurement of neutron spectra —— 175
4.61 Multivariate distribution function when rolling two dice —— 177
4.62 Positioning precision —— 177
4.63 Continuation of example 4.62 —— 180
4.64 Continuation of example 4.61 —— 182
4.65 Transformation to polar co-ordinates —— 183
4.66 Transformation to spherical co-ordinates —— 184
4.67 Transformation to cylindrical co-ordinates —— 184
4.68 General linear transformation —— 185
4.69 Marginal density of a two-dimensional probability density —— 186
4.70 Sum and difference of random quantities —— 187
4.71 Product of random quantities —— 188
4.72 Quotient of random quantities —— 189
4.73 Convolution of normal distributions —— 190
4.74 Convolution of uniform distributions —— 191

4.75 Oscillating random quantity —— **197**
4.76 Linear combination of two random quantities —— **199**
4.77 Uncorrelated, stochastically dependent random quantities —— **203**
4.78 Variance of a linear combination of random quantities —— **204**
4.79 Delay of flights —— **207**
4.80 Estimation of the constant k —— **208**

5.1   Estimating the parameters of a normal distribution —— **216**
5.2   FISHER information matrix of a normal distribution —— **220**
5.3   Method of moments for a normal distribution —— **222**
5.4   Method of moments for a uniform distribution —— **223**
5.5   Estimator of radioactive isotope mean lifetime —— **226**
5.6   Success parameter of a BERNOULLI experiment —— **227**
5.7   Parameters of a normally distributed random quantity —— **229**
5.8   Parameters of a uniformly distributed quantity —— **231**
5.9   Linear regression —— **237**
5.10 Repeated measurements —— **244**
5.11 Calibration of a measurement system —— **246**
5.12 Continuation of example 5.5 —— **248**
5.13 Continuation of example 5.6 —— **249**
5.14 Continuation of example 5.7 —— **250**
5.15 Continuation of example 5.8 —— **253**
5.16 Confidence interval for the mean of a normal distribution —— **256**
5.17 Continuation of example 5.16 —— **259**
5.18 Coverage interval for the mean of a normal distribution —— **262**
5.19 Coverage interval for a measurement system calibration —— **264**

6.1   Effect of correlations when using the E_n criterion —— **284**

A.1   Resistance and reactance —— **302**
A.2   Circle parameters —— **303**
A.3   Alignment —— **305**

B.1   Uniform distribution —— **314**
B.2   Exponential distribution —— **315**

List of Figures

2.1   Hierarchic subdivision of a quantity. —— 9
2.2   On the definition of measurement accuracy. —— 24
2.3   On the definition of measurement trueness. —— 25
2.4   On the definition of measurement precision. —— 25
2.5   Comparison of measurement trueness and measurement precision. —— 26

3.1   Typical I-U characteristic of a semiconductor diode. —— 43
3.2   Linearization of a non-linear equation. —— 56
3.3   Sensitivity of the output quantity to changes in the input quantity. —— 61
3.4   Cosine error in length measurement. —— 70
3.5   A typical directed graph. —— 75
3.6   Marking of a vertex. —— 78
3.7   A typical dependency graph. —— 79
3.8   Ohm's law. —— 80
3.9   Resistance of an incandescent lamp. —— 80
3.10 Electric power loss in a wire (after the first modelling step). —— 82
3.11 Electric power loss in a wire (after the second modelling step). —— 82
3.12 Electric power loss in a wire (after the third modelling step). —— 83
3.13 Electric power loss in a wire (after the last modelling step). —— 84
3.14 Weighted edges. —— 86
3.15 Electric power loss in a wire (dependency graph with weighted edges). —— 87

4.1   Outcomes after rolling an ideally symmetric die 100 times. —— 96
4.2   Relative frequencies when an ideally symmetric die is rolled. —— 98
4.3   Variations of the relative frequency when rolling a die. —— 99
4.4   The event $\mathfrak{E}$ can occur, if the outcome ω is contained in $\mathfrak{E}$. —— 115
4.5   Assignment of probabilities to the events contained in a set of events. —— 117
4.6   Two different random quantities when rolling a die. —— 137
4.7   Probability distribution of the sum of dots when rolling two dice. —— 141
4.8   Heaviside function. —— 144
4.9   Probability distribution function of a continuous random quantity. —— 146
4.10 Probability density function of a continuous random quantity. —— 152
4.11 Probability density of the sum of dots when rolling two dice. —— 155
4.12 Commutative diagram of the transformation of random quantities. —— 156
4.13 Creation of a two-dimensional random quantity. —— 173
4.14 Density function of the sum of two uniformly distributed random quantities. —— 194
4.15 Convolution of uniform distributions. —— 210

5.1   Estimation of a parameter of a population. —— 216
5.2   Confidence intervals of a normally distributed random quantity. —— 259
5.3   Illustration of the relations between sets. —— 268
5.4   Credible region of a bimodal posterior probability density function. —— 270

6.1   Systematic error and random error. —— 274

https://doi.org/10.1515/9783111453712-013

List of Tables

2.1   Names and symbols of the seven base units and their dimensions. —— 12

3.1   Types of vertices and the numbers assigned to them. —— 77

4.1   Basic set operations and their corresponding event relations. —— 110
4.2   Probabilities of the events $\mathfrak{E}_k$. —— 143

C.1   Data of the synthetic example obtained from the sampled data by applying the usual formulae for sample mean, sample variance and sample covariance. —— 343
C.2   Degrees of equivalence for the deviation from the nominal values and their associated standard uncertainties, as well as the key comparison reference values and their associated standard uncertainties for the groups A and B, respectively, obtained by the application of the proposed linking method for the synthetic example. —— 344
C.3   Results for the deviations from the nominal length and their associated standard uncertainties as taken from the CCL-K1 [120] and SIM.L-K1 [122] gauge block comparison reports (steel gauge block, nominal length 100 mm). —— 345
C.4   Degrees of equivalence for the deviation from the nominal values and their associated standard uncertainties, as well as the key comparison reference values and their associated standard uncertainties for the groups A and B respectively, obtained by applying the proposed linking method (results obtained for the original data in Tab. C.3). —— 346
C.5   Degrees of equivalence for the deviation from the nominal values and their associated standard uncertainties, as well as the main reference values for comparison and their associated standard uncertainties for the groups A and B, respectively, obtained by applying the proposed linking method after increasing the uncertainty of INMETRO1 from 4.0 nm to 11.2 nm. —— 347

https://doi.org/10.1515/9783111453712-014

List of Symbols

For each symbol the page number is given, where it is introduced or used for the first time.

$(\Omega, \mathcal{A})$	measurable space (p. 114)
$(\Omega, \mathcal{A}, \mu)$	measure space (p. 115)
$(\Omega, \mathcal{A}, P)$	probability space (p. 117)
$G_X(x)$	probability distribution function of the random quantity X (p. 141)
$G_{X_1,\ldots,X_n}(x_1, \ldots, x_n)$	multivariate (joint) probability distribution function of the random quantity $X = (X_1, \ldots, X_n)^\mathsf{T}$ (p. 176)
$G_X(x)$	multivariate (joint) probability distribution function of the random quantity $X = (X_1, \ldots, X_n)^\mathsf{T}$ (p. 176)
$H_n(\mathfrak{E})$	absolute frequency of the occurrence of the event $\mathfrak{E}$ in n experiments (p. 96)
$M_k(X)$	k-th moment of the sample vector $X = (X_1, \ldots, X_n)^\mathsf{T}$ (p. 222)
$R(X)$	remainder of the deviation of the linear approximation of the function $f(X)$ (p. 55)
R_{XY}	sample correlation coefficient (p. 214)
S^2	sample variance (p. 214)
S_{XY}	sample covariance (p. 214)
S_X^2	sample variance of the components of the random quantity X (p. 214)
X	input quantity (p. 46)
$X(\omega)$	random quantity X (function, assigning a real number to an outcome ω) (p. 136)
$X_{\max}$	maximum of $(X_1, \ldots, X_n)$ (p. 215)
X_{med}	sample median of $(X_1, \ldots, X_n)$ (p. 215)
$X_{\min}$	minimum of $(X_1, \ldots, X_n)$ (p. 215)
X_k	k-th input quantity (p. 46)
$X_{(k)}$	k-th order statistic of $(X_1, \ldots, X_n)$ (p. 214)
Y	output quantity (p. 46)
Y_{lin}	linear approximation of the output quantity of a model function (p. 58)
Y_{qu}	quadratic approximation of the output quantity of a model function (p. 72)
ΔX	change of the input quantity X (p. 56)
$\Theta(x)$	HEAVISIDE function (p. 144)
Ω	sample space (p. 106)
$\mathrm{Cov}(X_1, X_2)$	covariance of the random quantities X_1 and X_2 (p. 200)
$\det J$	determinant of the JACOBIAN matrix J (JACOBIAN determinant) (p. 183)
$\mathrm{D}(X; \xi)$	dispersion of the random quantity X about the reference value ξ (p. 167)
$\mathrm{E}[X]$	expectation of the random quantity X (p. 163)
$\mathrm{E}[X_i]$	expectation of the i-th component of a multivariate random quantity X (p. 195)
$\mathrm{E}[X]$	expectation of the multivariate random quantity X (p. 195)
$\mathrm{E}[X^k]$	k-th moment of a probability distribution function (p. 221)
$\mathrm{E}[f(X)]$	expectation of the random quantity X transformed by the function f (p. 165)
$\mathrm{E}[f(X_1, \ldots, X_n)]$	expectation of the function $f(X_1, \ldots, X_n)$ of the multivariate random quantity $X = (X_1, \ldots, X_n)^\mathsf{T}$ (p. 195)
$\hat{\Theta}$	estimator of the parameter Θ (p. 216)
$\hat{\theta}$	estimated value of the parameter Θ (p. 216)
$\mathbb{R}$	set of real numbers (p. 113)

https://doi.org/10.1515/9783111453712-015

$\mathbf{C}$	covariance matrix (p. 205)
$\mathbf{C}(\hat{\boldsymbol{\Theta}})$	covariance matrix of the estimator $\hat{\boldsymbol{\Theta}}$ (p. 219)
$\mathbf{I}(\theta)$	FISHER information matrix of the parameter vector θ (p. 219)
$\mathbf{M}$	matrix $\mathbf{M}$ (p. 185)
$\mathbf{P}$	correlation matrix (p. 206)
$\mu(\mathfrak{M})$	measure of the set $\mathfrak{M}$ (p. 115)
ω	outcome of an experiment (p. 108)
ω_k	k-th outcome of an experiment (p. 108)
$\overline{X}$	sample mean (p. 214)
$\overline{x}$	arithmetic mean of the values of the discrete random quantity X (p. 160)
$P(\mathfrak{E})$	probability of the event $\mathfrak{E}$ (p. 117)
$\rho(X_1, X_2)$	correlation coefficient of the random quantities X_1 and X_2 (p. 201)
$\mathcal{A}$	set of events (σ-algebra on the sample space Ω) (p. 114)
$\mathcal{P}(\mathfrak{M})$	power set of the set $\mathfrak{M}$ (p. 112)
$\mathfrak{E}$	event (p. 108)
$\mathfrak{E}^c$	complementary event (p. 112)
$\sigma(X)$	standard deviation of the random quantity X (p. 168)
$\sigma_r(X)$	relative standard deviation (coefficient of variation) of the random quantity X (p. 169)
$\mathrm{Var}(X)$	variance of the random quantity X (p. 168)
$X(\omega)$	multivariate random quantity $X(\omega) = (X_1, \ldots, X_n)^\mathsf{T}$ (vector function which assigns a vector of real numbers $x = (x_1, \ldots, x_n)^\mathsf{T}$ to the outcomes $\omega \in \Omega$) (p. 174)
θ	parameter (p. 216)
c_k	k-th sensitivity coefficient (p. 62)
$f'(X_0)$	derivative of the function $f(X)$ at X_0 (p. 56)
$f(X)$	model function of the input quantity X (p. 55)
$f(X_1, \ldots, X_n)$	model function of the input quantities $X_1, \ldots, X_n$ (p. 47)
$f(Y, X_1, \ldots, X_n)$	model function of the output quantity Y and the input quantities $X_1, \ldots, X_n$ (p. 48)
$f * g$	convolution of the functions f and g (p. 190)
$f^{-1}(Y)$	inverse function of the function $f(X)$ (p. 156)
$g_X(x)$	probability density function of the random quantity X (p. 151)
$g_{X_1, \ldots, X_n}(x_1, \ldots, x_n)$	multivariate (joint) probability density function of the random quantities $X_1, \ldots, X_n$ (p. 180)
$g_X(x)$	multivariate (joint) probability density function of the multivariate random quantity $X = (X_1, \ldots, X_n)^\mathsf{T}$ (p. 180)
$h_n(\mathfrak{E})$	relative frequency of the occurrence of the event $\mathfrak{E}$ in n experiments (p. 96)
$r(X)$	relative deviation, as function of the quantity X (p. 56)

Bibliography

[1] JCGM 200:2008. *International vocabulary of metrology — Basic and general concepts and associated terms (VIM), Corrigendum*. Paris: Bureau International des Poids et Mesures (BIPM).

[2] JCGM 200:2008. *International vocabulary of metrology — Basic and general concepts and associated terms (VIM)*. Paris: Bureau International des Poids et Mesures (BIPM).

[3] ISO/IEC Guide 99:2007. *International vocabulary of metrology — Basic and general concepts and associated terms (VIM)*. Genève: International Organization for Standardization (ISO).

[4] JCGM 100:2008. *Evaluation of measurement data — Guide to the expression of uncertainty in measurement*. Paris: Bureau International des Poids et Mesures (BIPM).

[5] ISO/IEC Guide 98:1995. *Guide to the expression of uncertainty in measurement (GUM)*. Genève: International Organization for Standardization (ISO).

[6] DIN 1319-1:1995. *Grundlagen der Messtechnik — Teil 1: Grundbegriffe*. Berlin: Beuth Verlag.

[7] M. Krystek. *Quantities and Units*. Berlin: De Gruyter Oldenbourg, 2023.

[8] M. P. Krystek. "The term 'dimension' in the international system of units". In: *Metrologia* 52 (2015), p. 297.

[9] DIN 55350-13:1987. *Begriffe der Qualitätssicherung und Statistik; Begriffe zur Genauigkeit von Ermittlungsverfahren und Ermittlungsergebnissen*. Berlin: Beuth Verlag.

[10] D. Sadeh. "Experimental Evidence for the Constancy of the Velocity of Gamma Rays, Using Annihilation in Flight". In: *Phys. Rev. Lett.* 10 (1963), pp. 271–273.

[11] P. J. Mohr, B. N. Taylor, and D. B. Newell. "CODATA recommended values of the fundamental physical constants: 2006". In: *Rev. Mod. Phys.* 80 (2008), pp. 633–730.

[12] ISO 3534-2:2006. *Statistics — Vocabulary and symbols — Part 2: Applied statistics*. Genève: International Organization for Standardization (ISO).

[13] DIN 55350-12:1989. *Begriffe der Qualitätssicherung und Statistik; Merkmalsbezogene Begriffe*. Berlin: Beuth Verlag.

[14] C. F. Gauss. *Theoria combinationis observationum erroribus minimum obnoxiae*. Göttingen: Commentationes societatis regiae scientiarum Gottingensis, 1821.

[15] D. Dörner. *The Logic of Failure — Recognizing and Avoiding Error in Complex Situations*. New York: Metropolitan Books, 1996.

[16] DKD-3-E1:1998. *Angabe der Messunsicherheit bei Kalibrierungen — Ergänzung 1 — Beispiele*. Braunschweig: Physikalisch-Technische Bundesanstalt (PTB).

[17] DKD-3-E2:2002. *Angabe der Messunsicherheit bei Kalibrierungen — Ergänzung 2 — Zusätzliche Beispiele*. Braunschweig: Physikalisch-Technische Bundesanstalt (PTB).

[18] T. M. Apostol. *Calculus*. 2nd ed. Vol. I. New York: John Wiley & Sons, Inc., 1967.

[19] L. B. Rall. *Automatic Differentiation: Techniques and Applications*. Berlin: Springer Verlag, 1981.

[20] B. Pascal. *Oeuvre de Blaise Pascal, Nouvelle Édition, Tome Quatrième*. French. Paris: Libraire chez Lefèvre, 1819.

[21] P. S. de Laplace. *Théorie Analytique des Probabilités*. French. Paris: Gauthier-Villars, 1812.

[22] B. Bertrand. *Calcul des probabilités*. French. Paris: Gauthier-Villars et fils, 1889.

[23] B. Pascal. *Traite du triangle arithmetique, avec quelques autres petits traitez sur la mesme matière*. French. Paris: Libraire chez Guillaume Despres, 1665.

[24] G. W. Leibniz. *Dissertatio de arte combinatoria*. Leipzig: Fick & Seubold, 1666.

[25] J. Bernoulli. *Ars conjectandi, opus posthumum. Accedit Tractatus de seriebus infinitis, et epistola gallicé scripta de ludo pilae reticularis*. Basel: Gebrüder Thurneysen, 1713.

[26] P. S. de Laplace. "Mémoire sur les approximations des formules qui sont fonctions de très grand nombres (suite)". French. In: *Mém. Académ. Roy. Scie.;Oeuvres complètes* X (1783;1786), pp. 295–338.

[27] R. von Mises. "Grundlagen der Wahrscheinlichkeitsrechnung". German. In: *Math. Zeitschrift* 5 (1919), pp. 52–99.

https://doi.org/10.1515/9783111453712-016

[28]   R. von Mises. *Wahrscheinlichkeit, Statistik und Wahrheit*. German. Wien: Springer Verlag, 1928.

[29]   R. von Mises and H. Geiringer. *The Mathematical Theory of Probability and Statistics*. German. New York, NY: Academic Press, 1964.

[30]   T. M. Apostol. *Mathematical Analysis*. Reading, MA: Addison-Wesley Publishing Co., 1981.

[31]   H. Reichenbach. "Axiomatik der Wahrscheinlichkeitsrechnung". German. In: *Mathematische Zeitschrift* 34 (1932), pp. 568–619.

[32]   H. Reichenbach. *The theory of probability, an inquiry into the logical and mathematical foundations of the calculus of probability*. Oakland, CA: University of California Press, 1948.

[33]   F. P. Ramsey. "Truth and Probability". In: *The Foundations of Mathematics and other Logical Essays*. Ed. by R. B. Braithwaite. London: Kegan, Paul, Trench, Trubner & Co., 1926. Chap. 7, pp. 156–198.

[34]   B. de Finetti. "La prévision: ses lois logiques, ses sources subjektives". Italian. In: *Annales de l'Institut Henri Poincaré* 7 (1937), p. 168.

[35]   L. J. Savage. *The foundations of statistics*. New York: John Wiley & Sons, 1954.

[36]   H. Jeffreys. *Theory of Probability*. Oxford: Oxford University Press, 1939.

[37]   R. T. Cox. *The Algebra of Probable Inference*. Baltimore: John Hopkins Press, 1961.

[38]   ISO 3534-1:2007. *Statistics — Vocabulary and symbols — Part 1: General statistical terms and terms used in probability (corrected version)*. Genève: International Organization for Standardization (ISO).

[39]   R. Carnap. *Logical Foundations of Probability*. Chicago: University of Chicago Press, 1950.

[40]   J. Venn. *Symbolic Logic*. London: Macmillan & Co., 1881.

[41]   G. Boole. *An investigation on the laws of thought*. London: Walton and Maberly, 1854.

[42]   G. Peano. *Arithmetices principia nova methodo exposita*. Italian. Roma: Bocca, 1889.

[43]   P. R. Halmos. *Naive Set Theory*. Princeton, NJ: D. von Nostrand Company, 1960.

[44]   E. Borel. *Leçons sur la théorie des fonctions*. French. Paris: Gauthier-Villars, 1898.

[45]   A. Kolmogoroff. *Grundbegriffe der Wahrscheinlichkeitsrechnung*. German. Ergebnisse der Mathematik und ihrer Grenzgebiete. Berlin: Springer Verlag, 1933.

[46]   A. N. Kolmogorov. *Foundations of the Theory of Probability*. New York, NY: Chelsea Publishing Company, 1956.

[47]   G. Vitali. *Sul problema della misura dei gruppi di punti di una retta*. Italian. Bologna: Gamberini and Parmeggiani, 1905.

[48]   A. Rényi. *Foundations of Probability*. San Francisco, CA: Holden Day, Inc., 1970.

[49]   T. Bayes. "An essay towards solving a problem in the doctrine of chances". In: *Phil. Trans. Roy. Soc. London* 53 (1763), pp. 370–418.

[50]   P. S. de Laplace. "Mémoire sur la probabilité des causes par les évènemens". French. In: *Mém. Académ. Roy. Scie.* VI (1774), pp. 621–656.

[51]   Th. Abelin. "Rauchen als Ursache von Lungenkrebs". German. In: *Z. Präventivmedizin* 6 (1961), pp. 349–366.

[52]   S. N. Bernstein. *Theory of Probability*. Russian. Moskau, Leningrad, 1927.

[53]   P. Erdős and A. Rényi. "On a new law of large numbers". In: *Jour. Analyse Mathematique* 23 (1970), pp. 103–111.

[54]   ISO/IEC Guide 98-3:2008/Suppl 1:2008:2008. *Evaluation of measurement data — Supplement 1 to the "Guide to the expression of uncertainty in measurement" — Propagation of distributions using a Monte Carlo method*. Genève: International Organization for Standardization (ISO).

[55]   JCGM 101:2008. *Evaluation of measurement data — Supplement 1 to the "Guide to the expression of uncertainty in measurement" — Propagation of distributions using a Monte Carlo method*. Paris: Bureau International des Poids et Mesures (BIPM).

[56]   O. Heaviside. *Electromagnetic Theory*. Vol. II. London: "The Electrician" Printing and Publishing Company Ltd., 1899.

[57]   P. Billingsley. *Convergence of Probability Measures*. New York: John Wiley & Sons, Inc., 1968.

[58]   R. S. Klessen. "One-Point Probability Distribution Functions of Supersonic Turbulent Flows in Self-gravitating Media". In: *The Astrophysical Journal* 535.2 (2000), p. 869.

[59]   H. Lebesgue. *Leçons sur l'intégration et la recherche des fonctions primitives*. French. Paris: Gauthier-Villars, 1904.

[60]   P. A. M. Dirac. *The Principles of Quantum Mechanics*. Oxford: Oxford University Press, 1930.

[61]   L. Schwartz. "Généralisation de la notion de fonction, de dérivation, de transformation de Fourier et applications mathématiques et physiques". French. In: *Annales de l'université de Grenoble* 21 (1945), pp. 57–74.

[62]   R. Strichartz, ed. *A Guide to Distribution Theory and Fourier Transforms*. Boca Raton, FL: CRC Press, 1994.

[63]   R. N. Bracewell. *The Fourier Transform and Its Applications*. New York: McGraw-Hill, 1965.

[64]   C. Huygens. "De ratiociniis in aleæ ludo". In: *Exercitationum mathematicarum libri quinque: Quibus accedit Christiani Hugenii tractatus De ratiociniis in aleæ ludo*. Ed. by F. à Schooten. Lugd. Batav.: Ex officina Johannis Elsevirii, 1657, pp. 517–534.

[65]   K. Knopp. *Theory and Application of Infinite Series*. London: Blackie & Son Ltd., 1951.

[66]   C. F. Gauss. *Theoria motus corporum coelestium in sectionibus conicis solem ambientium*. Hamburg: Perthes, 1809.

[67]   H. Lebesgue. "Sur l'intégration des fonctions discontinues". French. In: *Annales scientifiques de l'École Normale Supérieure, Sér. 3* 27 (1910), pp. 361–450.

[68]   C. G. J. Jacobi. "De determinantibus functionalibus". In: *Crelle's Journ.* 22 (1841), pp. 319–359.

[69]   K. Novianingsih and R. Hadianti. "Modeling Flight Departure Delay Distributions". In: *International Conference on Computer, Control, Informatics and Its Applications (IC3INA)*. Bandung, 2014, pp. 30–34.

[70]   M. Dwass. "On the Convolution of Cauchy Distributions". In: *American Mathematical Monthly* 92.1 (1985), pp. 55–57.

[71]   C. R. Rao. "Information and accuracy obtainable in an estimation of a statistical parameter". In: *Bull. Calcutta Math. Soc.* 37 (1945), p. 81.

[72]   H. Cramér. *Mathematical methods in statistics*. Princeton University Press, 1946.

[73]   R. A. Fisher. "On the mathematical foundations of theoretical statistics". In: *Philosophical Transactions of the Royal Society* A 222 (1922), pp. 309–368.

[74]   W. Chauvenet. *A Manual of Spherical and Practical Astronomy*. Vol. 2. London: Trübner & Co., 1864, pp. 481–482.

[75]   R. A. Fisher. "On an absolute criterion for fitting frequency curves". In: *Messenger of Mathematics* 41 (1912), pp. 155–160.

[76]   A. M. Legendre. *Nouvelles méthodes pour la détermination des orbites des comètes*. French. Paris: Courcier, 1805.

[77]   F. R. Helmert. *Die Ausgleichsrechnung nach der Methode der kleinsten Quadrate*. German. Leipzig: Teubner-Verlag, 1872.

[78]   W. W. Johnson. *Theory of errors and method of least squares*. London: Chapman & Hall Ltd., 1905.

[79]   M. Krystek and M. Anton. "A weighted total least-squares algorithm for fitting a straight line". In: *Measurement Science and Technology* 18 (2007), pp. 3438–3442.

[80]   M. Krystek and M. Anton. "A least-squares algorithm for fitting data points with mutually correlated coordinates to a straight line". In: *Measurement Science and Technology* 22 (2011), 035101 (9pp).

[81]   A. A. Markoff. *Wahrscheinlichkeitsrechnung*. German. Leipzig: B. G. Teubner Verlag, 1912.

[82]   H. Jeffreys. "An Invariant Form for the Prior Probability in Estimation Problems". In: *Proceedings of the Royal Society of London* A 186 (1946), pp. 453–46.

[83]   E. T. Jaynes. "Information Theory and Statistical Mechanics". In: *Physical Review* 106 (1957), pp. 620–630.

[84]   E. T. Jaynes. "Information Theory and Statistical Mechanics II". In: *Physical Review* 108 (1957), pp. 171–190.

[85] E. T. Jaynes. "Prior Probabilities". In: *IEEE Transactions On Systems Science and Cybernetics* sec-4, no. 3 (1968), pp. 227–241.

[86] C. E. Shannon. "A Mathematical Theory of Communication". In: *The Bell System Technical Journal* 27 (1948), pp. 623–656.

[87] H. Raiffa and R. Schlaifer. *Applied Statistical Decision Theory*. Tech. rep. Harvard: Division of Research, Graduate School of Business Administration, Harvard University, 1961.

[88] J. B. S. Haldane. "A note on inverse probability". In: *Mathem. Proc. Cambridge Phil. Soc.* 28 (1932), pp. 55–61.

[89] J. Neyman. "Outline of a Theory of Statistical Estimation Based on the Classical Theory of Probability". In: *Phil. Trans. R. Soc. Lond.* A 236 (1937), pp. 333–380.

[90] A. Erdélyi et al. *Higher Transcendental functions*. Vol. 2. New York: MacGraw-Hill, 1953.

[91] Student. "The Probable Error of a Mean". In: *Biometrika* 6(1) (1908). Author: Gosset, W. S., pp. 1–25.

[92] E. T. Jaynes. "Confidence intervals vs Bayesian intervals". In: *Foundations of Probability Theory, Statistical Inference, and Statistical Theories of Science*. Ed. by W. L. Harper and C. A. Hooker. Vol. II. Dordrecht-Holland: D. Reidel Pub. Co., 1976, pp. 175–257.

[93] M. Krystek. "Zuverlässigkeitsbereiche als Maße der Messgenauigkeit". German. In: *Technisches Messen* 81 (2014), pp. 605–617.

[94] G. E. P. Box and G. C. Tiao. "Multiparameter problems from a Bayesian point of view". In: *Ann. Math. Statist.* 36 (1965), pp. 1468–1482.

[95] J. H. Lambert. "Theorie der Zuverlässigkeit der Beobachtungen und Versuche". German. In: *Beyträge zum Gebrauche der Mathematik und deren Anwendung*. Vol. I. Berlin: Verlag des Buchladens der Realschule, 1765.

[96] C. Eisenhart. "Realistic evaluation of the precision and accuracy of instrument calibration". In: *Journal of Research of the National Bureau of Standards* 67C (1963), pp. 161–187.

[97] K. Weise and W. Wöger. *Eine Bayessche Theorie der Messunsicherheit*. German. PTB-Bericht N-11. Braunschweig: Physikalisch-Technische Bundesanstalt, 1992.

[98] K. Weise and W. Wöger. "A Bayesian theory of measurement uncertainty". In: *Measurement Science and Technologie* 4.1 (1993), pp. 1–11.

[99] K. Weise and W. Wöger. *Meßunsicherheit und Meßdatenauswertung*. German. Weinheim: Wiley-VCH Verlag GmbH, 1999.

[100] W. Wöger. "Remarks on the E_n-Criterion Used in Measurement Comparisons". In: *PTB-Mitteilungen* 1 (1999), pp. 24–27.

[101] B. L. Welch. "Specification of rules for rejecting too variable a product, with particular reference to an electric lamp problem". In: *J. Roy. Statist. Soc. Suppl.* 3 (1936), pp. 29–48.

[102] B. L. Welch. "The significance of the difference between two means when the population variances are unequal". In: *Biometrica* 29 (1938), pp. 350–362.

[103] B. L. Welch. "The Generalization of 'Student's' Problem when Several Different Population Variances are Involved". In: *Biometrika* 34 (1947), pp. 28–35.

[104] F. E. Satterthwaite. "An Approximate Distribution of Estimates of Variance Components". In: *Biometrics Bulletin* 2 (1946), pp. 110–114.

[105] R. N. Kacker. "Bayesian alternative to the ISO-GUM's use of the Welch–Satterthwaite formula". In: *Metrologia* 43 (2006), pp. 1–11.

[106] JCGM 102:2008. *Evaluation of measurement data - Supplement 2 to the "Guide to the expression of uncertainty in measurement" — Extension to any number of output quantities*. Paris: Bureau International des Poids et Mesures (BIPM).

[107] A. Griewank. *Evaluating Derivatives: Principles and Techniques of Algorithmic Differentiation*. Philadelphia: SIAM, 2000.

[108] I. Lira and W. Wöger. "Evaluation of the uncertainty associated with a measurement result not corrected for systematic effects". In: *Meas. Sci. Technol.* 9.6 (1998), pp. 1010–1011.

[109]   F. Härtig and M. Krystek. "Correct treatment of systematic errors in the evaluation of measurement uncertainty". In: *Proc. IX. Int. Symp. on Measurement Technology and Intelligent Instruments*. 2009.

[110]   B. O. Koopman. "On Distributions Admitting a Sufficient Statistic". In: *Transactions of the American Mathematical Society* 39.3 (1936), pp. 399–409.

[111]   E. J. G. Pitman. "Sufficient statistics and intrinsic accuracy". In: *Mathematical Proceedings of the Cambridge Philosophical Society* 32.4 (1936), pp. 567–579.

[112]   S. Kullback and Leibler R. A. "On Information and Sufficiency". In: *Ann. Math. Statist.* 22.1 (1951), pp. 79–86.

[113]   M. G. Cox and B. R. L. Siebert. "The use of a Monte Carlo method for evaluating uncertainty and expanded uncertainty". In: *Metrologia* 43.4 (2006), pp. 178–188.

[114]   J. S. Bendat and A. G. Piersol. *Random Data Analysis and Measurement Procedures*. New York: Wiley & Sons, 2000.

[115]   Elster2003 C, Link A, and Wöger W. "Proposal for linking the results of CIPM and RMO key comparisons". In: *Metrologia* 40 (2003), pp. 189–194.

[116]   Sutton2004 C M. "Analysis and linking of international measurement comparisons". In: *Metrologia* 41 (2004), pp. 272–277.

[117]   White2004 D R. "On the analysis of measurement comparisons". In: *Metrologia* 41 (2004), pp. 122–131.

[118]   Steele2005 A G, Wood B M, and Douglas R J 2005. "Linking key comparison data to appendix B". In: *TEMPMEKO 2004 : 9th International Symposium on Temperature and Thermal Measurements in Industry and Science* (Dubrovnik, Croatia). Vol. 2. June 22, 2004, pp. 1087–1092.

[119]   Nielsen2000 L. *Evaluation of measurement intercomparisons by the method of least squares*. Technical Report DFM-99-R39. Danish Institute of Fundamental Metrology. 2000.

[120]   Thalmann R. "Analysis and linking of international measurement comparisons". In: *Metrologia* 39 (2002), pp. 165–177.

[121]   Decker J E et. al. "Measurement science and the linking of CIPM and regional key comparisons". In: *Metrologia* 45 (2008), pp. 223–232.

[122]   Decker J E et al. Report on SIM.L-K1 regional comparison: stage one calibration of gauge blocks by optical interferometry. In: *Metrologia* vol. 44 (2008).

[123]   Acko B. Final report on EUROMET Key Comparison EUROMET.L-K7: Calibration of line scales. In: *Metrologia* vol. 49 (2012).

Index

All **bold** printed page numbers refer to pages that contain the definition of the respective term, while normally printed page numbers refer to pages on which the respective term is used.

A

absolute zero point **17**

abstraction 40

accuracy

    measurement ~ **23**

actualization

    ~ of the mean value 252

    ~ of the variance 252

addition theorem **126**

algebra

    Boolean ~ 109

    ~ of events 109

    ~ of sets 109

almost

    ~ never **147**

    ~ surely **147**

amount

    ~ of substance **12**

ampere **12**

amplification factor 62

analogue

    ~ computer 53

approach

    error ~ **19**

    systematic ~ 54

    uncertainty ~ **19**

approximation

    linear ~ **55**, 62

    quadratic ~ **72**

        of an explicit function **72**

    ~ quality **40**

average 16, 24, 30

axioms

    measure ~ **116**

    ~ of KOLMOGOROV **117**

    ~ of conditional probability **123**

B

base

    ~ quantity **12**

    ~ unit **12**

BAYESian

    ~ correction **246**

    ~ estimator **242**

        expectation of a ~ **243**

    ~ method **240**

BAYES-LAPLACE

    ~ theorem 242

BERNOULLI

    ~ experiment **227**

BERTRAND'S

    ~ paradox 94

best-fit

    ~ straight line **236**

bias **32**, 217

    ~ of an estimator **217**

bimodal 271

black-box

    ~ modelling **41**

Boolean

    ~ algebra 109

bound

    CRAMÉR-RAO ~ **218**

branch

    ~ of an inverse function **156**

C

calculation

    ~ of a measurement error 29

    ~ of errors **30**

calibration 29

    ~ certificate 31, 246, 247

    ~ of a measurement system 246, 264

candela **12**

cardinality **113**, **138**

carrier

    ~ of a characteristic **211**

    ~ of a property **7**

    ~ of the quantity **14**

CAUCHY

    ~ distribution 221

cause-effect

    ~ interrelation 132

    ~ relation 53, 54

certainty 105

certificate

    calibration ~ 31, 246, 247

https://doi.org/10.1515/9783111453712-017

characteristic
    carrier of a ∼ **211**
    quantitative ∼ 18, **212**
closeness 40
co-ordinates
    cylindrical ∼ **184**
    polar ∼ **184**
    spherical ∼ **184**
coefficient
    confidence ∼ **255**
    correlation ∼ **201, 239**
      properties of the ∼ **201**
    ∼ of variation 25, **169**
    sample correlation ∼ **214**
    sensitivity ∼ 85
coefficients
    sensitivity ∼ 279
collection
    data ∼ 212
combinatorics 95
comparison
    ∼ of measurement standards **29**
compatibility
    metrological ∼ **20**
    ∼ of measurement results **20**
compensation **32**
complement
    set ∼ **110**
complexity
    mathematical ∼ 40
computability 41
computer
    analogue ∼ 53
    ∼ simulation 54
condition
    measurability ∼ **139, 174**
    regularity ∼ **139**, 140, **174**, 175
conditional ∼ 133
confidence
    ∼ coefficient **255**
    ∼ interval **255**
    ∼ level **255**
    level of ∼ 286
confidence level 34
constant
    expectation of a ∼ **166**
    fundamental ∼ **20**
    mathematical ∼ 17

constraints
    mathematical ∼ 41
continuum **113**
convolution
    ∼ of Gaussian distributions **191**
    ∼ of normal distributions **191**
    ∼ of probability density functions **190**
    ∼ of uniform distributions **193**
correction 15, **31**, 32, 34
    Bayesian ∼ **246**
correlation
    ∼ and stochastic dependence 203
    ∼ coefficient **201, 239**
      properties of the ∼ **201**
    ∼ matrix **206**
covariance **200**
    ∼ matrix **205**
      ∼ of an estimator 219
    properties of the ∼ **201**
    sample ∼ **214**
coverage
    ∼ factor 285
    ∼ interval **261**
      symmetric ∼ **266**
    ∼ probability **261**, 286
Cramér-Rao
    ∼ bound **218**
credible region 267, **271**
current
    electric ∼ **12**
cycle **76**

D
data
    ∼ collection 212
    measurement ∼ 19
decomposition
    ∼ of models 52
definition
    circular ∼ 93, 100
definitional uncertainty 20, 34
degree
    effective ∼ **286**
degrees of freedom
    number of ∼ **239**
density **266**
    ∼ function **266**
      bimodal ∼ **266**
      conjugated ∼ **248**

marginal ~ 185
posterior probability ~ **242**
prior probability ~ **241**
density function 151
   conditional ~ **155**
   conditions of a ~ **153**
   joint ~ **180**
   marginalization of a ~ **187**
   multivariate ~ **180**, 182
      conditions of a ~ **181**
      transformation of a ~ **183**
   transformation of a ~ **157**
dependence
   stochastic ~
      correlation and ~ 203
      ~ of disjoint events **134**
dependency
   ~ graph 78
derivative
   partial ~ **60**
description
   mathematical ~ 14, **40**
   system ~ 41
detailing 40
deviation
   mean ~ **234**
   mean square ~ **167**, **234**, 239
   relative ~ 55, 60
   sample standard ~ **214**
   standard ~ 212
device
   displaying ~ 28
diagram
   EULER-VENN ~ **109**
difference
   ~ of random quantities **188**
   set ~ **110**
differentiability 40
differentiation
   automatic ~ **88**
dimension
   ~ one **10**
   symbol of ~ **12**
δ-function
   DIRAC's **153**
discontinuity 40
   jump ~ 143, 147, 179
dispersion
   measure of ~ **166**

~ of a random quantity **167**
distribution *see* probability distribution function
   beta ~ **250**
   CAUCHY ~ 221
   exponential ~ 226
   frequency ~ 212
   ~ function
      central moment of a ~ **221**
      generalized moment of a ~ **221**
   inverse gamma ~ **249**, **251**
   marginal ~ 185
   normal ~ 222, 229, 246
   normal gamma ~ **251**
   PARETO ~
      modified ~ **253**
   probability ~ 34
   RAYLEIGH ~ 180
   rectangular ~ 152, 180
   statistical ~ 34
   uniform ~ 152, 180, 223, 232, 246
distribution function
   conditional ~ **149**
   continuous ~ **141**, 145, 147
   continuous multivariate ~ 179
   discrete ~ 143, **145**
      normalisation condition 145
   empirical ~ 145
   exponential ~ 162
   geometric ~ **162**
   joint ~ **176**
   marginalization of a ~ **186**
   multivariate ~ **176**
   properties of a ~ 146
   properties of a multivariate ~ **179**
   rules of a ~ **148**
   univariate ~ **176**
drift 31
duration **12**

E
edge **74**
entropy
   information ~ **246**
   principle of maximum ~ **246**
equation
   model ~ 14, **47**, **48**, 51, 236
   quantity ~ **13**, 41
error
   ~ approach **19**

measurement ~ **29**
observational ~ 28
random ~ **274**
systematic ~ 170, **274**
estimate 19, 31, 34
~ of a mean 16
~ of a median 16
estimation
~ of a parameter 216
estimator **216**
asymptotically unbiased ~ **218**
BAYESIAN ~ **242**
bias of an ~ **217**
biased ~ **217**
covariance matrix of an ~ 219
efficient ~ **219**
interval ~ **221**
moment ~ **222**
point ~ **221, 255**
region ~ **221**
unbiased ~ **217**
EULER-VENN
~ diagram **109**
evaluation
~ model 19
objective of the ~ **19**
type A ~ **282**
type B ~ **282**
event **108**
certain ~ **109**
complementary ~
conditional probability of the ~ **124**
elementary ~ **109**
impossible ~ **109**
conditional probability of the ~ **123**
probability of the ~ **117**
~ relation **109**
events
mutually exclusive ~ **111**
set of ~ **114**
existence
~ of an expectation 163
expectation 30
existence of an ~ 163
linearity of an ~ **165**
linearity of the ~ **196**
monotonicity of an ~ **166**
~ of a BAYESIAN estimator **243**
~ of a constant **166**

~ of a continuous random quantity **163**
~ of a discrete random quantity **160**
~ of a function
~ of a random quantity **165**
~ of a function of a multivariate random
quantity **195**
~ of a linear function **166**
~ of a measured quantity 166
~ of a multivariate random quantity **195**
~ of a product of independent random
quantities **197**
~ of a sum of random quantities **196**
experience 34
experiment
BERNOULLI ~ **227**
random ~ **96**, 97
exponential
~ distribution 226

F
factor
coverage ~ 285
feasibility
numerical ~ 40
feedback **54**
FISHER
~ information 218
~ information matrix 219
formula
BIENAYMÉ **205**
frequency
absolute ~ **96**
~ distribution 212
relative ~ **96**, 99
friction 28
function
HEAVISIDE **144**
HEAVISIDE ~ 153
implicit ~ **63**
inverse ~ **156**
branch of an ~ **156**
likelihood ~ **225**
linear ~
expectation of a ~ **166**
linearisation of an explicit ~ **63**
linearization of an implicit ~ **65**
log-likelihood ~ **225**
loss ~ **218**
quadratic ~ **218**

probability density ~ 15
risk ~ **218**
step ~ 143, 144
transformation ~
    measurable ~ **157**
unit step ~ **144**
~ function
    multimodal ~ **266**

G
GAUSS-MARKOV
    ~ theorem **240**
graph
    connected ~ **76**
    dependency ~ 78, **78**
    directed ~ **74**
    directed acyclic ~ **76**
    edge-weighted ~ **85**
    named ~ **75**
    named, coloured vertex ~ **77**
    named, marked vertex ~ **77**
GUM **19**

H
height
    step ~ 143
HESSE
    ~ matrix **229**
hyperparameters **248**
hypothesis 105
hysteresis 31

I
impossibility 105
independence
    stochastic ~ **133**
        ~ of complementary events **134**
        total ~ **135**
indication 16, 28
indifference
    ~ principle 243
inequality
    CHEBYSHEV'S **208**
    MARKOV'S **207**
influence quantity 19
information 19, 34, 103
    available ~ 104
    ~ entropy **246**
    FISHER ~ 218

relevant ~ **15**
input
    ~ quantity 45, 59, 79, 281
    ~ vertex 79
intensity
    luminous ~ **12**
interrelation
    cause-effect ~ 132
intersection of sets **110**
interval
    confidence ~ **255**
    coverage ~ **261**
        symmetric ~ **266**
    statistical coverage ~ **261**
    statistical tolerance ~ **261**

J
jump discontinuity 143, 147, 179

K
kelvin **12**
kilogram **12**
kind
    ~ of quantity 14
knowledge
    complete ~ 17
    prior ~ 41

L
Last Squares Estimator **236**
law
    ~ of large numbers 97
Laws
    ~ of Mathematics 148
laws
    ~ of mathematics 17
learning process **252**
length **12**
level
    confidence ~ **255**
    ~ of confidence 34, 286
    ~ of significance **255**
likelihood **225**, 242
    ~ function **225**
limit 99
    stochastic ~ 100
linear
    ~ function
        expectation of a ~ **166**

linearisation
~ of an explicit function **63**
~ of an implicit model equation 67
linearity
~ of an expectation **165**
~ of the expectation **196**
linearization 59, 64
~ of an implicit function **65**
~ of the model equation **55**
log-likelihood
~ function **225**
loop **74**, 76
loss
~ function **218**
luminous
~ intensity **12**

M
marginalization **185**
marking
of a vertex 78
mass **12**
material
reference ~ 9, 10, 29
mathematical
~ description 14
Mathematics
Laws of ~ 148
mathematics
laws of ~ 17
matrix
correlation ~ **206**
covariance ~ **205**
~ of an estimator 219
FISHER information ~ 219
HESSE ~ **229**
uncertainty ~ **205**
variance-covariance ~ **205**
maximum
~ likelihood method **224**
sample ~ **215**
mean
arithmetic ~ **160**, 191
sample ~ **214**
~ square deviation 239
mean square
~ deviation **167**
mean value
actualization of the ~ 252

measurability
~ condition **139**, **174**
measurand **14**, 18
smallest change of a ~ 28
specification of a ~ **14**
measure **116**
~ axioms **116**
normalized ~ **116**
~ of dispersion **166**
problem of ~ 146
~ space **115**, 117
measured data
tabulated ~ 40
measurement **14**
~ accuracy **23**
~ bias **32**
detailed description of a ~ **23**
~ error 23, **29**
calculation of a ~ 29
random ~ 24
systematic ~ 24, 26, 28
ideal ~ 17, 18
intermediate precision condition of ~ **27**
~ method **21**
differential ~ **22**
direct ~ **22**
indirect ~ **22**
null ~ **22**
substitution ~ **22**
model of the ~ 23
~ operation **15**
operations used in a ~ 21
~ precision **25**, 28
precision of a ~ 23
~ principle **21**
~ procedure 9, 10, **23**, 27
repeatability condition of ~ **26**
replicate ~ 25, 27
reproducibility condition of ~ **27**
~ result **15**, 23
~ series **213**
~ standard 31
~ trueness 24
~ uncertainty 15, 23, 28
target ~ **23**
~ unit 9, 10, **11**, 12, 14
measurement data 19
measurement error
component of the ~ 30, 31

random ~ **30**
systematic ~ **31**, 32, **235**
    unknown ~ 32
measurement procedure
    reference ~ 29
measurement result
    quality of a ~ 32
measurement standard 34
measurement system
    calibration of a ~ 246, 264
measurement uncertainty 32, **34**
    associated ~ 29
    calculation of the ~ 32
    contribution to the ~ 29
    standard ~ 34
    Type A evaluation of ~ 34
    Type B evaluation of ~ 34
measuring
    ~ system 14
measuring system 27
median 16
    sample ~ **215**
method
    BAYESIAN ~ **240**
    maximum likelihood ~ **224**
    pivot ~ **259**
    stochastic ~ 132
    traditional ~ **275**
metre **12**
metrological compatibility **20**
metrology 16
minimum
    sample ~ **215**
model 19, **39**, 51, 79
    complex ~ 50
    ~ equation 14, **47**, **48**, 51, 236
    evaluation ~ 19
    experimental ~ **40**
    hybrid ~ 44
    ideal ~ 53
    implicit ~ **48**
    linear ~ **46**
    mathematical ~ 78
    non-linear ~ **47**
    ~ of reality 55
    ~ of the measurement 23
    ~ order **40**
    ~ parameter **40**
    physical ~ **41**, 44

SHOCKLEY ~ **42**
    theoretical ~ **41**
model equation
    implicit ~ 50
    linearisation of an implicit ~ 67
    linearization of the ~ **55**
    linearized ~ 55
    solution of an implicit ~ **48**
model parameter
    empirical ~ 42
modelling 19, 39
    black-box ~ **41**
    experimental ~ **40**
    grey-box ~ **41**
    method of ~ **40**
    modularization during ~ 52
    physical ~ **41**
    stochastic ~ **237**
    ~ strategy 52
    unambiguous ~ 40
    white-box ~ **41**
modularization
    ~ during modelling 52
mol **12**
moment
    ~ estimator **222**
    ~ of a probability distribution function
        central ~ **221**
        generalized ~ **221**
    sample ~ **222**
monotonicity
    ~ of an expectation **166**
monotonicity relation **125**
multimodal 271
multiplication theorem **127**

N
noise 28
normal
    ~ distribution 222, 229
null set 147
number
    ~ of experiments 97
numerical
    ~ quantity value **12**
    ~ value 11

O
operating condition 27

operating point **55**, 58, 59
operation
    linear ~ 165
    measurement ~ 15
    set ~ 109
optimization
    system ~ 42
order
    ~ statistic **214**
output
    ~ quantity 45, 59, 79, 281
    ~ vertex 79

P

paradox
    BERTRAND'S ~ 94
parameter
    estimation of a ~ 216
    model ~ **40**
    non-negative ~ 34
    population ~ **212**
    success ~ **227**
    unknown ~ 19
parametrization 212
PARETO
    ~ distribution
        modified ~ **253**
partition
    disjoint ~ **111**, **127**
path **75**, 79
performance
    dynamic ~ 40
    static ~ 40
perturbation 40
phenomenon **8**, 14, 21
    natural ~ 21
pivot
    ~ method **259**
pivotal
    ~ quantity **258**
point
    ~ estimator **255**
population **211**
    ~ parameter **212**
posterior **242**
    ~ probability density **242**
power set **111**
precision 212
    measurement ~ **25**, 28

    ~ of a measurement 23
predecessor
    direct ~ **76**, 79
prevalence 130
principle
    indifference ~ 102, 243
    measurement ~ **21**
    ~ of insufficient reason 102
    ~ of maximum entropy **246**
prior **241**
    conjugated ~ **248**
    improper ~ **244**
    ~ probability density **241**
    uninformative ~ **243**
probability 89, 105, 133
    a posteriori ~ **132**
    a priori ~ **132**
    classical ~ **91**, 104
    combinatorial ~ **95**
    conditional ~ **121**
        addition theorem **126**
        axioms of ~ **123**
        boundedness **125**
        monotonicity relation **125**
        multiplication theorem **127**
        ~ of conjoint events **124**
        ~ of the complementary event **124**
        ~ of the impossible event **123**
    coverage ~ 34, **261**, 286
    ~ density
        bimodal ~ **266**
        conjugated ~ **248**
        multimodal ~ **266**
        unimodal ~ **266**
    ~ density function 15, 34, 151
        bimodal ~ **266**
        conditional ~ **155**
        conditions of a ~ **153**
        conjugated ~ **248**
        joint ~ **180**
        marginalization of a ~ **187**
        multimodal ~ **266**
        multivariate ~ **180**, 182
        transformation of a ~ **157**
        unimodal ~ **266**
    ~ density functions
        convolution of ~ **190**
    ~ distribution 34
        central moment of a ~ **221**

generalized moment of a ~ **221**
~ distribution function
  central moment of a ~ **221**
  conditional ~ **149**
  continuous ~ **141**, 145, 147
  continuous multivariate ~ 179
  discrete ~ 143, **145**
  empirical ~ 145
  generalized moment of a ~ **221**
  joint ~ **176**
  marginalization of a ~ **186**
  multivariate ~ **176**
  properties of a ~ 146
  properties of a multivariate ~ **179**
  rules of a ~ **148**
  univariate ~ **176**
frequentist ~ **99**, 104
mathematical ~ **116**
~ of the impossible event **117**
~ space **117**
~ statement 101, 105
subjective ~ **103**, 104, 105
total ~ **127**
~ value 102
probability density 151
  conditional ~ **155**
  conditions of a ~ **153**
  joint ~ **180**
  marginalization of a ~ **187**
  multivariate ~ **180**, 182
    conditions of a ~ **181**
    transformation of a ~ **183**
  transformation of a ~ **157**
probability distribution
  conditional ~ **149**
  continuous ~ **141**, 145, 147
  continuous multivariate ~ 179
  discrete ~ 143, **145**
    normalisation condition **145**
  empirical ~ 145
  geometric ~ **162**
  joint ~ **176**
  marginalization of a ~ **186**
  multivariate ~ **176**
  one-point ~ **150**
  properties of a ~ 146
  properties of a multivariate ~ **179**
  rules of a ~ **148**
  univariate ~ **176**

probability distribution function
  geometric ~ **162**
problem
  ~ of measure 146
procedure
  measurement ~ 9, 10, **23**, 27
product
  ~ of random quantities **189**
product rule **122**
properties
  ~ of the standard deviation **170**
  ~ of the variance **170**
property
  ~ carrier **7**
  perfectly defined ~ **18**
  quantifiable ~ **7**
  quantitative ~ **18**
proposition **106**
  logically meaningful ~ 106

Q
quality
  ~ of a measurement result 32
  ~ of approximation **40**
  ~ of the **239**
quantity **8**, 10, 14, 18, 139
  base ~ **12**
  carrier of the ~ **14**
  characteristic ~ **41**
  ~ equation **13**, 41
  influence ~ 19, 54
  input ~ 45, 59, 79, 281
  intermediate ~ 79
  kind of ~ **8**, 14
  measured ~
    expectation of a ~ 166
  ~ of dimension one **10**
  ~ of the same kind 11
  output ~ 45, 59, 79, 281
  perfectly defined ~ **18**
  pivotal ~ **258**
  random ~ **136**
    continuous ~ **138**
    continuous multivariate ~ **174**
    discrete ~ **138**
    discrete multivariate ~ **174**
    mixed multivariate ~ **174**
    multivariate ~ **174**
    stochastically independent ~ 177, 182

subdivision of a ~ **8**
~ symbol **8**
true ~ **19**
quantity value **8**, 10, 12, 14
conventional ~ 29–31
measured ~ 16
real scalar ~ 11
reference ~ 24, 26, **29**
~ true 16
true ~ 29, 31
quotient
~ of random quantities **189**

R
random quantities
difference of ~ **188**
product of ~ **189**
quotient of ~ **189**
sum of ~ **188**
uncorrelated ~ **200**
random quantity
continuous ~
expectation of a ~ **163**
discrete ~
expectation of a ~ **160**
dispersion of a ~ **167**
function of a ~
expectation of a ~ **165**
relative standard deviation of a ~ **169**
scaling of a ~ **166**
shifting of a ~ **166**
standard deviation of a ~ **168**
transformation of a ~ **156**
linear ~ **157**
logarithmic ~ **159**
quadratic ~ **158**
variance of a ~ **168**
random quantity ~
expectation ~
of a function of a multivariate ~ **195**
of a multivariate ~ **195**
range **30**
sample ~ **215**
ratio 8
reality
model of ~ 55
realization **139**
~ of a sample vector **213**

reference
~ material 9, 10
reference standard 34
regression
linear ~ **236**
regularity
~ condition **139**, 140, **174**, 175
relation
causal ~ 79
cause-effect ~ 53, 54
event ~ **109**
functional ~ 79
monotonicity ~ **125**
subset ~ 109
reliability theory 150
remainder **55**, 58, 59
resolution **27**, 28
result
corrected ~ 15
measurement ~ **15**, 23
uncorrected ~ 15
risk
~ function **218**
rounding error 17

S
sample **212**
~ coefficient of variation **214**
~ correlation coefficient **214**
~ covariance **214**
empirical ~ **213**
~ maximum **215**
~ mean **214**
~ median **215**
~ minimum **215**
~ moment **222**
observed value of a ~ **212**
~ range **215**
~ size **213**
~ standard deviation **214**
~ variance **214**
~ vector **212**
realization of a ~ **213**
sample space
continuous ~ 111
finite ~ 111
subset of the ~ 109
scalar **8**

scale
  transformation of ~ **171**
scaling
  ~ of a random quantity **166**
scatter 26
secant 56, 60
  slope of the ~ **57**
second **12**
sensitivity 130
  ~ coefficient 85
  ~ coefficients 279
sensitivity coefficient **62**, 69
series
  measurement ~ **213**
set
  basic ~ 111
  ~ complement **110**
  ~ difference **110**
  null ~ **116**
  ~ of events **114**
  ~ of vertex marks 78
  ~ of vertex names 77, 78
  ~ of vertices 75
  ~ operation 109
  power ~ **111**
  ~ theory 109
shifting
  ~ of a random quantity **166**
SHOCKLEY
  ~ model **42**
σ-algebra **113**
significance
  level of ~ **255**
simulation 39
  computer ~ 54
singularity **154**
slope
  ~ of the secant **57**
  ~ of the tangent 56, 60, 62
solution
  ~ of an implicit model equation **48**
solvability
  unique ~ **50**, 65
space
  measurable ~ **114**
  measure ~ **115**, 117
  probability ~ **117**
  sample ~ **106**

specification
  ~ of a measurand **14**
specificity 130
speed of light **17**
stability
  system ~ 40
standard
  ~ deviation 212
  measurement ~ 31
standard deviation 25, 34
  experimental ~ 34
  ~ of a random quantity **168**
  properties of the ~ **170**
  relative ~
    ~ of a random quantity **169**
start
  ~ vertex **74**
statement
  probability ~ 101, 105
statistic **213**
  order ~ **214**
statistical
  ~ coverage interval **261**
  ~ tolerance interval **261**
statistics 145
  mathematical ~ 104
step
  ~ function 143, 144
  ~ height 143
stochastic
  ~ dependence
    ~ of disjoint events **134**
  ~ independence **133**
    ~ of complementary events **134**
  total ~ **135**
  ~ method 132
straight line
  best-fit ~ **236**
submodel 54
  reusability 52
submodels **50**
subset **110**
  ~ of the sample space 109
  ~ relation 109
substance
  amount of ~ **12**
subtasks 50
success
  ~ parameter **227**

successor
    direct ~ **76**
sum
    ~ of random quantities **188**
symbol
    ~ of a quantity 11
    ~ of a quantity value 11
    ~ of dimension **12**
    ~ of unit **12**
    quantity ~ **8**
syntax tree
    abstract ~ **87**
System
    ~ of Units **11**
system
    ~ description **40**, 41
        physical reasonable ~ **41**
    measuring ~ 14
    ~ optimization 42
    ~ stability 40
systematic error
    ~ of the result 17

T
tangent 56, 60
    slope of the ~ 56, 60, 62
Taylor expansion 69
temperature
    thermodynamic ~ **12**
tensor **8**
theorem
    addition ~ **126**
    Bayes-Laplace ~ 242
    central limit ~ **209**
    Gauss-Markov ~ **240**
    multiplication ~ **127**
    ~ of Bayes-Laplace **129**
theory
    ~ of distributions **153**
    reliability ~ 150
time **12**
transformation
    affine ~ **185**
    general linear ~ **185**
    linear ~
        ~ of a random quantity **157**
    logarithmic ~
        ~ of a random quantity **159**
    ~ of a density function **157**

~ of a multivariate probability density
    function **183**
    ~ of a probability density **157**
    ~ of a probability density function **157**
    ~ of a random quantity **156**
    ~ of scale **171**
    quadratic ~
        ~ of a random quantity **158**
    ~ to cylindrical co-ordinates **184**
    ~ to polar co-ordinates **184**
    ~ to spherical co-ordinates **184**
treatment
    numerical ~ 41
true quantity 19
trueness
    measurement ~ **24**

U
uncertainty
    ~ approach **19**
    definitional ~ 20, **36**
    expanded ~ **285**
    ~ matrix **205**
    measurement ~ 15, 28
        target ~ **23**
    ~ of measurement **34**
    standard ~
        combined ~ **279**, 283
        type A ~ **282**
        type B ~ **282**
uniform
    ~ distribution 223, 232
unimodal 271
unimodal ~ **266**
union of sets **110**
unit 11
    base ~ **12**
    measurement ~ 9, 10, **11**, 12, 14
    ~ name 11
    symbol of ~ **12**
unit step
    ~ function **144**
update 252

V
validation 39, 50
value
    determination of the ~ 21
    estimated ~ **216**

exact ∼ **17**
measured ∼ **15**
numerical ∼ 11
numerical quantity ∼ **12**
observed ∼
    ∼ of a sample **212**
∼ of a theoretical (mathematical) charac-
    teristic **17**
∼ of the speed of light **17**
probability ∼ 102
true ∼ **18, 19**
variance 25, 30, 212
    actualization of the ∼ 252
    ∼ of a linear combination **205**
    ∼ of a random quantity **168**
    properties of the ∼ **170**
    sample ∼ **214**
variance-covariance
    ∼ matrix **205**

variation 99
    coefficient of ∼ 25, **169**
    sample coefficient of ∼ **214**
vector **8**
    column ∼ **172**
vertex **74**
    end ∼ **74**
    final ∼ **75**
    initial ∼ **75**
    inner ∼ **77**, 79
    input ∼ **77**, 79
    input degree of a ∼ **76**
    isolated ∼ **76**
    ∼ mark **77**
    marking of a ∼ 78
    ∼ name **77**
    output ∼ **77**, 79
    output degree of a ∼ **76**
    start ∼ **74**
VIM **19**

www.ingramcontent.com/pod-product-compliance
Lightning Source LLC
Chambersburg PA
CBHW080704220326
41598CB00033B/5301